Y0-CDB-757

06/24
STAND PRICE
$ 5.00

McGRAW-HILL TECHNICAL EDUCATION SERIES

Norman C. Harris, Series Editor

Introduction to Electron Tubes and Semiconductors • Alvarez and Fleckles

Data Processing: A Text and Project Manual • Cashman and Keys

Architectural Drawing and Planning • Goodban and Hayslett

Calculus with Analytic Geometry • Green

Introductory Applied Physics, 2d ed. • Harris and Hemmerling

Experiments in Applied Physics • Harris

Elementary Mathematics: Arithmetic, Algebra, & Geometry • Hemmerling

Pulse and Switching Circuits • Ketchum and Alvarez

Introduction to Microwave Theory and Measurements • Lance

Microwave Experiments • Lance, Considine, & Rose

Electronics Mathematics • Nunz and Shaw
Volume I: Arithmetic and Algebra
Volume II: Algebra, Trigonometry, and Calculus

Specifications Writings for Architects and Engineers • Watson

(Other volumes in preparation.)

ALGIE L. LANCE

Hughes Aircraft Company Microwave Standards Laboratory
Instructor, Santa Monica City College
Los Angeles Trade Technical College

INTRODUCTION TO MICROWAVE THEORY AND MEASUREMENTS

McGRAW-HILL BOOK COMPANY

New York

St. Louis

San Francisco

Toronto

London

Sydney

Mexico

Panama

INTRODUCTION TO MICROWAVE THEORY AND MEASUREMENTS

Library of Congress Catalog Card Number: 63-23468

07-036104-5

15 16 17 18 19 20 KPKP 7 8 3 2 10 9 8

PREFACE

This book was written as a basic text for use in introductory courses in microwave theory and measurements and to provide a reference for engineers and technicians whose work is related to microwave measurements and microwave systems or components. There is an increasing need for a text which covers basic microwave theory and techniques and the applications of these techniques to measurement problems. This text was developed as a result of teaching microwave courses and it is based on practical experience in the development and applications of precision microwave measurement systems and techniques.

The book presents a compact, logical description of physical concepts, mathematical formulations, measurement systems, and illustrative examples of ideas and measurement procedures. In the organization of this material, particular attention has been given to fundamental principles and applications. The physical significance of the mathematical formulations is presented by use of pictorial diagrams and/or descriptive explanations. Even though this is basically a qualitative approach to the microwave technique, one must realize that an adequate treatment of fundamental principles requires certain mathematical details.

Most of the problems are used to illustrate or amplify points discussed in the text. Some problems outline details omitted from the text in order to shorten the presentation. In general, the problems are designed to help the student gain an understanding of the concepts and techniques and to help form the foundation for understanding more advanced microwave techniques.

The final judgment of relative emphasis on text material is based upon various discussions with my colleagues, especially J. M. Considine, whose suggestions were very helpful in organizing the original outline.

I wish to acknowledge my indebtedness to the many scientists and engineers upon whose work this text is based. Also, I would like to express my appreciation for the continued interest and cooperative efforts on the part of my colleagues. Special thanks are due T. Mukaihata, Head, Microwave Standards Laboratory, Hughes Aircraft Company, for his continued support and encouragement.

Algie L. Lance

CONTENTS

INTRODUCTION

Microwave theory deals with electromagnetic phenomena occurring in the wavelength range from 30 centimeters to a few millimeters. This is a general consideration since boundaries of the microwave region are not specifically defined.

The transition between the specialized point of view of lumped-constant circuits and the fundamental approach to electromagnetic theory occurs in the region between 50 and 100 Mc, which corresponds to wavelengths of 6 and 3 m. In this region the lumped-constant analysis is largely replaced by wave theory associated with the conventional transmission line. It is proper to regard transmission of power as taking place through the space between conductors and not through the conductors themselves. This concept seems to hold regardless of the frequency range. The power is specified by the intensities of the electric and magnetic fields and the velocity with which the configuration is propagated along the line.

It is possible to introduce some of the most fundamental concepts of electromagnetic wave theory without becoming involved in all the complications of vector field theory. This approach to the subject begins in Chap. 2.

The distribution of the electric and magnetic fields must be measured in order to explain the behavior of the given microwave system. Precise information concerning the configuration of these fields can be obtained by proper application of the impedance concept in one form or another.

The importance of measurement techniques in the microwave field cannot be overemphasized since progress in the application of microwaves demands the development of measurement techniques.

Microwaves have a broad range of application in the numerous forms of "communication of information." The overall trend in expansion is towards microwaves, thus placing increasing demands for knowledge of the essentials of this subject.

CHAPTER

I

INTRODUCTION TO ELEMENTARY FIELD THEORY

The *current* aspect of electricity meets most needs and requirements at low frequencies; only occasionally is there a need to discuss lines of electric and magnetic force. There is a tendency to regard the electric and magnetic fields as almost unrelated quantities since their roles are so different at the lower frequencies. At higher frequencies these fields are so intimately related that they may be regarded, at times, as different aspects of the same thing.

We are concerned with phenomena which are described in terms of the motions of charges. An introduction to the basic nature of charges leads to a description of the fundamental properties of electric and magnetic lines of force and serves as a background for the basic definitions of transmission line parameters. Therefore, an understanding of the basic nature of charges and of the relationship of these charges to electric and magnetic fields and to the basic definitions of transmission line parameters is necessary. The manner in which conductors and dielectrics affect the charge distribution and the electric and magnetic field distribution must be known.

1·1 Basic phenomena

A fundamental property of electricity is that every particle of it exerts a force on every other particle. The magnitude of this force depends on the electric charges of the particles, the medium in which the charges are placed, and their locations in space and time.

1·2 Electrostatic fields

Electricity has a dual aspect in that it consists of an electric charge q, which exists at a point (infinitesimal volume), and an electric flux ψ, which occupies the rest of space but with intensity diminishing with distance. *Force exists by the interaction of the flux of one particle upon the center of source of flux of another particle; the region in space in which the force can be detected is called the "field" of the charge.*

1·3 Force between charges

The unit charge of electricity, the coulomb, is defined in terms of the experimental law of Coulomb, which gives the force between electric charges and includes the following information:

1. Like charges repel and opposite charges attract.
2. Force is dependent upon the medium in which the charges are located and acts in a line joining the charges.
3. Force is proportional to the charge magnitudes.
4. Force is inversely proportional to the square of the distance between the charges.

$$F = \frac{q_1 q_2}{4\pi\epsilon r^2} \qquad \text{mks system of units} \tag{1·1}$$

F is the force in newtons (1 newton $= 10^5$ dynes or 102 g weight) that q_1 plus all the polarized particles of the dielectric exerts upon q_2. ϵ is a property of the medium which may be called the permittivity (farads per meter). q_1 and q_2 (coulombs represent the magnitude and the sign of the charges. The distance between the charges is r. The dyne is the force necessary to accelerate a mass of one gram at the rate of one centimeter per second per second. The coulomb is a charge of 6.25×10^{18} electrons. The quantity $1/4\pi$ is a constant of proportionality peculiar to the mks system of units.

The advantage of the mks system of units is that the units of the electric quantities are those actually measured. *Length* is in *meters*, *mass* is in *kilograms*, *time* is in *seconds*, *current* is in *amperes*, *potential* is in *volts*, *impedance* is in *ohms*, and *power* is in *watts*.

1·4 Fields in dielectrics

If the charges $-q$ and $+q$ are immersed in a dielectric medium, the particles polarized by these charges will line up as shown in Fig. 1·1. The resultant effects cancel out in the medium, and, since the polarized particles are of opposite sign, they tend to reduce the effective magnitudes of the charges.

$$F_f = \epsilon' F \tag{1·2}$$

F_f is the force due to free charges, i.e., charges in free space with no polarized

Fig. 1·1 Polarized particles in a dielectric between charges.

charges. ϵ' is the relative permittivity and is a function that F must be multiplied by to obtain F_f. The permittivity of the dielectric is defined by

$$\epsilon = \epsilon' \epsilon_0 \tag{1·3}$$

ϵ_0 is the *permittivity of free space* or *specific inductive capacity*. If the medium is free space, without polarized charges, then

$$\epsilon = \epsilon_0 \qquad \text{and} \qquad F = F_f$$

The nature of permittivity ϵ, or relative permittivity ϵ', depends upon the medium in which the field exists. If the polarization properties of the medium are such that, throughout the entire field, the medium is *linear* (ϵ' does not depend upon the magnitude of the flux density), *isotropic* (ϵ' does not depend upon the direction of flux density), and *homogeneous* (ϵ' does not depend upon the location of the flux density), then ϵ' is a scalar constant called the *dielectric constant.*

Free space, having no polarized particles, is linear, homogeneous, and isotropic. Therefore, the *permittivity of free space* ϵ_0 is a universal constant.

$$\epsilon_0 = \frac{1}{36\pi} \times 10^{-9} = 8.854 \times 10^{-12} \text{ farad per m} \tag{1·4}$$

1·5 Electric field strength

The *electric field strength* or *electric field intensity* **E** at a point is defined as the *force per unit charge* exerted upon a test charge placed at a point.

$$\mathbf{E} = \frac{F}{q} \tag{1·5}$$

where F is the vector force acting upon the infinitesimal test charge q.

The electric field arising from the test charge is

$$\mathbf{E} = \frac{q}{4\pi\epsilon r^2}(\bar{a}_r) \qquad \text{volts per m} \tag{1·6}$$

Since $\bar{a}_r$ is the unit vector directed from the point in a direction away from the charge, the electric field vector points away from positive charges and toward negative charges. The electric field strength has magnitude and direction and

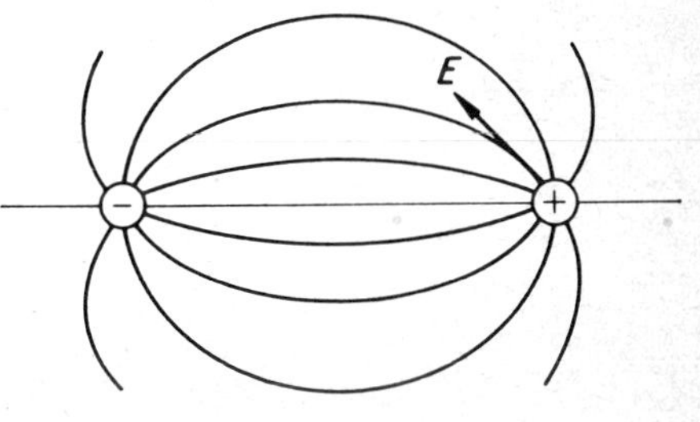

Fig. 1·2 Electrostatic lines of force in the region between two oppositely charged spheres.

is therefore a vector quantity. If the positive direction is taken, a positive test charge (proton) is displaced, as shown in Fig. 1·2. The electric field strength is measured in *volts per meter*.

1·6 Magnetic field strength

The force exerted on a unit magnetic pole is a measure of the *magnetic field intensity* **H**. It is a vector quantity measured in *amperes per meter*. The direction of force on the unit north pole is shown in Fig. 1·3. The force

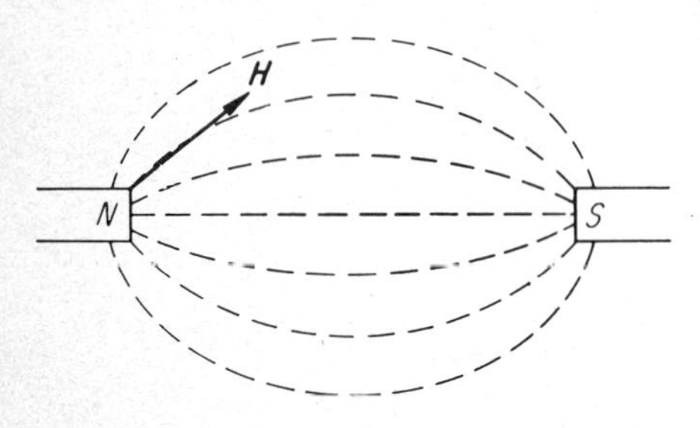

Fig. 1·3 Direction of force on a unit north pole in the region between two oppositely magnetized poles.

between magnetic charges (pole strength) m_1 and m_2, measured in webers, is expressed by

$$F = \frac{m_1 m_2}{4\pi\mu r^2} \qquad \text{newtons} \tag{1·7}$$

B is the magnetic flux density in webers (1 weber is 10^8 maxwells or "lines"). This choice is convenient in time-varying fields since a rate of change of 1 weber per second generates an electromotive force of 1 volt. This measure of volt-seconds per square meter equals henrys per meter from the definition of inductance.

$$\mathbf{B} = \mu_0 \mathbf{H} \tag{1·8}$$

where μ_0 is the permeability of free space. Permeability is the unit of measurement which indicates the ease with which a magnetic field may be set up in a material. It represents the ability of the medium to support tubes of magnetic force.

The permeability of the medium

$$\mu = \mu' \mu_0 \tag{1·9}$$

where μ' is the relative permeability. μ_0 is the permeability of free space in *henrys per meter*.

$$\mu_0 = 4\pi \times 10^{-7} \qquad \text{henry per m} \tag{1·10}$$

1·7 Resistance and conductance

From Ohm's law, the resistance $R = E/I$.

When two equipotential surfaces (for example, two conductors) have a potential difference between them, a certain amount of current will flow because of the finite resistance of the insulation. This leakage is called *conductance* and is determined by the dielectric medium between the surfaces or conductors. *Conductance* G is the factor by which the potential between the two equipotential surfaces, at any instant, must be multiplied to give the current flowing between the two surfaces.

$$I = GE \qquad \text{or} \qquad G = \frac{I}{E}$$

from the definitions $R = 1/G$ and $G = 1/R$.

1·8 Capacitance and inductance

If a potential E exists between two equipotential surfaces (such as two conductors), an electric charge $+q$ will be set up on the surface at a positive potential. An equal charge of $-q$ will, from the principle of conservation, be set up on the other surface. The *charge per unit difference of potential* is called *capacitance* C and is measured in farads.

Inductance. *Inductance* is the property of a circuit by which it opposes any change in current. It manifests itself in a back emf (electromotive force) that is developed when current is changed. *The inductance of a circuit is the back emf induced in it by a unit time rate of change of current.* The unit of inductance is the *henry*. Inductance can also be defined as *flux linkages per unit current.*

1·9 Properties of electric and magnetic fields

The properties of electric and magnetic fields which explain numerous phenomena of electrical transmission are as follows:

1. When lines of electric force are displaced laterally, lines of magnetic force are induced in the immediate adjacent space. This resultant magnetic force has an intensity **H** which is proportional to the velocity **v** of displacement and to the intensity **E** of the electric force. The direction of the induced magnetic force is normal to the direction of the original electric force. This property of the electric and magnetic fields is expressed by the vector notation

$$\mathbf{H} = \epsilon(\mathbf{v} \times \mathbf{E}) \tag{1·11}$$

The above notation is pronounced "v cross E." The physical significance of this notation concerning the direction of **H** is given by the right-hand rule and is applied as follows:

> Let the curled fingers of the right hand lie in the plane of **v** and **E** and point in the direction from **v** to **E** through the smaller angle; the thumb will point in the direction of **H**.

This rule is used for the vector or cross product of any two vectors. It can be seen that in Fig. 1·4, $\mathbf{v} \times \mathbf{E}$ gives the direction of **H** to the left.

2. When lines of magnetic force are displaced laterally, lines of electric force are induced in the immediate adjacent space. This resultant electric force has an intensity **E** which is proportional to the velocity **v** of displacement and the intensity **H** of the magnetic force. The direction of the induced electric force is normal to the direction of motion and also normal to the direction of the original magnetic force. This property of the electric and magnetic fields is expressed by the vector notation

$$\mathbf{E} = -\mu(\mathbf{v} \times \mathbf{H}) \tag{1·12}$$

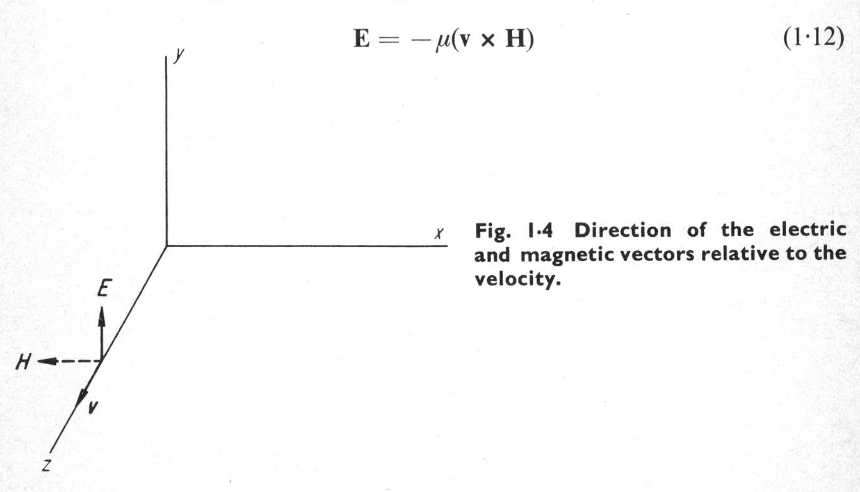

Fig. 1·4 Direction of the electric and magnetic vectors relative to the velocity.

In applying the right-hand rule to $\mathbf{v} \times \mathbf{H}$ in Fig. 1·4 it is noted that the thumb points in the direction *opposite* to the direction of **E** (downward). The minus (—) sign in the equation indicates that the direction of **E** is opposite the direction as determined by $\mathbf{v} \times \mathbf{H}$.

1·10 Energy flow

The amount of energy transferred from one point to another depends upon the magnitudes, distribution, and phases of the electric and magnetic fields. The Poynting concept specifies that the magnitude of the energy flow per unit volume across a unit area measured perpendicular to **v** is proportional to the product of **E** and **H** and is in a direction normal to both **E** and **H**. The vector notation for the Poynting concept is

$$\mathbf{P} - \mathbf{E} \times \mathbf{H} \tag{1·13}$$

The relative directions of **P**, **E**, and **H** are shown in Fig. 1·5. The energy moves at a velocity defined by

$$\mathbf{v} = \frac{1}{\sqrt{\mu\epsilon}} = \frac{1}{\sqrt{\mu'\mu_0\epsilon'\epsilon_0}} = \frac{1}{\sqrt{\epsilon'}}\frac{1}{\sqrt{\mu_0\epsilon_0}} \tag{1·14}$$

Since μ' is 1.

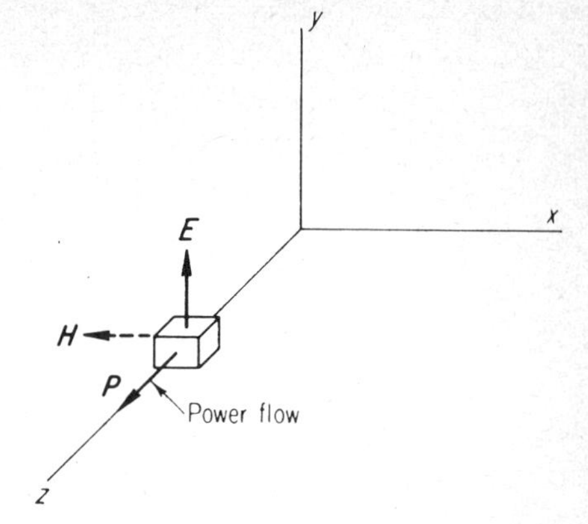

Fig. 1·5 Direction of the electric and magnetic vectors relative to the Poynting vector.

1·11 Electric current

The current per unit area is called the *current surface density* **J**, a vector measured in amperes per square meter. By vector notation, $\mathbf{J} = \mathbf{n} \times \mathbf{H}$ and is illustrated in the diagram of Fig. 1·6. **n** represents a unit vector perpendicular to the surface, and **H** is the magnetic field tangent to the surface. It is noted that the direction of the current flow is normal to the magnetic field.

1·12 Boundary conditions

The tangential components of the electric and magnetic fields must be continuous in traversing the interface between physically real media. That is, the tangential components of the electric and magnetic fields must be equal on the two sides of the boundary. Therefore, the amplitudes of the tangential components of the incident and reflected waves at the interface must equal the amplitude of the transmitted components of the transmitted wave.

No conductor is perfect, but in many practical problems it is desirable to neglect the finite electric field along the conductor. *In the case of an ideal perfect conductor, there can be no component of electric field tangential to the surface and no component of time-varying magnetic field perpendicular to the surface.*

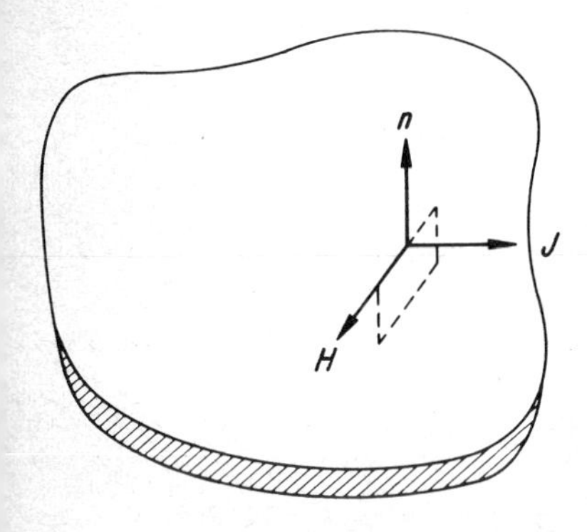

Fig. 1·6 Direction of current surface density relative to the magnetic field and the unit vector normal to the surface.

PROBLEMS

1·1 Calculate $\sqrt{\mu_0/\epsilon_0}$. What is the significance of this value?

1·2 Calculate $1/\sqrt{\mu_0\epsilon_0}$. What is the significance of this value?

1·3 If $q_1 = q_2 = \frac{1}{3} \times 10^{-9}$ coul ~~per m~~, $F = 1$ *dyne*, and q_1 and q_2 are spaced 1 cm apart (10^{-2} m), calculate ϵ.

1·4 Calculate the velocity in a medium which has a dielectric constant of 2. Repeat for dielectric constants of 6 and 12.

1·5 Consider a sheet of paper as the plane of the following diagrams and indicate the direction of the third vector given in the parentheses.

a. **E** points up and **v** points left. Show the direction of (**H**).
b. **v** points to the left and **H** points up. Show the direction of (**E**).
c. **E** points left and **H** points down. Show the direction of (**P**).
d. **P** points left and **H** points up. Show the direction of (**E**).

1·6 Draw a diagram showing an electric and magnetic field approaching a perfect conducting metal sheet. Draw a diagram alongside this sheet to indicate the magnitude of the electric and magnetic fields at the surface of this perfect conductor.

1·7 Draw an end view of a single conductor showing concentric lines of magnetic field directed clockwise around the conductor. Show electric lines of force normal to the surface and directed away from the conductor.

a. What is the direction of power flow?
b. If there is a component of electric field at the surface of the conductor directed away from the observer (parallel to the wire), what is the direction of power flow due to this component of electric field?
c. What is indicated by part *b*?

CHAPTER

2

TRANSMISSION LINES

Introduction. A transmission line is any structure used to guide the flow of energy from one point to another. This general definition is required as evidenced by the cross sections of various types of guiding structures shown in Fig. 2·1.

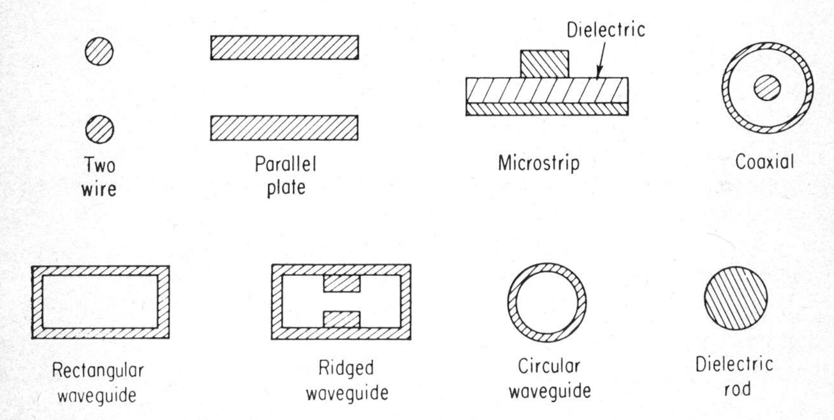

Fig. 2·1 Cross-sectional configurations of various types of guiding structures.

This chapter presents an introduction to basic transmission line theory involving the behavior of voltages and currents applied to the basic two-wire transmission line. The basic theory of reflections is the starting point for the major parts of this text. A treatment of the familiar two-wire transmission line is given, since it exhibits certain properties which are common to all types of transmission lines. The fundamental assumption is that the uniform spacing of the wires is so close that the effect in one wire of a change in current in the other wire is instantaneous. Therefore, the conductor spacing must be small compared to wavelength, and the length of the line must be long compared to the spacing when a description of the behavior of sinusoidal waves of voltage and current is considered.

2·1 The two-wire transmission line

The two-wire transmission line consists of two parallel conductors properly supported and insulated from each other.

The concept of movement of charge along a two-wire transmission line from a current-forcing source is illustrated in Fig. 2·2. For simplicity, it is customary to say that the charge flows into the line or that a charge moves

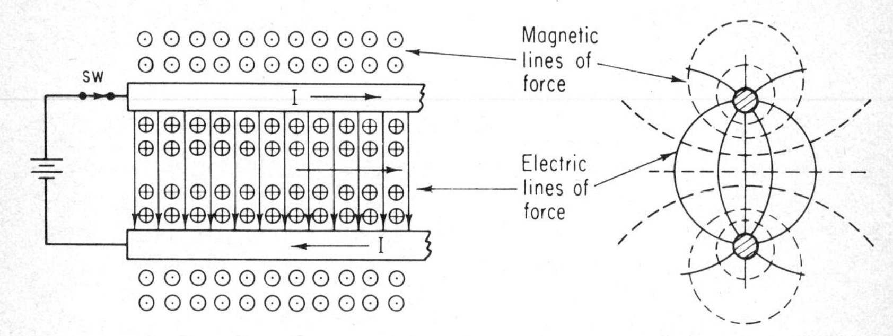

Fig. 2·2 Transmission of d-c power along a two-wire line.

down the line. It should be noted, however, that individual charges experience only small changes in their positions. The movement of electrons constitutes a current that produces an equivalent movement of charge. Also, no individual electron travels any great distance from its original position on the line.

It is necessary to describe the behavior of the charges on the two-conductor transmission line of Fig. 2·2. When the source voltage is connected to the two wires, a current flow exists because of the flow of charges into the line. There is a magnetic field, proportional to this current, surrounding the conductors. The associated flux linkages per unit current I are called *inductance L*. Therefore, there is an inductance per unit length of line when a unit current is flowing. The charge on the conductor is proportional to the potential difference (voltage). Therefore, the line has *shunt capacitance C*, previously defined as the charge per unit potential difference. If the dielectric medium between the conductors is not perfect, a conductive element must be assumed between the lines to account for this loss. This is the *conductance G* per unit length of line. In addition to the above parameters, there is a *series resistance R* associated with the conductors, since the perfect conductor does not exist in practice. The resistance depends upon the resistivity of the material, the length and cross section of the conductor, and the distribution of currents in the cross section.

If a given length of this transmission line were divided into more and more sections, the ultimate case would be an infinitesimal section of the basic elements, resistance R, conductance G, inductance L, and capacitance C. The equivalent circuit that results is shown in Fig. 2·3. When the source is applied to the line, it "sees" this first section of line which is made up of the basic circuit parameters.

The voltage generated by the flow of charge is referred to as a *voltage wave*, and the current induced in the line is referred to as the *current wave*. The relationship of the voltage wave to the current wave, the relationship of the electric field to the magnetic field, and the velocity with which these waves travel down the transmission line are determined by the values of the basic

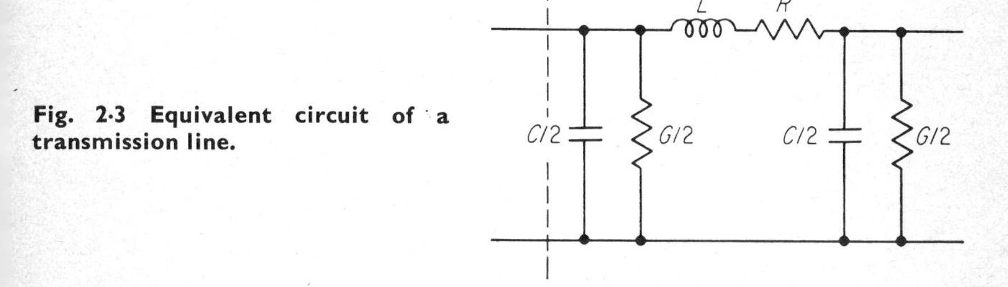

Fig. 2·3 Equivalent circuit of a transmission line.

circuit parameters. This is a physical concept of the electromagnetic theory set forth in Chap. 1, where it was pointed out that the motion of electric field lines of force gives rise to magnetic lines of force in the immediate vicinity, and the two fields together give rise to component Poynting vectors representing power flow. The series resistance and inductance are shown in one line only, but they are actually in both lines and can be distributed either equally or unequally. If the basic transmission line parameters are spread evenly along the entire length of the line, the constants are said to be *distributed*. The capacitance of the line is spread out along the line; the effect of this capacitance is not the same that would be obtained if it were all centered at one point.

2·2 Displacement current

In Fig. 2·2 the current is shown as flowing to the right in the upper conductor and to the left in the lower conductor. The direction of force and current flow is given for conventional current flow; therefore electron flow is in the opposite direction. There seems to be an inconsistency since the current path is not complete. This apparent inconsistency can be resolved by taking into account the *displacement current*, which is defined as the *time rate of change of electric flux through a surface*. The electric field arises from the moving charge and is varying with time, thus producing a magnetic field proportional to the electric field. The current, other than leakage conductance current, from one conductor to the other is $C\,dv/dt$, where dv/dt is the change in voltage with respect to time. Another example of displacement current is the capacitor in an electronic circuit. The completed current path is by means of a displacement current between the capacitor plates.

2·3 Traveling waves on a lossless transmission line

In most microwave transmission lines the losses are extremely small, and in most practical cases the line can be considered lossless. If the loss parameters R and G are zero, then the equivalent circuit of Fig. 2·3 is reduced to series sections of inductance and shunt sections of capacitance.

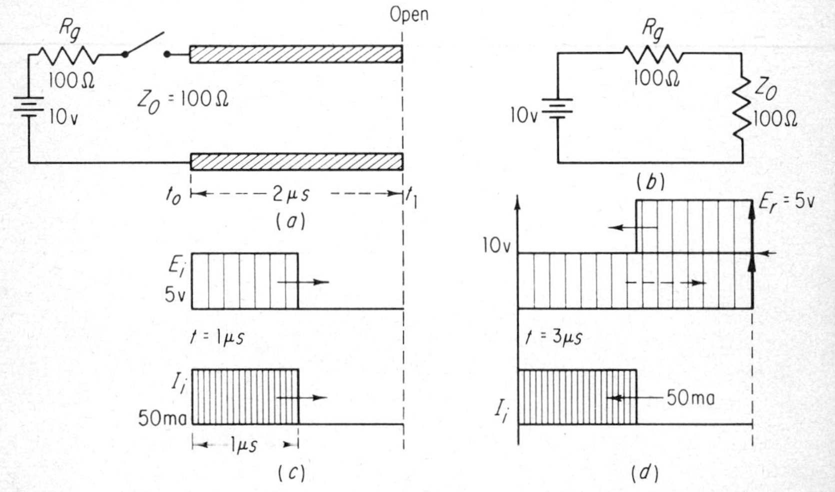

Fig. 2·4 Traveling waves on a transmission line terminated in an open circuit.

When the switch is closed (Fig. 2·2), the battery "sees" the first infinitesimal section of inductance and capacitance. The voltage cannot appear instantaneously at all points on the line because the series inductance of the line, which is associated with the magnetic flux, opposes a change in current, and the shunt capacitance, which is associated with the electric charge, opposes a change in voltage. A finite time is required to charge the capacitance of each small section of the line.

The voltage wave can progress down the line only as fast as the current can carry the necessary charge to the wavefront to produce the change in voltage. Also, the current can travel down the line only as fast as the voltage that is required, at the wavefront, to force the current through each short section of line inductance. *Therefore, the voltage and current must travel together along the line.* In Fig. 2·4 the current wave and voltage wave are shown in step at the wavefront with a current flowing away from the battery in the top conductor and a current flowing toward the battery in the lower conductor. The current path is completed by the displacement current between the conductors as explained in Sec. 2·2.

Define τ as the time required for waves of voltage and current to travel a unit length along the line. The velocity of travel is $v = 1/\tau$.

During the time interval τ, $I\tau$ is the charge that flows into the line, and CE is the charge accumulated on the line. Therefore,

$$I\tau = CE \tag{2·1}$$

LI is the increase in flux linkages encircling the conductor. $E\tau$ is the rate at which flux linkages are being produced at the wavefront. Therefore,

$$E\tau = LI \tag{2·2}$$

Multiply Eq. (2·1) by Eq. (2·2) and solve for τ.

$$\tau = \sqrt{LC} \qquad \text{and} \qquad v = \frac{1}{\sqrt{LC}}$$

Divide Eq. (2·2) by Eq. (2·1) and multiply by E/I.

$$\frac{E}{I} = \sqrt{\frac{L}{C}} \qquad \text{or} \qquad Z_0 = \sqrt{\frac{L}{C}} \tag{2·3}$$

The ratio E/I is called the *characteristic impedance* Z_0 of the transmission line. Z_0, *the characteristic impedance, is the ratio of the voltage to the current traveling in a particular direction.* This means that Z_0 is the *ratio of voltage to current traveling together in one direction or the other on the line.* By this definition it can be seen that any change in voltage and current on a transmission line has a constant of proportionality which is the characteristic impedance Z_0 of the line. This also indicates that regardless of the initial current and voltage conditions, if a wave of voltage and current is sent down the line, the voltage and current waves still travel at the velocity determined by the line parameters, and the ratio of the voltage to current is still Z_0 because the transmission line is a linear device.

Waves launched on a uniform two-conductor line that is infinitely long are assumed to be propagated to infinity. If the transmission line is terminated in a pure resistance equal to the characteristic impedance, a steady state would exist, since this value of resistance would exactly support the voltage and current of the traveling wave. This is, in effect, a transmission line of infinite length. The importance of this particular value of termination will become more apparent in subsequent discussions.

2·4 Reflections at an open-line termination

Reflections are caused by discontinuities in the transmission line structure. These discontinuities may be regarded as changes in the characteristic impedance of the transmission line or guiding structure. The circuit in Fig. 2·4 with a charging resistor R_g equal to the characteristic impedance Z_0 simplifies the initial discussion of the behavior of the open-circuit termination. When the switch is closed, the line appears as shown in the equivalent circuit of Fig. 2·4*b* because the battery sees Z_0 in series with the charging resistor R_g. Waves of voltage and current start from the source toward the open termination. These waves are designated *incident* waves and are labeled E_i (incident voltage wave) and I_i (incident current wave).

Assume that the velocity of propagation is such that a voltage wave and current wave can travel the length of the line in 2 μsec, as indicated in Fig. 2·4. At t_0, 5 volts is dropped across R_g and a 5-volt wave of voltage accompanied by a 50-ma current wave starts down the transmission line. These wavefronts travel *together* down the line at a velocity equal to $1/\sqrt{LC}$ and charge each element of line capacitance to 5 volts. No capacitance remains to be charged when the wave reaches the end of the line after 2 μsec of travel. The collapsing magnetic field, associated with the inductance element at the end of the line, tries to maintain the original current flow into the capacitance. This additional current flow charges the end capacitance to 10 volts ($2E_i$), and the current at the end of the line has decreased to zero. This action continues from section to section back toward the battery, as shown in Fig. 2·4*d*. This corresponds to a *reflected* wave of voltage E_r equal to the incident wave E_i, and a *reflected* wave of current I_r equal to the incident current I_i traveling back toward the battery. The reflected waves travel at the same velocity ($v = 1/\sqrt{LC}$) as the incident waves, and the ratio of E_r to I_r is equal to the characteristic impedance Z_0. The entire line is charged to 10 volts when the reflected wave arrives at the battery after a total travel time of 4 μsec. The steady-state condition is reached; a voltage of 10 volts exists between the conductors, and no current flows in the conductors.

Note: At an open-circuit termination, the *reflected voltage* E_r is *equal* in magnitude and is *in phase* with the incident voltage E_i. The reflected current I_r is *equal to* and 180° *out of phase* with the incident current I_i.

2·5 Reflections at a short-circuit termination

Another simple form of reflection occurs when the transmission line of Fig. 2·4 is terminated with a transverse sheet of metal which is assumed to be a perfect conductor (short circuit). At the instant the incident waves of voltage and current reach the termination, the boundary conditions of the perfect conductor indicate that there can be no tangential component of electric field, i.e., there can be no voltage drop across zero resistance.

The resultant zero voltage at the surface can be accounted for if it is assumed that the reflecting surface merely reverses the direction of the electric lines of force, thereby giving rise to a reflected wave which cancels the incident wave at the surface of the conducting plate or short circuit.

The behavior at the end of the line can also be described in terms of the inductance and capacitance of the line as considered for the open circuit. When the 5-volt, 50-ma incident waves reach the short circuit, a reflected voltage of 5 volts with reversed polarity starts back toward the battery. The line capacitance at the wavefront is discharged by a net current of 50 ma, and the voltage between the conductors drops from 5 volts to zero. This 50 ma of current adds to the 50-ma incident current, and the total line current flow is 100 ma at the wavefront. When the reflected wave reaches the battery, the voltage between the conductors is zero and the current flow is 100 ma. This is the steady state since the 100 ma of current is exactly the current drawn by the 100-ohm resistor R_g across a 10-volt battery. At a short circuit E_r is *equal to* E_i but has *opposite polarity*, while I_r is *equal to* and *in phase* with I_i.

2·6 Reflections from resistive terminations

If the transmission line is terminated in a resistance equal to Z_0, the line behaves as though it has infinite length. Since there are no reflected waves on the line, it is said to be match terminated.

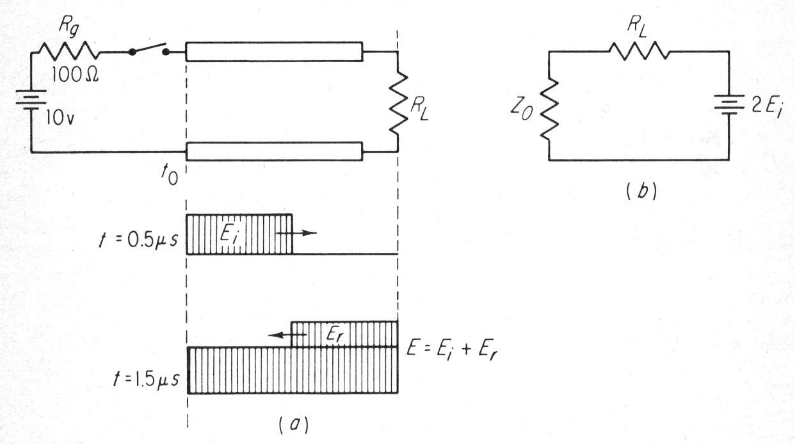

Fig. 2·5 Traveling waves on a transmission line terminated in a resistance greater than Z_0.

If the line is terminated in any resistance other than Z_0, there will be reflected waves from the termination. The amplitude and polarity of these reflected waves will be determined by the particular value of load and the value of characteristic impedance of the line.

Consider the line in Fig. 2·5 which is terminated with a pure resistance R_L somewhat greater than Z_0. After 1 μsec of time, E_i and I_i reach the resistive load termination R_L. The voltage cannot jump to $2E_i$ and it cannot go to zero as in the open-circuit or short-circuit considerations. The voltage at the load E_L can be determined by calculating E_L for an open circuit and then using Thévenin's theorem to find the value of E_L; then calculate the value of reflected voltage E_r.

Assume that at the instant the voltage on the open line jumps to $2E_i$, the load resistance Z_L is connected to the line. The equivalent circuit is shown in Fig. 2·5*b*. Since Thévenin's theorem for this circuit states that the current through Z_L is the ratio of the open-circuit voltage and the sum of Z_L and Z_0 of the line,

$$I_L = \frac{2E_i}{R_L + Z_0}$$

$$E_L = \frac{(R_L)2E_i}{R_L + Z_0} = 2E_i \frac{R_L}{R_L + Z_0} \tag{2·4}$$

E_L is the *total* voltage at the end of the line. The reflected voltage is obtained by subtracting the incident voltage E_i from this value.

$$E_r = 2E_i \frac{R_L}{R_L + Z_0} - E_i = E_i \frac{R_L - Z_0}{R_L + Z_0} \tag{2·5}$$

The ratio of the reflected voltage E_r to incident voltage E_i is called the *reflection coefficient* Γ (gamma).

$$\Gamma = \frac{E_r}{E_i} = \frac{R_L - Z_0}{R_L + Z_0} \tag{2·6}$$

The reflection coefficient is calculated from the above equation *provided the load is a pure resistance*. The reflection coefficient for sinusoidal waves and complex impedance is considered in a subsequent chapter.

The initial voltage and current wave for the lines discussed depends upon the value of R_g and Z_0 as shown by the equivalent circuit in Fig. 2·5*b*. In each of the previous discussions the steady-state condition was reached after the reflection from the load reached the battery. If the value of R_g had not been equal to the characteristic impedance, there would have been a re-reflection from the source end of the line. The re-reflected wave can be determined in the same way that the reflection from the load was determined. The resistance R_g is now the load for the reflected wave, and the reflection coefficient is calculated from the values of R_g and Z_0. Therefore, when there is a mismatched load and a mismatched source, there may be several round-trip reflections before the steady state is reached. As an example, 90 volts is applied to the line where $R_g = 2Z_0$. The voltage across R_g is 60 volts, and a 30-volt wave of voltage travels down the line to the open circuit illustrated in Fig. 2·4. At the end of the line the 30 volts is reflected in phase and travels back toward the source. When the 30-volt wave arrives at R_g, there is 60 volts on the line. The 30-volt reflected wave encounters the mismatch of $2Z_0$, and part of the 30-volt wave is re-reflected. From the reflection coefficient equation,

$$\Gamma = \frac{Z_L - Z_0}{Z_L + Z_0} = \frac{2Z_0 - Z_0}{2Z_0 + Z_0} = \frac{1}{3}$$

Therefore a new incident wave of $(\frac{1}{3})30 = 10$ volts travels from the source toward the open end of the transmission line. This 10-volt wave is reflected from the open circuit, and upon arriving back at the input, one-third of this voltage is re-reflected. This process continues until there is 90 volts on the transmission line.

Conclusions: Any change on a transmission line takes the form of voltage and current waves traveling in one direction or the other or of pairs of voltage and current waves traveling in opposite directions on the transmission line. The characteristic impedance is the ratio of the voltage wave to the current wave traveling together in a particular direction. Z_0 is determined by the physical characteristics of the line which in turn determine $\sqrt{L/C}$. The waves of voltage and current travel at a velocity of $1/\sqrt{LC}$.

If a transmission line is terminated in a pure resistance Z_L *greater than* the characteristic impedance Z_0, the reflected voltage wave E_r is *in phase* with the incident voltage wave E_i, and the reflected current wave I_r is 180° *out of phase* with the incident current wave I_i.

If a transmission line is terminated in a pure resistance Z_L *less than* the characteristic impedance Z_0, the reflected voltage wave E_r is 180° *out of phase* with the incident voltage wave E_i, and the reflected current wave is *in phase* with the incident current wave I_i.

If the source impedance (R_g in this discussion) is not equal to the characteristic impedance Z_0 of the line, there will be reflections from the source end of the line back toward the load.

PROBLEMS

2·1 An artificial transmission line can be formed using lumped L and C. Calculate the delay of an artificial line composed of eight sections of inductance $L = 4$ mh per section, and capacitance $C = 40$ $\mu\mu$f per section.

2·2 Draw a diagram illustrating the sinusoidal electric and magnetic fields on a two-wire line and show the complete loops of current flow including the displacement current. Refer to the following diagram when working Probs. 2·3 to 2·10.

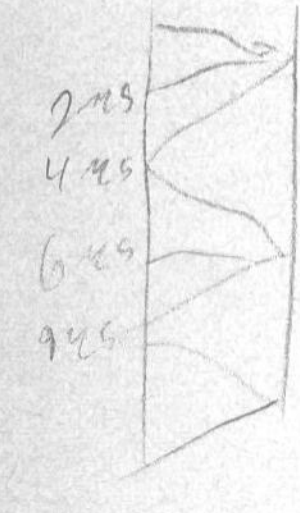

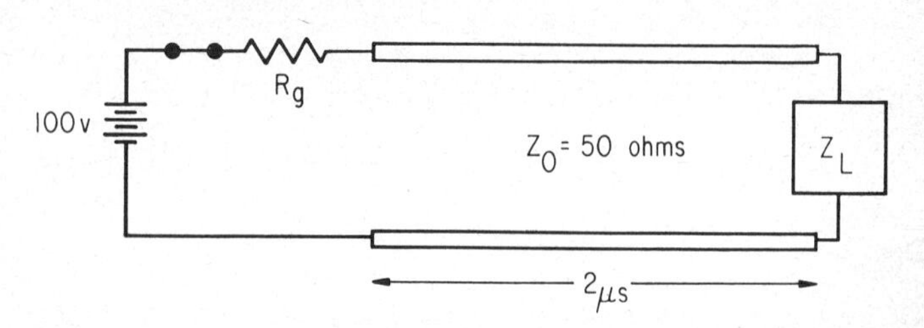

2·3 In the above diagram, R_g is 50 ohms and Z_L is an open circuit. Calculate the total voltage at the points indicated by the following time intervals: (*a*) 1 μsec, (*b*) 2 μsec, (*c*) 3 μsec, (*d*) 4 μsec, and (*e*) 6 μsec.

2·4 Repeat Prob. 2·3 if Z_L is a short circuit.

2·5 Repeat the calculations in Prob. 2·3 if R_g is 450 ohms and Z_L is an open circuit.

2·6 Repeat the calculations in Prob. 2·3 if R_g is 450 ohms and Z_L is a short circuit.

2·7 Repeat the calculations in Prob. 2·3 if R_g and Z_L are 450 ohms.

2·8 Repeat the calculations in Prob. 2·3 if R_g is 25 ohms and Z_L is 50 ohms.

2·9 If R_g is 50 ohms and the reflected voltage is 20 volts, (*a*) What is the value of load resistance? (*b*) What is the value of load resistance if the reflected voltage is −20 volts?

2·10 If R_g is 50 ohms and Z_L is 150 ohms, calculate (*a*) reflected voltage, (*b*) incident power, and (*c*) reflected power. (*d*) Express the ratio of reflected to incident power in terms of the voltage reflection coefficient. (*e*) If the ratio calculated in *d* is multiplied by 100, what does the resulting value indicate?

CHAPTER

3

TRANSMISSION LINES AT MICROWAVE FREQUENCIES

Introduction. A variety of phenomena may take place when an alternating electromotive force is connected to the transmission line. The concepts of resonant lines, traveling waves, standing waves, propagation constant, and impedance are presented for dissipationless transmission lines.

The useful methods of analysis gained from this study of transmission lines propagating the *transverse electromagnetic* (TEM) wave can also be applied to problems involving such guiding structures as waveguides. In the TEM wave, both the electric and magnetic fields are transverse (at right angles) to the direction of wave travel. There are no electric or magnetic components in the direction of wave travel for lossless lines. TEM waves exist in free space and two-conductor transmission lines such as two-wire, parallel-plate, microstrip, and coaxial lines and resonators.

All microwave phenomena can ultimately be expressed in terms of frequency, wavelength, and power. This chapter is devoted to the presentation of analytical expressions involving these quantities, although power is not used directly in the analysis. Instead, the original electric and magnetic field problems are reduced to an impedance problem in which discontinuities are expressed in terms of impedances rather than field intensities. The generalized impedance concept (the complex ratio of voltage to current) forms the foundation of transmission line theory and will be considered in detail.

3·1 Transmission of power at high frequencies

If the frequency is very high, the initial lines of force sent out by the source will not travel far before the emf at the source reverses direction. Figure 3·1*b* shows that this second group of lines of force is exactly like the first group except that the lines are directed in the opposite direction. Alternate groups are observed to be identical. Figure 3·1*a* shows the lines of electric and magnetic force on the transmission line. Figure 3·1*b* shows the space relationship between the electric and magnetic fields. Most of the total power flow takes place in the immediate vicinity of the wire, as shown in the diagram

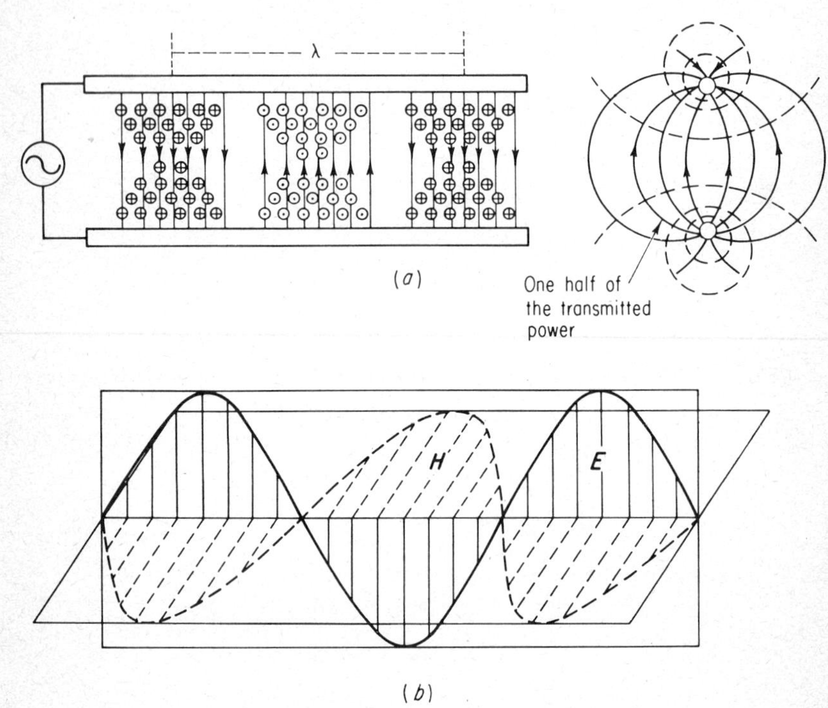

Fig. 3·1 (a) Electric and magnetic lines of force in the longitudinal and transverse sections of an infinitely long transmission line. (b) Space relationships of electric and magnetic vectors.

by the circle enclosing one-half of the transmitted power. The remaining half of the power extends to infinity.

3·2 Traveling waves

The voltage is always accompanied by a current wave of similar shape on a uniform lossless transmission line, and, regardless of their shape, these waves will be propagated without any change in magnitude or shape.

Figure 3·2 shows a sinusoidal traveling wave of voltage on a transmission line. This line has a *physical length* l measured in meters. It also has *electrical length* which is measured in wavelengths λ. The *wavelength* λ is defined as the distance between successive points of the same electrical phase in a wave. It depends on the frequency f of alternation, the velocity of propagation, and the nature of the medium through which the wave travels. For free space, the velocity is substantially 300,000,000 m per sec (186,000 miles per sec). If the medium has a relative dielectric constant, the wavelength will be reduced by the factor $1/\sqrt{\epsilon'}$.

$$\lambda = \frac{v}{f} = \frac{1}{\sqrt{\epsilon'}}\frac{v}{f} \tag{3·1}$$

Wavelength is also defined as the distance in which the phase changes by 2π rad (1 rad $= 180°/\pi = 57.3°$).

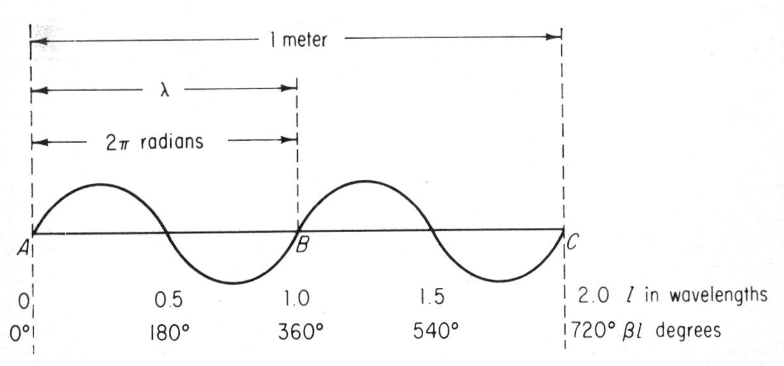

Fig. 3·2 Traveling wave of voltage on a lossless transmission line.

3·3 Propagation constant and characteristic impedance

The incident waves of voltage and current *decrease* in magnitude and vary in phase as one goes toward the receiving end of the transmission line which has losses. The *propagation constant* γ is a measure of the phase shift and attenuation along the line.

$$\gamma = \alpha + j\beta = \sqrt{(R + j\omega L)(G + j\omega C)} \tag{3·2}$$

α is the attenuation per unit length of line and is called the *attenuation constant*. The attenuation constant may be expressed in decibels per unit length or *nepers* per unit length, where 1 neper equals 8.686 db.

β is the phase shift per unit length, called the *phase constant*, and is measured in radians per unit length.

ω is $2\pi f$ and is called the *angular frequency*, in radians per second.

$(R + j\omega L)$ is the complex series impedance per unit length of transmission line, as shown in Fig. 2·3.

$(G + j\omega C)$ is the complex shunt admittance per unit length of line as shown in Fig. 2·3.

Since there is attenuation associated with the transmission line, the actual characteristic impedance of a transmission line (Fig. 2·2) is given by

$$Z_0 = \sqrt{\frac{R + j\omega L}{G + j\omega C}} \tag{3·3}$$

It is evident that the characteristic impedance Z_0 of a lossy line is a pure resistance only if $RC = GL$. As explained in Chap. 2, most microwave lines have very small losses and Z_0 can be considered to be a pure resistance $Z_0 = \sqrt{L/C}$. Also, the attenuation constant is zero, and the propagation constant becomes $\gamma = j\omega\sqrt{LC}$.

The characteristic impedance of the two-wire transmission line ranges from 100 to 300 ohms according to

$$Z_0 = 276 \log \frac{2b}{a} \tag{3·4}$$

where a is the diameter of the conductors and b is the distance between centers of conductors.

The *phase constant* is defined as the rate of change of phase with distance for fixed values of time and is given by

$$\beta = \omega\sqrt{LC} = \frac{2\pi f}{v} = \frac{2\pi}{\lambda} \tag{3·5}$$

Thus β determines the wavelength and for this reason is sometimes called the *wavelength constant*. At the end of each unit length of line, the voltage and current will lag the voltage and current at the beginning of that unit length by an angle of β rad. In other words, in traveling a distance l, in either direction on the line, the phase is retarded by βl rad. As an example, at the point B in Fig. 3·2,

$$\beta l = \frac{2\pi l}{\lambda} = \frac{2\pi(0.5)}{0.5} = 360°$$

At C,

$$\beta l = \frac{2\pi l}{\lambda} = \frac{2\pi(1)}{0.5} = 720°$$

The *phase velocity* is the *velocity of a point* denoting the location of a definite phase of the periodic disturbance *in space*. Phase velocity in meters per second is given by

$$v_p = f\lambda$$

where f is the frequency in cycles per second and λ is the wavelength in meters.

The time delay t_d of a transmission line is the time it takes *a point* to travel the length of the line. If the phase velocity is independent of frequency, this time is the time it takes a pulse or signal to travel the length of the line.

$$t_d = \frac{l}{v_p} = \frac{\beta l}{\omega} = \frac{lT}{\lambda} \tag{3·6}$$

The time delay in terms of the period T is equal to the number of wavelengths in the line.

3·4 Standing waves on the lossless transmission line

The basic theory of reflections as set forth in Chap. 2 will be applied to the case of sinusoidal waves of voltage and current, *in phase*, traveling along a transmission line which is terminated in an impedance different than the characteristic impedance of the line. *Reflected waves of voltage and current exist owing to the impedance mismatch.* The instantaneous total voltage or total current at any point on the line is the sum of the incident and reflected voltages or currents at that point.

The behavior of the sinusoidal voltage wave on a line terminated in a short circuit is considered for simplicity of explanation. The diagrams in Fig. 3·3 represent the traveling wave of voltage at different time intervals. The dotted sine waves to the right of the short circuit in each diagram indicate the position and distance that the wave would have traveled in the absence of the short circuit. With the short circuit placed at X, the wave travels the same distance back toward the generator. In order to satisfy the boundary conditions, the voltage at the short circuit must be zero at all times. This is accomplished by a reflected voltage wave which is equal in magnitude and reversed in polarity (shown by the superimposed reflected wave and the resultant total voltage on the line, as indicated by the dark line). In Fig. 3·3*a*, a time t_0 is selected after the wave has been reflected from the short circuit. The total voltage is $2E_i$ at a distance of one-quarter of a wavelength back toward the generator, and the total voltage is zero at a distance of one-half wavelength from the short. This point is labeled (*b*).

Figure 3·3*b* indicates a time t_1 when the wave has traveled (at the speed of light) one-quarter of a wavelength to the right. At this instant the reflected and incident waves cancel at all points on the line, resulting in zero voltage along the line.

At time t_2 the wave has traveled to the right a total of one-half wavelength. The incident and reflected waves add to $2E_i$ in a negative direction at a distance of one-quarter wavelength from the short. Again it is noted that the zero voltage point is at b and also at a, which is one-half wavelength from b.

Figure 3·3*d* shows the instantaneous voltage on the line as a function of position at successive intervals of time starting at t_0 which is labeled (1). As the wave moves to the right, the instantaneous voltages labeled (2) and (3) indicate a decrease in total voltage as the wave progresses to the right. When the wave has reached time t_1, which is one-quarter of a wavelength as shown in Fig. 3·3*b*, the voltage is zero along the line as indicated (4). As the wave travels from t_1 to t_2, a distance of one-quarter of a wavelength, the instantaneous voltages reach a total negative value of $2E_i$ in successive time intervals indicated by (5), (6), and (7). The corresponding current magnitude and the relative positions of maxima and minima points are indicated in Fig. 3·3*e*.

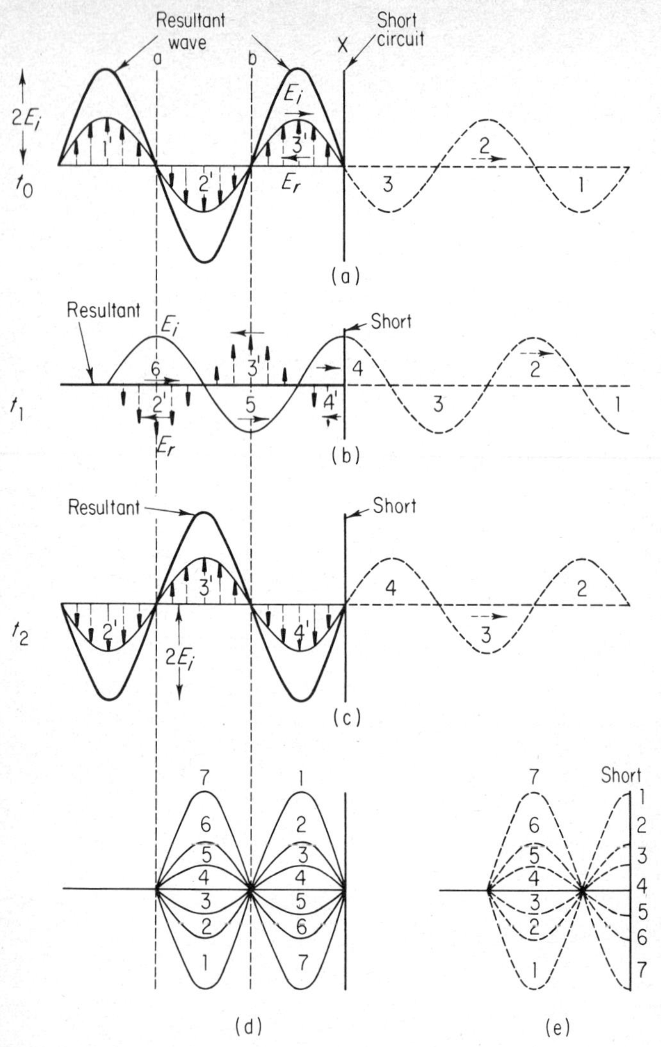

Fig. 3·3 Generation of standing waves on a shorted transmission line. Dotted lines to the right of the short circuit represent the distance the wave would have traveled in absence of the short. Dotted vectors represent the *reflected wave*. The heavy solid line represents the vector sum of the incident and reflected waves. (*d*) and (*e*) represent instantaneous voltages and currents at different intervals of time.

The total voltage pattern is called a *standing wave*. Standing waves exist as the result of two waves of the *same frequency* traveling in *opposite directions* on a transmission line.

If the signal source is assumed to have an internal impedance equal to Z_0, the steady-state conditions are reached when the reflected waves arrive at the input terminals. If the source is not matched, the re-reflections will result in an overall incident wave and an overall reflected wave with the same characteristics as previously discussed.

Figure 3·3*d* shows that the total voltage at any instant has a sine-wave distribution along the line with zero voltage at the short and zero points at half-wave intervals from the short circuit. The points of zero voltage are called voltage *nodes*, and the points of maximum voltage halfway between these nodes are called *antinodes*. By superimposing the current standing-wave diagram on the voltage standing-wave diagram, it can be seen that the current and voltage nodes are one-quarter of a wavelength apart and also that the antinode points are one-quarter of a wavelength apart.

At a distance of one-quarter of a wavelength from the short, the voltage is found to be $2E_i$, which is equivalent to an *open* circuit. Therefore, this same distribution would be obtained if an open circuit were placed $\lambda/4$ from the short. In this case, it is found that the zero points (nodes) are $\lambda/4$ from the open end of the line.

3·5 Standing-wave ratio

The *voltage-standing-wave ratio* is denoted by the symbol ρ and is defined as the ratio of the maximum voltage to the minimum voltage on a transmission line. This ratio is most frequently referred to as VSWR. The voltage-standing-wave ratio referred to in this text is as defined above. If the standing-wave ratio is measured in terms of the square of the voltage, the ratio is called the *power-standing-wave ratio* (PSWR) and is designated ρ^2.

$$\text{VSWR} = \rho = \frac{E_{\text{max}}}{E_{\text{min}}} = \frac{E_i + E_r}{E_i - E_r} = \frac{1 + |\Gamma|}{1 - |\Gamma|} \tag{3·7}$$

Also
$$\text{VSWR} = \frac{I_{\text{max}}}{I_{\text{min}}}$$

If the equation for VSWR is solved for Γ, it is found that

$$|\Gamma| = \frac{\rho - 1}{\rho + 1} \tag{3·8}$$

If the transmission line is terminated in a short or open circuit, the reflected voltage E_r is equal to the incident voltage E_i. From the above equation, the reflection coefficient is 1.0, and the voltage-standing-wave ratio (VSWR) is infinite. If a matched termination is connected to the line, the reflected wave is zero, the reflection coefficient is zero, and the VSWR is 1.0.

3·6 Transmission line impedance

The *input impedance* Z_i of a transmission line is defined as the ratio of *total voltage* to *total current* at a point on the line looking toward the load.

$$Z_i = \frac{E_{\text{total}}}{I_{\text{total}}} \tag{3·9}$$

The incident and reflected voltage waves may be shown graphically as indicated in Fig. 3·4*b*. At the short circuit, E_r is equal to E_i and is reversed in polarity as shown. If the incident wave is rotated counterclockwise and the reflected wave is rotated clockwise as shown in the diagram, the resultant amplitude of $2E_i$ is the first $E_{\max}$ on the standing-wave diagram. Following this procedure of rotating the two waves, the complete standing-wave diagram can be plotted.

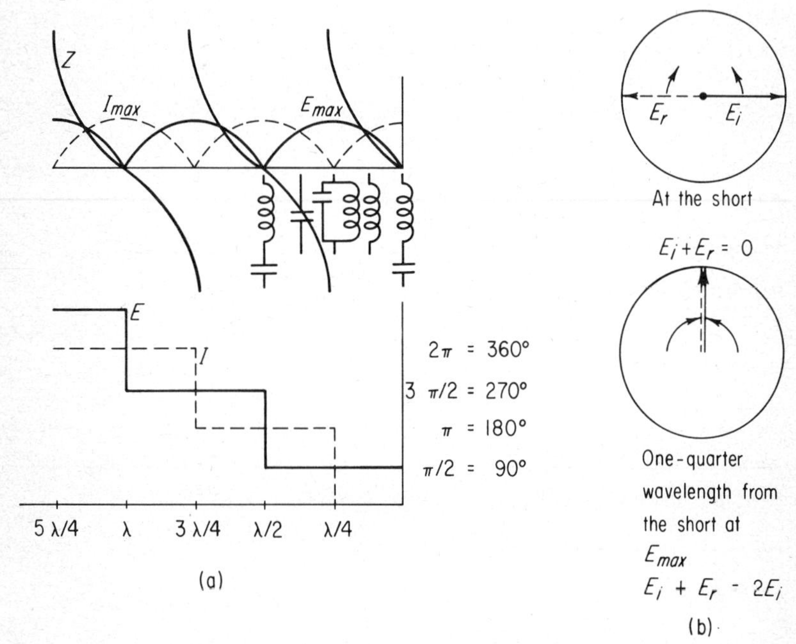

Fig. 3·4 Variations in impedance and the magnitude and phase of the voltage and current on a transmission line terminated in a short circuit.

During the first quarter wave of travel from the short toward the generator, the voltage leads the current by 90° and the transmission line input impedance is *inductive*, as shown on the diagram in Fig. 3·4*a*. The impedance is zero at the short circuit and increases, according to the tangent function, to positive infinity. At this point the impedance changes to a minus infinity value since here the current begins to lead the voltage, and the circuit becomes *capacitive*. It is noted that the voltage and current are always $\pm 90°$ apart when the incident and reflected voltages are equal. This condition exists when the transmission line is terminated in a pure open, short, inductance, or capacitance.

For the ideal dissipationless case, the input impedance of a short-circuited line is infinite for a line one-quarter of a wavelength long and is zero for a line one-half of a wavelength long. This is somewhat analogous to series and parallel resonance in a dissipationless LC circuit, and the corresponding series and parallel points are shown on the diagram of Fig. 3·4*a*. The actual impedance at a short circuit is a low pure resistance. Also, the actual imput impedance of a low loss quarter-wave shorted line is not infinite but is a pure resistance in the range of 400 kilohms.

The equation for the input impedance of a transmission line is given by

$$Z_i = Z_0 \frac{Z_L + jZ_0 \tan \beta l}{Z_0 + jZ_L \tan \beta l} = Z_0 \frac{Z_L \cos \frac{2\pi l}{\lambda} + jZ_0 \sin \frac{2\pi l}{\lambda}}{Z_0 \cos \frac{2\pi l}{\lambda} + jZ_L \sin \frac{2\pi l}{\lambda}} \tag{3·10}$$

3·7 Voltage and current relationships on the lossless transmission line

The properties of the voltage and current waves on the lossless transmission line are shown in Fig. 3·5 for the various terminating impedances. The angle of the reflection coefficient is denoted by ψ and represents the angle between the incident and reflected voltage waves. The angle of the reflection coefficient

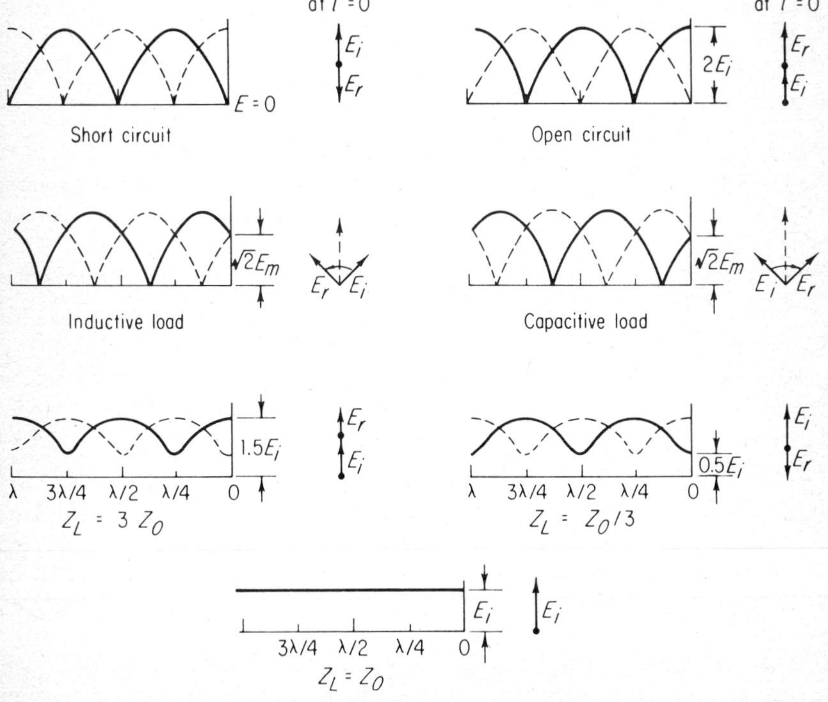

Fig. 3·5 Voltage and current distributions and relations on a lossless transmission line.

is zero when the incident and reflected voltage waves are in phase. Therefore, the angle of the reflection coefficient is always zero at E_{max} points on the standing wave.

Compare the following relationships and properties of voltage and current with the corresponding voltage and current distributions for the lossless line shown in Fig. 3·5.

Open-circuit Termination

1. The incident and reflected voltages are in phase at the open circuit and at one-half wavelength intervals from the open.
2. The angle of the reflection coefficient is zero at the open circuit and at intervals of one-half wavelength from the open.
3. The reflected current is equal in amplitude and 180° out of phase with the incident current at the open and at one-half wavelength intervals from the load.
4. The magnitude of the reflection coefficient is 1.0.
5. The VSWR is infinite.
6. The first voltage minimum along the line is located one-quarter of a wavelength from the open-circuit termination.
7. The first current minimum is located one-half wavelength from the open-circuit termination.

Short-circuit Termination

1. The incident and reflected currents are equal in amplitude and in phase at the short circuit and at one-half wavelength intervals along the line from the short circuit.
2. The VSWR is infinite.
3. The reflection coefficient is 1.0, and the angle of the reflection coefficient is 180°.
4. The first voltage minimum is located one-half wavelength along the line from the short circuit.
5. The first current minimum is located one-quarter wavelength from the short.
6. The input impedance of the line is a function of the line length.

Matched Termination

1. The reflected wave is zero.
2. There are no standing waves.
3. The VSWR is 1.0:1.
4. The reflection coefficient is zero.
5. The input impedance of the line is independent of the length of the line.

Pure Resistance Termination Greater Than Z_0

1. The incident and reflected waves of voltage are in phase at the load and at one-half wavelength intervals from the load.
2. A voltage maximum exists at the load and at one-half wavelength intervals from the load.
3. The angle of the reflection coefficient is zero at the load and at one-half wavelength intervals from the load.
4. The amplitude of the reflected wave, the magnitude of the reflection coefficient, and the VSWR depend upon the values of Z_0 and Z_L.
5. The reflected current is 180° out of phase with the incident current at the load and at one-half wavelength intervals from the load.
6. The wavelength locations of voltage and current maxima and minima follow the same pattern as for the open circuit except for amplitude variations.

Pure Resistance Termination Less Than Z_0

1. The incident and reflected currents are in phase at the load and at one-half wavelength intervals from the load.
2. The incident and reflected voltages are 180° out of phase at the load and at one-half wavelength intervals from the load.
3. A voltage minimum is located at the termination.

Pure Reactance Termination

1. The incident and reflected voltages are out of phase except at E_{max} and E_{min} points where they are either in phase or 180° out of phase, respectively.
2. The VSWR is infinite.
3. The reflection coefficient is 1.0.

3·8 Transmission line resistance

At high frequencies, current does not penetrate deeply into metal. When a conductor is carrying a current which is uniformly distributed throughout the cross section and this current begins to increase in value, a current path near the axis is encircled by more flux than is a current path near the surface. Therefore, there is a larger inductance associated with the internal path, and an easier path for current will lie toward the surface of the conductor. This phenomenon is called *skin effect*. An increase in current in the outer portions of the conductor will cause a greater heat loss than the reduction due to a decrease of current near the center. This redistribution of current results in an increase of total conductor resistance.

The current density decreases exponentially with distance beneath the surface, and the phase changes linearly with distance. At a depth δ, the skin depth, the current density decreases to $1/e$ times its surface value. The current at this depth lags the surface current by 1 rad. If the curvature of the conductor is large compared to the skin depth, the resistance of the conductor may be calculated by assuming that the current density is uniform and confined to a surface of thickness

$$\delta = \sqrt{\frac{\text{resistivity}}{\pi \mu_0 f}}$$

There is some current within the metal at a depth greater than the skin depth.

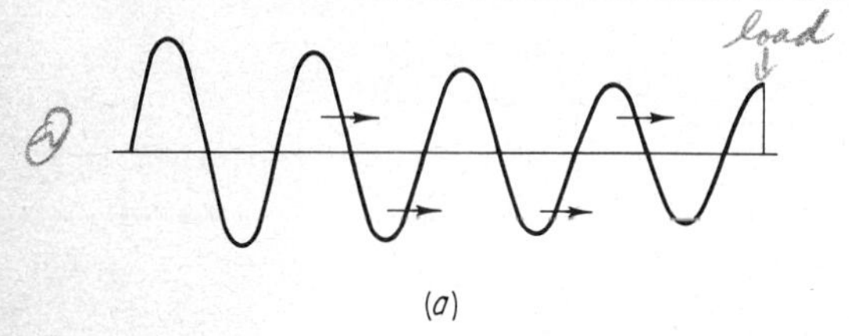

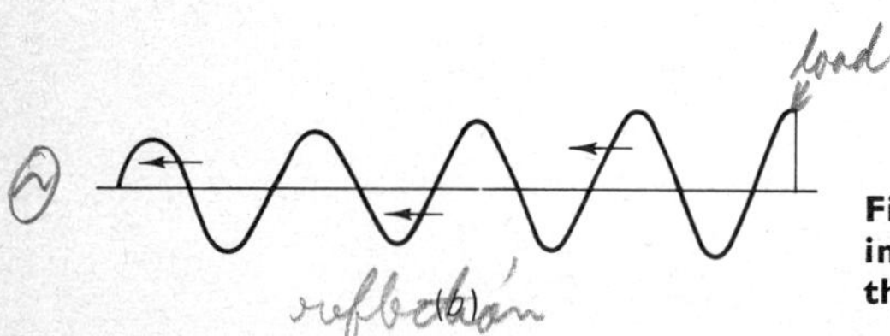

Fig. 3·6 Effects of losses on the incident and reflected voltage and the resulting standing-wave pattern.

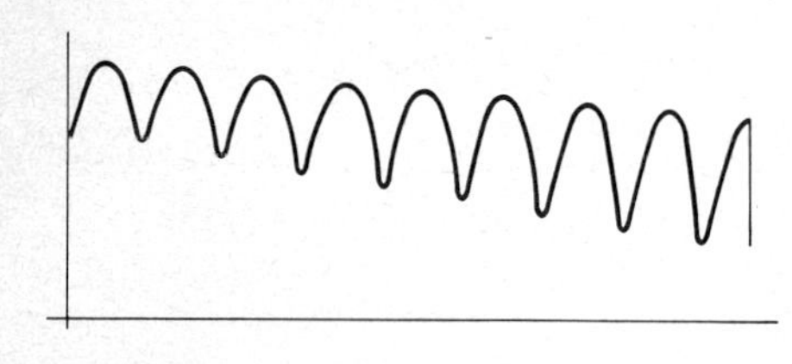

For example, at a depth of three times the skin depth, the current density is 5 per cent of the value at the surface. The total loss, however, is the same as it would be if the current were uniformly distributed over a surface layer of depth equal to the skin depth.

Figure 3·6 represents a traveling wave of voltage on a lossy transmission line and the corresponding reflection from a termination greater than Z_0. The incident and reflected waves are attenuated as shown. The following facts concerning these waves are of special significance.

1. At the load, the ratio of the reflected to the incident wave is a maximum and indicates that the reflection coefficient and VSWR are greatest *at the mismatch.*

2. As the point of observation is moved from the mismatch toward the generator, the VSWR and reflection coefficient decrease. Therefore, the maxima and minima of the standing waves of voltage and current are not constant on the lossy transmission line. A detailed discussion of the properties of the lossy transmission line is given in the chapter dealing with the radial scaled parameters of the Smith chart.

PROBLEMS

3·1 The frequency of the signal applied to a two-wire transmission line is 3 Gc.

a. What is the wavelength if the dielectric in the medium is air?

b. What is the wavelength if the dielectric constant of the medium is 3.6?

3·2 A signal is applied to a quarter-wave section of transmission line which is terminated in a short circuit. Draw a diagram illustrating your answer to each of the following questions.

a. What is the theoretical value of input impedance?

b. What is Z_i if the frequency is lowered and the line length is not changed?

c. What is Z_i of the line if the section of line is shortened but the frequency of the applied wave remains fixed?

d. What is Z_i if the frequency is increased slightly and the line length is not changed?

3·3 Repeat Prob. 3·2 for a quarter-wave section of line terminated in an open circuit.

3·4 What is the characteristic impedance of a two-wire line if the distance between centers of the conductors is 0.8 cm and the diameter of the conductor is 0.15 cm?

3·5 Plot a graph of voltage-standing-wave ratio versus reflection coefficient in steps of 0.05 from zero to 0.2 and in steps of 0.1 from 0.2 to 1.0.

3·6 A transmission line has a characteristic impedance of 50 ohms and is terminated in an open circuit. Calculate the input impedance at the following wavelength intervals and compare your results with the diagrams in Fig. 3·5.

a. One-quarter wavelength from the load

b. One-half wavelength from the load

c. One-eighth wavelength from the load

3·7 Repeat Prob. 3·6 if the load impedance is a short circuit.

3·8 Repeat Prob. 3·6 if the load impedance is equal to the characteristic impedance of the line.

3·9 Explain the relationships of the various transmission line parameters for a lossy transmission line terminated in a mismatched load.

3·10 The resistivity of copper is 1.724×10^{-6} ohms per cm.

a. Calculate the skin effect at frequencies of 100 Mc, 1 Gc, and 3 Gc.

b. Silver has a conductivity of 6.17×10^{7} mhos per m and aluminum has a conductivity of 3.72×10^{7} mhos per m. Repeat part *a*.

3·11 If the incident voltage is 100 volts and the reflected voltage is 50 volts, calculate the reflected power in per cent, the voltage and power reflection coefficients, the voltage- and power-standing-wave ratios, and the transmitted power in per cent.

3·12 The characteristic impedance of a transmission line is 100 ohms and the load impedance is 200 ohms. If the incident voltage is 50 volts, calculate the reflected power in per cent, the VSWR, and the voltage reflection coefficient.

CHAPTER

4

GRAPHICAL REPRESENTATION OF TRANSMISSION LINE CHARACTERISTICS

Introduction. A graphical treatment of the impedance properties of the lossless transmission line is given in this chapter. The presentation consists of a description of incident and reflected waves in terms of complex numbers and is a step-by-step method of approach to the Smith chart.[1] The Smith chart is the most useful of the many graphical aids which have been presented for use in performing transmission line computations.

4·1 Application of complex exponentials

When a voltage is written as a complex exponential, a quantity is used which can be analyzed into real and imaginary parts. The real part is a cosine function, and the imaginary part is a sine function multiplied by $\sqrt{-1}$, written j. The voltage represented as a rotating vector may be written in the exponential form $Ee^{j\theta}$ where $e^{j\theta} = \cos\theta + j\sin\theta$.

In the study of transmission lines, complex numbers are frequently used to represent rms values and phase angles of sinusoidal functions of time.

If the instantaneous reference voltage at the input terminals of a line is represented by the complex number $E_P = (E/\sqrt{2})e^{j0°}$, a point 90° from the input terminals is represented by $(E/\sqrt{2})e^{j90°}$. The wave has the same amplitude but is delayed 90° in phase.

In the following analysis, the complex number notation is used to describe positions on a transmission line at a distance measured from the *receiving end* of the line.

Figure 4·1 is a circuit diagram of a shorted transmission line with a point P located a distance l from the short-circuit load termination. E_{iL} is defined as the incident voltage *at the load*, and E_{rL} is the reflected voltage *at the load.* The incident voltage E_i and the reflected voltage E_r are to be represented at P, measured from the load, in terms of the incident and reflected voltages at the load.

The incident wave is delayed in phase by βl in traveling from P to the short circuit. The incident wave at P can be expressed as

$$E_i = E_{iL}e^{j\beta l} \tag{4·1}$$

The reflected wave is also delayed in phase by βl in traveling from the

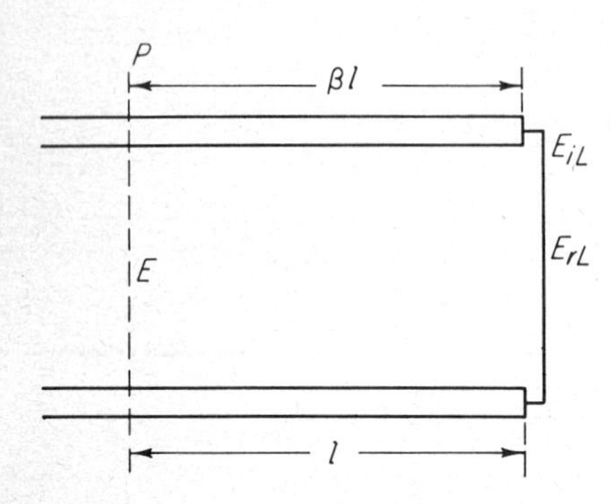

Fig. 4·1 The incident wave is delayed βl degrees in traveling from P to the short circuit while the reflected wave is delayed $-\beta l$ in traveling from the short to the point P.

short to the point P and is expressed as

$$E_r = E_{rL}e^{-j\beta l} \tag{4·2}$$

$$e^{j\beta l} = 1\underline{/\beta l} = \cos \beta l + j \sin \beta l$$

$$e^{-j\beta l} = 1\underline{/-\beta l} = \cos \beta l - j \sin \beta l$$

Since E_{iL} equals E_{rL} at the short circuit, the total voltage at P is

$$E_t = E_{iL}e^{j\beta l} + E_{rL}e^{-j\beta l} = E_{iL}(e^{j\beta l} - e^{-j\beta l})$$

$$E_t = j2E_{iL} \sin \beta l \tag{4·3}$$

This is true since the $\sin \beta l = (e^{j\beta l} - e^{-j\beta l})/2j$. The total current can be derived in the same manner and is found to be

$$I_t = 2\frac{E_{iL}}{Z_0} \cos \beta l \tag{4·4}$$

The input impedance is the ratio of these two complex numbers and is found to be

$$Z_i = \frac{E_t}{I_t} = jZ_0 \tan \beta l \tag{4·5}$$

The voltage and current for the open circuit can be derived in the same manner as for the shorted line, and the results are

$$E_t = 2E_{iL} \cos \beta l \qquad I_t = j2\frac{E_{iL}}{Z_0} \sin \beta l \qquad Z_i = -jZ_0 \cot \beta l \tag{4·6}$$

Compare Eqs. (4·3) to (4·5) with Fig. 3·4, which shows the total voltage and current relationships.

4·2 Graphical representation of propagation characteristics

The circle diagram in Fig. 3·4 served its purpose for indicating the reflection coefficient and the plot of standing waves when the incident and reflected waves were the same amplitude. If the reflected wave is smaller than the incident wave, the resulting elliptical diagram would not be so simple or informative. In Fig. 3·4 the reflected and incident waves are rotated at the same speed but in opposite directions. A new diagram is to be constructed using E_i and I_i as unit vectors. These vectors are made to remain stationary while the vectors representing reflected voltage and current waves are rotated at *twice* the previous rate. The same results will be obtained with this diagram as were obtained with the diagram of Fig. 3·4.

The vector representation is illustrated in Fig. 4·2. The circle diagram represents the voltage on the transmission line terminated in a *resistive load* which has a reflection coefficient of 0.5. The unit vector representing the incident voltage wave remains fixed, and the vector representing the reflected wave is rotated *clockwise* an angle $-2\beta l$ or 45°, as shown at *b*. The reflected wave on the diagram is rotated 90° at *c*, and this point corresponds to 45° travel on the transmission line, as indicated at *C* on the standing-wave diagram. When the reflected wave has traveled 180° on the circle diagram as indicated at *e*, the actual distance traveled on the line is 90°, as indicated at *E* on the standing-wave diagram.

The diagram can be described by the following mathematical presentation. The equation for total voltage on the transmission line is divided by the incident voltage and modified to obtain the following results:

$$\frac{E}{E_i} = 1 + \frac{E_r}{E_i} \tag{4·7}$$

$$= 1 + \frac{E_{rL}e^{-j\beta l}}{E_{iL}e^{j\beta l}} = 1 + |\Gamma|\, e^{-j2\beta l}$$

$$= 1 + |\Gamma| \underline{/-2\beta l}$$

$$= 1 + |\Gamma| \underline{/\psi - 2\beta l} \tag{4·8}$$

$$\phi = \beta l - \psi/2$$

$$-2\phi = \psi - 2\beta l$$

$$E/E_i = 1 + |\Gamma| \underline{/-2\phi}$$

$$\frac{I}{I_i} = 1 - |\Gamma| \underline{/\psi - 2\beta l} \tag{4.9}$$

ψ is the angle of the reflection coefficient, and -2ϕ is the angle measured from the point where ψ equals zero. The angle of the reflection coefficient ψ is zero degrees when the load is a pure resistance and the incident and reflected voltages are *in phase* as shown at *A*.

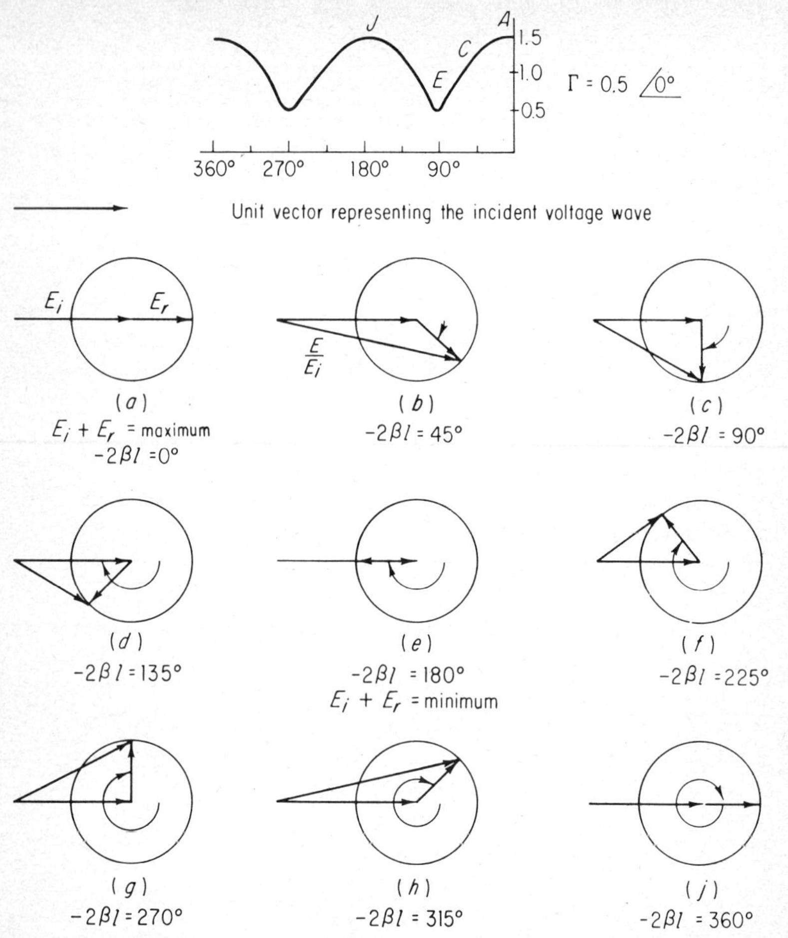

Fig. 4·2 Circle-diagram representation of the voltage on a transmission line terminated in a *resistive* load. The reflection coefficient is 0.5, which corresponds to a VSWR of 3.0:1.

A vector diagram representing Eqs. (4·8) and (4·9) is shown in Fig. 4·3 for a complex load impedance which has a reflection coefficient of 0.5 at an angle of 60°.

The vectors labeled E_i and I_i are the unit vectors representing the incident voltage and current and represent the first term on the right-hand side of Eqs. (4·8) and (4·9). The vector labeled E/E_i represents the total voltage on the line and is the left-hand term of Eq. (4·8). The vectors labeled $\Gamma\underline{/-2\beta l}$ and $-\Gamma\underline{/-2\beta l}$ are the remaining terms of Eqs. (4·8) and (4·9) and represent the reflected voltage and reflected current, respectively. I/I_i is the total current represented by the first term of Eq. (4·9).

The reflected voltage and current waves are drawn 180° out of phase on the circle diagram since all angles on the diagram are *twice* the angles on the transmission line.

Positive angles of reflection coefficient are measured *counterclockwise* from A, and negative angles are measured in a clockwise direction from A.

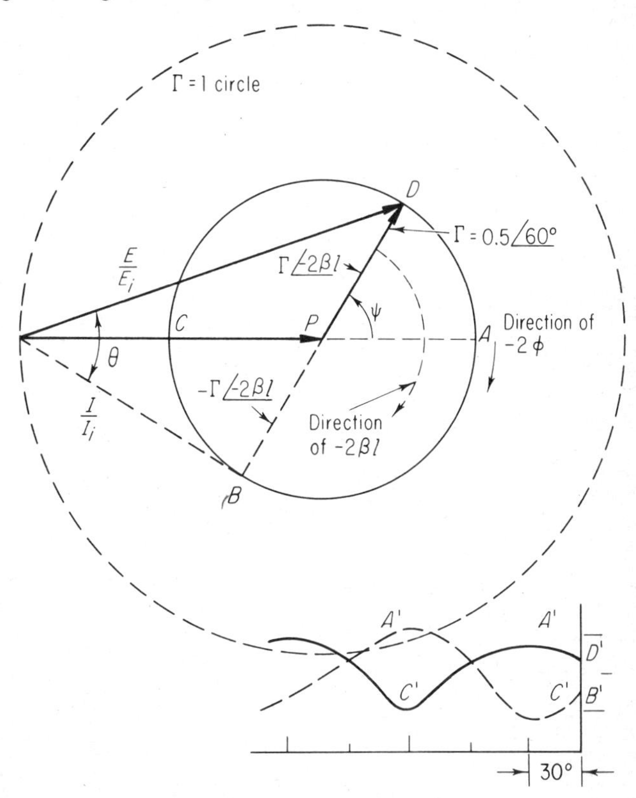

Fig. 4·3 Vector representation of voltage and current on a transmission line terminated in a complex impedance.

For this particular load, the angle of ψ is 60° on the circle diagram and corresponds to an angle of $2\beta l$ measured CCW from A. The corresponding point on the standing-wave diagram A' is found to be 30°, which is the distance of βl on the transmission line. In comparing the distance from A to C on the circle diagram to A' and C' on the standing-wave diagram, it is noted that a 90° distance on the transmission line is represented by 180° rotation of the reflected voltage and current vectors on the circle diagram. The point D on the diagram represents the voltage at the load, and the angle of the reflection coefficient is 60° as shown.

Traveling from D' to A' on the transmission line *toward the generator* corresponds to the distance D to A on the circle diagram. Therefore, clockwise rotation of the diagram represents moving along the transmission line *toward the generator* and *counterclockwise* rotation represents movement along the line *toward the load.*

The $\Gamma = 1$ circle is obtained when the incident and reflected waves are equal in amplitude. This circle is obtained when the line is terminated in a *short circuit*, *open circuit*, or *pure reactance*. The phase angle between E_t and I_t in this case is always $\pm 90°$. A family of reflection coefficient circles can be constructed with their centers at P. The reflection coefficient is zero at P and represents the matched condition. At this point the load impedance is equal to the characteristic impedance of the transmission line, and no standing waves exist on the line.

The angle θ represents the angle between the total current and total voltage on the transmission line. When the reflection coefficient approaches zero, the total voltage and total current approach the in-phase condition, which exists when only incident voltage and incident current appear on the line.

The transmission line characteristics which can be described on this diagram are referred to as the *propagation characteristics*. In addition to the family of reflection coefficient circles, a family of VSWR circles can be drawn on the diagram. The VSWR is *unity* at the center of the circle and *infinite* at the outside rim of the circle. Since 180° on the diagram represents 90° or one-quarter of a wavelength on the transmission line, it is convenient to label the 180° rotation in terms of fractional wavelength. In this case, one-half the distance around the diagram represents 0.25 wavelength. In this way, radii of constant phase in terms of fractional wavelengths (l/λ) are obtained. It was shown in Fig. 3·6 that the VSWR on a lossy line decreases as one travels toward the generator. This indicates that the radial distance on this diagram can also indicate attenuation along the line. A complete discussion of this characteristic will be covered in the discussion of the Smith chart.

4·3 Impedance and admittance coordinates

It is not necessary to understand the mathematical derivation in order to use the circle diagram. The approach to the *propagation grid* of the Smith chart as given in Sec. 4·2 is sufficient. Also, the mathematical proof that the loci of constant resistance and reactance are circles and the mathematical proof of the location of the centers of those circles will not be covered. However, a description of the construction of those circles will be presented.

The impedance or admittance at a point on a uniform lossless transmission line is defined by the amplitude and phase angle of the reflection coefficient. As previously shown, the angle of the reflection coefficient is directly related to the impedance and is indicated on the scale around the rim of the circle diagram.

The chart is constructed on a per unit basis. *The normalized or per unit load impedance is obtained by dividing the load impedance by the characteristic impedance* (Z_L/Z_0). In the case of complex impedances the *per unit* impedance is

$$z = \frac{R}{Z_0} \pm \frac{jX}{Z_0}$$

The construction line for the constant resistance and constant reactance circles is shown in Fig. 4·4 tangent to the circle at the point R/Z_0 = infinity,

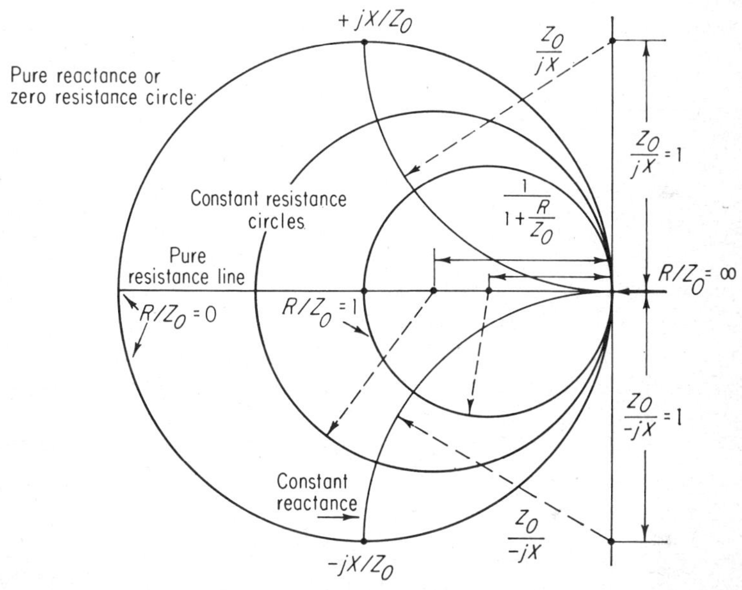

Fig. 4·4 Construction of constant resistance circles R/Z_0 and constant reactance curves $\pm jX/Z_0$.

located at the right-hand side of the diagram. The centers of the positive and negative reactance circles are located at a distance of Z_0/jX and $Z_0/-jX$ measured from the point (R/Z_0 = infinity), as shown on the diagram. The centers of the resistance circles are located on the pure resistance line R/Z_0 with the radii calculated from the formula shown on the diagram. Again it is important to note that the outside circle is the *pure reactance* circle and that the center line on the diagram represents *pure resistance*.

This diagram is intimately related to the diagram of Fig. 4·3. For example, the outside circle of Fig. 4·3 was generated when the incident and reflected voltages were equal in magnitude, a condition that exists when the line is terminated in a short, open, or pure reactance. The outside circle of Fig. 4·4 is also the pure reactance circle (zero resistance). The centers of the diagrams correspond since each represents the point where the line is terminated in its characteristic impedance and there are no reflections on the line. The extreme left in Fig. 4·3 is the zero voltage point, and the same point on Fig. 4·4 is R/Z_0, the zero resistance point which must also be the zero voltage point. The right side of Fig. 4·3 shows that E_i and E_r are in phase and a maximum when E_r equals E_i at an open circuit. This point corresponds to $R/Z_0 =$ infinity on the diagram of Fig. 4·4.

4·4 The Smith chart

By superimposing the propagation grid of Sec. 4·2, and the impedance or admittance coordinate grid of Sec. 4·3, the complete Smith chart of Fig. 4·5 is obtained. Several of the electrical parameters are not shown on the diagram proper because it would be quite confusing if all the properties of each coordinate system were included on one chart. Therefore, the parameters left off the chart proper are given in the form of radial-scaled parameters as shown in Fig. 4·5.

VSWR (*Voltage-Standing-wave Ratio*). The standing-wave ratio has been defined as the ratio of the maximum to minimum voltage or the ratio of maximum to minimum current on a transmission line.

$$\rho = \frac{E_{max}}{E_{min}} \quad \text{or} \quad \frac{I_{max}}{I_{min}} \tag{4·10}$$

From the above equation

$$\rho = \frac{I_{max}}{I_{min}} = \frac{I_i + I_r}{I_{min}} = \frac{E_i/Z_0 + E_r/Z_0}{I_{min}} = \frac{E_{max}/Z_0}{I_{min}} = \frac{Z_{max}}{Z_0} \tag{4·11}$$

from which $Z_{max} = Z_0\rho$ (a pure resistance). Also,

$$\rho = \frac{E_{max}}{E_{min}} = \frac{(I_i + I_r)Z_0}{E_{min}} = \frac{Z_0}{Z_{min}} \tag{4·12}$$

and $Z_{min} = Z_0/\rho$ (a pure resistance).

The normalized value of $Z_{max} = R + j0$ is the VSWR and is read on the right-hand axis of the chart as the normalized value of the constant resistance circles. As an example, a VSWR of 5.0:1 is read at point A, Fig. 4·5. The impedance at this point is 250 ohms for Z_0 of 50 ohms. The corresponding minimum impedance is indicated at B and is 10 ohms. *Note that the reciprocal of the normalized impedance is located diametrically opposite the given impedance value on the same VSWR circle.*

The VSWR circle is a circle with a radius equal to ρ drawn with the center at 1.0 on the chart as indicated in Fig. 4·5. For a normalized load impedance of $1 + j2$ located at C, the VSWR circle is drawn with a radius PC centered at P. The resulting VSWR is read as 5.9 where the VSWR circle crosses the Z_{max} axis at D. A load of $35 + j20$ ohms normalized on a 50-ohm line is $0.7 + j0.4$ located at E. The VSWR circle is drawn with the radius PE, and a VSWR of 1.8 is read where the circle crosses the R/Z_0 axis.

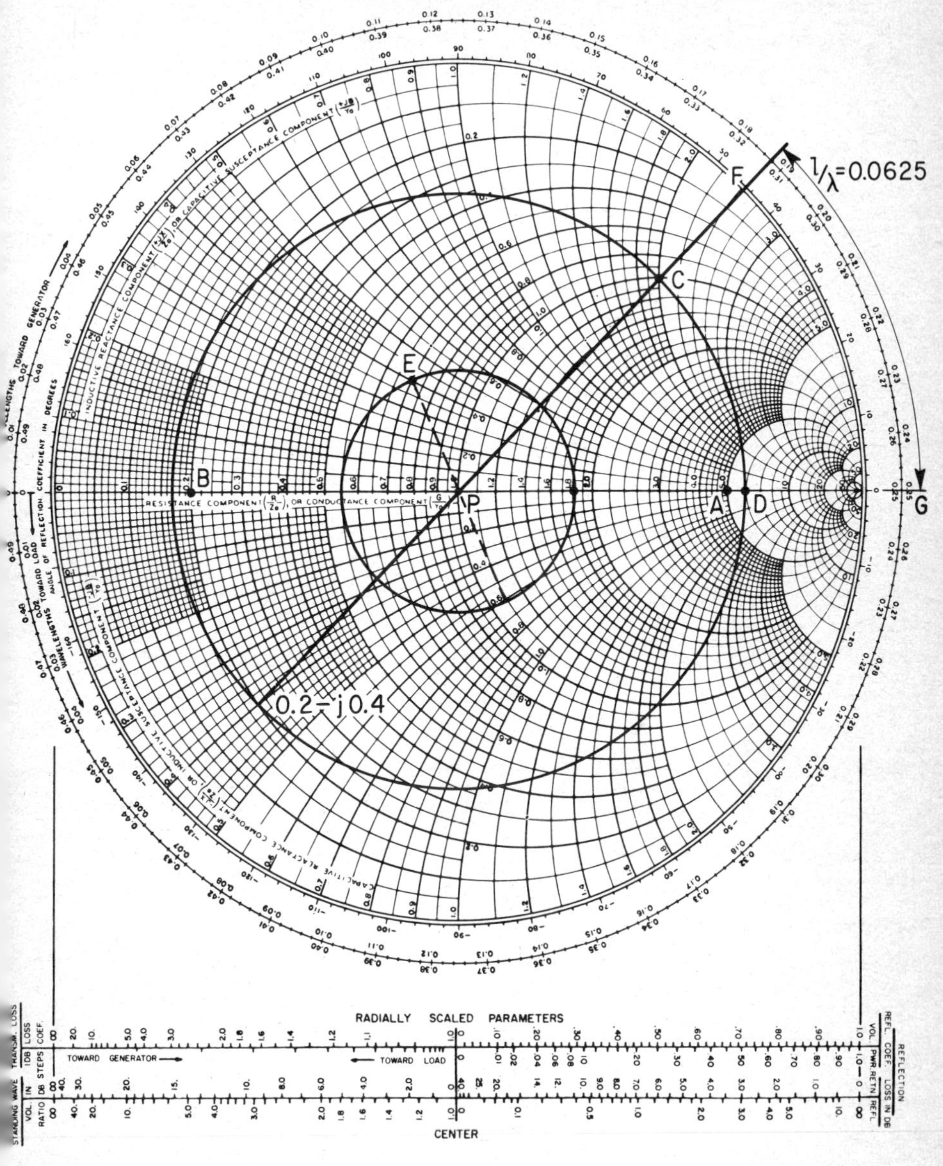

Fig. 4·5 Smith chart. *(From Irving L. Kosow (ed.), "Microwave Theory and Measurements," by the Engineering Staff of the Microwave Division, Hewlett-Packard Company, © 1962, by permission of Prentice-Hall, Inc., Englewood Cliffs, N.J.)*

Impedance and Admittance. Simplified schematic representations of the conventional forms of impedance and admittance are shown in Fig. 4·6. The impedance representation is a *series* one, as indicated at *a*, and the admittance is a *shunt* case, as indicated at *b*. The equation for impedance is

$$Z = R \pm jX \tag{4·13}$$

where R represents the resistive component and X represents the reactive

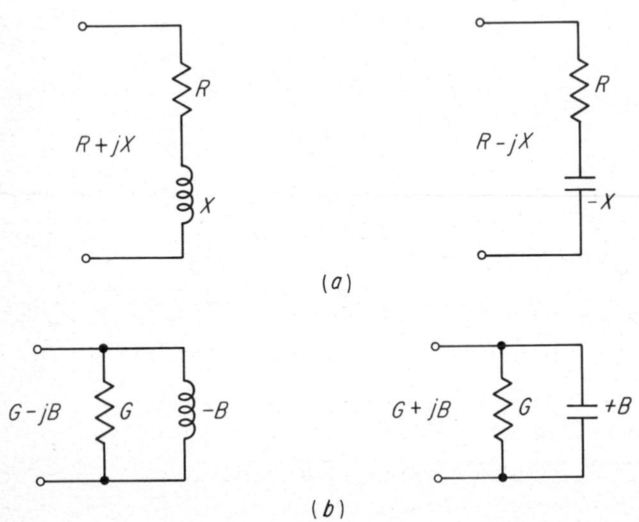

Fig. 4·6 Simplified schematic representation of the conventional forms of impedance and admittance.

component. The term j is the $\sqrt{-1}$ and indicates that R and X are perpendicular vectors and that the magnitude of Z is

$$Z = \sqrt{R^2 + X^2}$$

Admittance is expressed by

$$Y = G \pm jB \tag{4·14}$$

where G represents the conductance and B represents the susceptance. G is real and B is imaginary.

$$G = \frac{R}{R^2 + X^2} \qquad \text{and} \qquad B = \frac{X}{R^2 + X^2}$$

The chart provides a means of converting from impedance to admittance, and this conversion is complete even to the point of specifying resistance R in terms of conductance G and specifying reactance X in terms of susceptance B. This conversion is accomplished on the Smith chart by moving to a point diametrically opposite the known impedance or admittance and reading the *normalized value* of the desired admittance or impedance. It was pointed out in a previous section that the *normalized resistance* value of 5.0 in Fig. 4·5 could be used to find the reciprocal of the impedance by obtaining the value of normalized impedance on the pure resistance line at a distance equal to PA. The located point was determined as 0.2 at B. This principle will be applied to normalized values of complex impedance and admittance since it is known that

$$Z = \frac{1}{Y} \qquad \text{and} \qquad Y = \frac{1}{Z}$$

Let r and x represent the normalized resistance and reactance, respectively. Then

$$\frac{Y}{Y_0} = \frac{1}{Z/Z_0} = \frac{1}{r + jx} = \frac{1}{r + jx}\frac{r - jx}{r - jx} = \frac{r - jx}{r^2 - jrx + jrx - (jx)^2}$$

$$\frac{Y}{Y_0} = \frac{r - jx}{r^2 + x^2} = \frac{r}{r^2 + x^2} - \frac{jx}{r^2 + x^2}$$

As an example, assume that the impedance $Z = 50 + j100$ ohms. The normalized value becomes $1 + j2$. Substituting these values in the equation

$$\frac{Y}{Y_0} = \frac{1}{1 + (2)^2} - \frac{2}{1 + (2)^2} = 0.2 - j0.4$$

Compare the above results to the plotted points on the Smith chart of Fig. 4·5. The normalized impedance of $(1 + j2)$ is plotted at the point C, and the VSWR circle is drawn as shown. A straight line is drawn from the point C through the center P to the VSWR circle diametrically opposite, and the point of intersection with the VSWR circle is at $(0.2 - j0.4)$. This agrees with the calculated value. Therefore, the normalized admittance is read on the diagram at a point on the VSWR circle diametrically opposite the *normalized impedance* value for a lossless line. Also, the normalized impedance can be found by the same procedure when the normalized admittance is known.

The reciprocal properties at quarter-wave intervals are applied in quarter-wave impedance-matching problems. Suppose that it is desired to connect two lines which have characteristic impedance values Z_{01} and Z_{02}. This can be accomplished with a quarter-wave line as indicated by the schematic diagram of Fig. 4·7. Neglecting discontinuity capacitances at the steps

$$\frac{Z_{02}}{Z_0} = \frac{Z_0}{Z_{01}} \qquad \text{or} \qquad Z_{02} = \frac{Z_0^2}{Z_{01}}$$

From the above equation, the characteristic impedance of a quarter-wave line used to match two lines which have different characteristic impedances is

$$Z_0 = \sqrt{Z_{01}Z_{02}} \tag{4·15}$$

A numerical example is shown in Fig. 4·7*b*.

Line Length. The wavelength scale around the rim of the Smith chart calculator is linear. The values on the chart are fractional wavelengths l/λ. Fractional wavelengths on the chart correspond to fractional wavelengths on the transmission line. Wavelength measurements can be started from any point radially in line with any known impedance point on the coordinates and can be measured in either direction, clockwise toward the generator or counterclockwise toward the load.

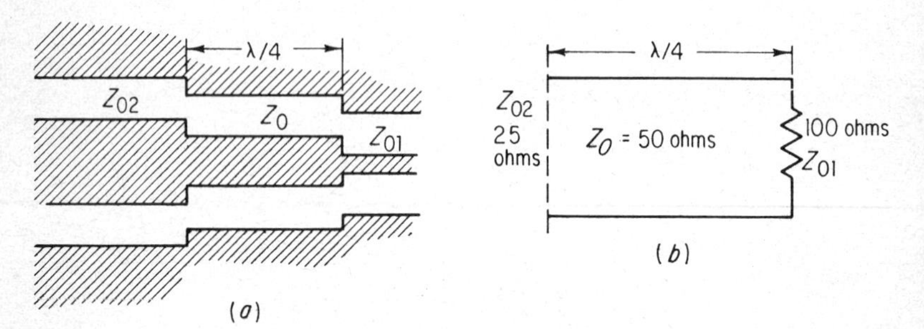

Fig. 4·7 A quarter-wavelength section of line with a characteristic impedance Z_0 used to connect two lines which have different values of characteristic impedance.

The radius vector *PC* of Fig. 4·5 is extended to *F*. The fractional wavelength readings are approximately 0.1875 on the outside scale and 0.3125 on the inside scale. The distance from *F* to *G* is 0.0625. Assuming an operating frequency of 3 Gc, the wavelength is 10 cm, and the distance on the transmission line from the load at *C* to E_{max} at *D* is 0.625 cm.

4·5 The decibel

The *decibel*, abbreviated db, is one-tenth of the international transmission unit known as the *bel.* The origin of the bel is the logarithm to the base 10 of the power ratio. The logarithm to the base 10 is the *common logarithm.* It is that power to which the number 10 must be raised in order to equal the given number. The number 10 is raised to the second power, or *squared*, in order to get 100. Therefore the log of 100 is 2.

The decibel is expressed mathematically by the equation

$$\text{db} = 10 \log \frac{P_2 \text{ (larger power)}}{P_1 \text{ (smaller power)}} \tag{4·16}$$

Throughout this text, the abbreviation log refers to the logarithm to the base 10 *unless otherwise specified.*

The complicated negative characteristics of the logarithm of the ratio can be avoided by always placing the larger power value in the numerator of the equation. In each problem it is known whether there is a gain or loss.

If the resistance level is the same at the points where both power levels are measured, the relative currents or voltages are expressed in decibels as

$$\text{db} = 20 \log R \tag{4·17}$$

where R is the ratio of voltages or currents.

The relationships of power ratio, decibels, and powers of 10 are given in Table 4·1. The table can be used to obtain approximate decibel or ratio values. By comparing the decibel column with the power of 10 column, it is noted that the number corresponding to the power of 10 also appears in the decibel column. The ratio of 1,000 is 10^3 or 30 db.

Table 4·1 Relationship of ratios to decibels

Power ratio	Power of 10	db	Power referred to 0 dbm = 1 mw
1	10^0	$0 = 10 \log 1$	1 mw
1.259	$10^{0.1}$	$1 = 10 \log 1.259$	1.259 mw
10	10^1	$10 = 10 \log 10$	10 mw
100	10^2	$20 = 10 \log 100$	100 mw
1,000	10^3	$30 = 10 \log 1{,}000$	1 watt
10,000	10^4	$40 = 10 \log 10{,}000$	10 watts
100,000	10^5	$50 = 10 \log 100{,}000$	100 watts
0.1	10^{-1}	$-10 = 10 \log 0.1$	100 μw
0.01	10^{-2}	$-20 = 10 \log 0.01$	10 μw
0.001	10^{-3}	$-30 = 10 \log 0.001$	1 μw

If it is desired to find the ratio corresponding to 27 db,

$$27 \text{ db} = 10 \log R$$

$$\frac{27}{10} = 2.7 = \log R$$

The ratio is found by taking the antilog of 2.7. The 2 represents the characteristic, and 0.7 is the mantissa. The number corresponding to the mantissa of 0.7 is obtained from a table of logarithms or from a slide rule. The number 2 indicates the number of zeros to add. From the table of mantissas, the number is 5.01. Two zeros are added to obtain the ratio of 501. From the table it is noted that the ratio must be between 100 and 1,000 or 20 db + x db.

The decibel value corresponding to a power ratio of 1,920 is db = 10 log 1,920. From Table 4·1 it is noted that the decibel value is 30 db + x db. The decibel value to be added is the mantissa of 1,920 multiplied by 10 or 10(0.283). The decibel value is 30 + 2.83 = 32.83 db.

The use of log tables can be avoided in practical applications where exact values of the power ratio are not required. A power ratio of 2 corresponds to 3.01 db. If the power ratio is expressed as 2^n,

$$\text{db} = n(3.01) \tag{4·18}$$

or

$$n = \frac{\text{db}}{3.01} \tag{4·19}$$

These equations can be used in conjunction with Table 4·1 to obtain any decibel value. For simplicity of calculations, the denominator in Eq. (4·19) is considered to be 3.0, that is, a change in power of ½ or 2:1 corresponds to 3 db.

If 3, 6, and 9 db values are substituted in Eqs. (4·18) and (4·19) and the power is expressed as 2^n, the following results are obtained:

$$3 \text{ db} = 2^n = 2 \tag{4·20a}$$

$$6 \text{ db} = 2^n = 4 \tag{4·20b}$$

$$9 \text{ db} = 2^n = 8 \tag{4·20c}$$

This technique is based on the fact that 3, 6, and/or 9 db can be added or subtracted (in some combination) to the decibel values in Table 4·1 to obtain any decibel value.

Example

1. 17 db = 20 db − 3 db
   The ratio of 20 db is 100 (Table 4·1), and the ratio of 3 db expressed as 2^n is 2 [Eq. (4·20*a*)]. Therefore, 100/2 = 50.
2. 36 db = 30 db + 6 db = 1,000 (Table 4·1) × 4 [Eq. (4·20*b*)] = 1,000(4) = 4,000.
3. 25 db = 10 db + 9 db + 6 db = 10(8)(4) = 320.
4. 5 db = 30 db − 9 db − 10 db − 6 db = 1,000/(8)(10)(4) = 3.16.

An alternate unit called the *neper* is defined in terms of the logarithm to the base *e*. $e = 2.718$.

$$1 \text{ neper} = 8.686 \text{ db}$$
$$1 \text{ db} = 0.1151 \text{ neper}$$

The decibel is not a unit of power. The unit of power in our exponential or logarithmic system of numbers is represented by dbm, where the *m* is the *unit*, meaning *above or below one milliwatt*. Since 1 mw is neither above nor below 1 mw, 1 mw = 0 dbm. Relative values are tabulated in Table 4·1. It should be noted that

$$\text{db} \pm \text{db} = \text{db}$$
$$\text{dbm} \pm \text{db} = \text{dbm}$$
$$\text{dbm} \pm \text{dbm} = \text{db}$$

PROBLEMS

4·1 The angle of the load reflection coefficient is 115° and the VSWR is 6.4. Plot the following points on a Smith chart: Z_L, Z_{min}, Z_{max}, Y_L, and the point where $2\beta l - \psi = 0$.

4·2 The VSWR on a 50-ohm transmission line is 4.0. If the load impedance is a pure resistance greater than the characteristic impedance, what are the values of Z_L, Z_{min}, Z_{max}, Y_L, and the angle of the reflection coefficient? Record the distance in wavelengths to the first E_{max} and E_{min} points.

4·3 The VSWR on a 100-ohm transmission line is 2.5. If the load impedance is a pure resistance smaller than the characteristic impedance, what are the values of Z_L, Z_{min}, Z_{max}, Y_L, and the angle of the reflection coefficient? What is the input impedance at a distance 0.14 wavelength from the load?

4·4 A 50-ohm transmission line is terminated in a load impedance $Z_L = 50 - j60$ ohms. Use a Smith chart to obtain the VSWR, Y_L, the magnitude and angle of the reflection coefficient, and the distance in wavelengths to the first E_{max} point. Record the input impedance and input admittance at a point 6.18 wavelengths from the load.

4·5 The VSWR on a 50-ohm transmission line is 2.0 and the first E_{min} is located 0.2 wavelength from the load. Find Z_L, Y_L, Z_{min}, Z_{max}, the magnitude and angle of the reflection coefficient, and the reflected power in per cent.

4·6 If the frequency is 1,000 Mc in Prob. 4·5, what is the distance in centimeters from the load to the first E_{min}?

4·7 The applied frequency is 3 Gc, Z_0 is 50 ohms, and the load impedance is $(120 + j85)$ ohms. Find the VSWR, the values of Z_{max} and Z_{min}, and the distance in centimeters to the first E_{max} and E_{min} points.

4·8 The reflection coefficient is 0.6 and βl from the load to the first E_{max} is 0.1 wavelength. Find the VSWR, Z_L, Y_L, Z_{min}, Z_{max}, and the angle of the reflection coefficient if Z_0 is 50 ohms.

4·9 Z_L is 100 ohms and Z_0 is 50 ohms. Show mathematically and explain why the load admittance as read on the Smith chart is the reciprocal of the normalized impedance.

4·10 It is desired to use a quarter-wave matching section of transmission line between two lines which have characteristic impedance values of 60 and 50 ohms, respectively. What is the characteristic impedance of the matching section?

4·11 Calculate the power ratios for the following decibel values: 7, 24, 46, 32, 66, and 18 db.

4·12 Calculate the decibel values corresponding to the power ratios 1,620; 145; 1,840; 188; 16,700; 560; and 58,600.

4·13 The signal to a three-stage system is −35 dbm. The first stage has a gain of 28 db, the second stage has a loss of 3 db, and the third stage has a gain of 76 db. What is the power output at each stage?

4·14 Express the following powers in dbm: 160 $\mu\mu$w, 280 $\mu\mu$w, 25 mw, 56 watts, 20 kw, 2 megawatts, 720 $\mu\mu$w, and 18 mμw.

4·15 Convert the following dbm values to microwatts: −20, −28, −92, 35, 8, and 87 dbm.

REFERENCE

1. P. H. Smith, Transmission Line Calculator, *Electronics*, vol. 12, pp. 29–31, January, 1939. An Improved Transmission Line Calculator, *Electronics*, vol. 17, p. 130, January, 1944.

CHAPTER

5

USING THE SMITH CHART

Introduction. This chapter is concerned with examples which illustrate how the Smith chart can be used for practical microwave computations. A considerable amount of microwave technique is concerned with impedance transformation produced by a length of guiding structure. The usual time-consuming calculations are eliminated by using the Smith chart, and, for this reason, the chart is invaluable to the microwave engineer.

The discussions of the chart include basic impedance matching problems, a brief discussion of each of the radial scaled parameters, and sample problems which outline the methods used to evaluate the radial scaled parameters in practical applications. The chart is used to evaluate the unique relationships of impedance, reflection coefficient, voltage-standing-wave ratio, and the position of a voltage minimum on the line.

5·1 Determination of input impedance and admittance

In the example of Fig. 5·1 the load impedance of $40 + j25$ ohms terminates a 50-ohm transmission line. The normalized impedance $z_L = (40 + j25)/50 = 0.8 + j0.5$ is located at the point A on the diagram. The angle of the reflection coefficient is 96°, and the reference wavelength point is 0.116λ, as indicated on the wavelength toward generator scale. The VSWR read at point B is 1.79. The point B is also the maximum voltage and maximum impedance (pure resistance of 89.5 ohms) point and is located $0.25 - 0.116 = 0.134\lambda$ from the load. The normalized load admittance is read diametrically opposite the normalized impedance z_L and is located at point C, where $y_L = 0.89 - j0.56$.

Practical measurements of load impedance usually require the location of the load impedance point at a particular distance from a voltage minimum point when the VSWR and the position of the voltage minimum have been determined. As an example, suppose that the load is located a distance of

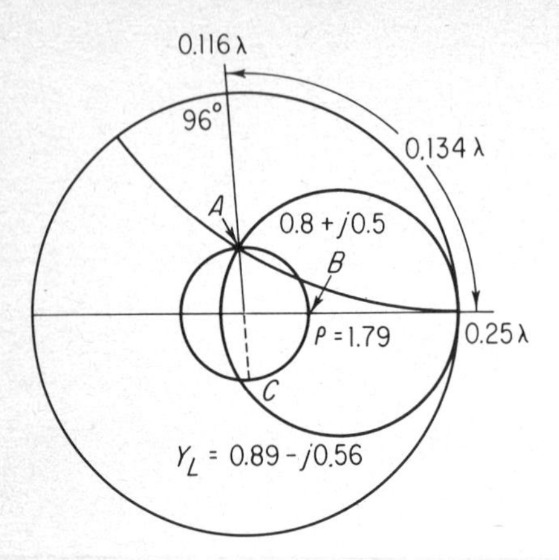

Fig. 5·1 Using the Smith chart to determine VSWR and admittance.

0.313λ from the first voltage minimum and the VSWR is 2.4. Figure 5·2 illustrates this problem in which the VSWR circle is shown, and the distance from $Z_{\min}$ to the load (0.313λ), measured CCW from 0 at the $E_{\min}$ point, is located at point A. A straight line from the center of the chart to A intersects the ρ circle at B, and the normalized impedance is read as $1.4 + j1.0$. The load impedance is therefore $50(1.4 + j1.0) = 70 + j50$ ohms. The normalized input admittance y_L is indicated at point C, where

$$y_L = (0.475 - j0.34)$$

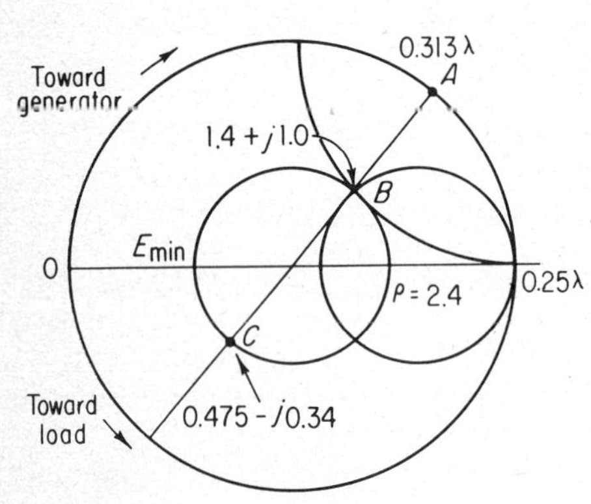

Fig. 5·2 Using the Smith chart to locate the load impedance when the VSWR and position of a voltage minimum have been determined.

5·2 Impedance matching

The principles of impedance matching are, in general, applicable to both waveguides and transmission lines, although the physical form and the behavior of a given matching structure as a function of frequency may be different.

Suppose it is desired to match a 50-ohm transmission line terminated with a load $Z_L = 100 + j50$ ohms. The normalized load impedance is $2 + j1$ ohms and is plotted as point A in Fig. 5·3. If the reactive component of the load is eliminated by placing a capacitive reactance $-j50$ ohms in series with the load, only the resistive component remains. This is shown by moving along the constant resistance circle to point B. *There is still a VSWR of 2.0, so it is obvious that elimination of the reactive component of the load is not the only requirement for matching the transmission line.* For a matched condition to exist on a transmission line, the normalized input impedance must be $1 + j0$. Therefore, the reactive component must be eliminated on the $R/Z_0 = 1$ circle. A line is drawn from the center of the chart through the point A intersecting the wavelength scale at 0.213λ at D. The $R/Z_0 = 1$ circle is located by traveling clockwise around the constant VSWR circle to point C, where the normalized input impedance of the line is $1 - j1$. The distance along the line to this point is $0.338\lambda - 0.213\lambda = 0.125\lambda$, as indicated on the wavelengths toward generator scale. A *series* reactance of $+j50$ ohms placed at this point eliminates the capacitive reactance, and the input impedance point moves to the center of the chart, where $R/Z_0 = 1$ and $X/Z_0 = 0$. The transmission line is now matched. In other words, a VSWR equal in amplitude and opposite in phase has been introduced at point C.

The requirement that the input impedance for a match must be $1 + j0$ can also be obtained by traveling around the constant VSWR circle to the point P. The impedance is $1 + j1$, and a capacitive reactance must be added in *series* with the line to cancel the inductive reactance at this point. The distance from the load at A to this new point at P is 0.449λ as indicated.

5·3 Single-stub transformer

Since admittance values are additive at parallel junctions, transmission line problems are considerably simplified if they are considered in terms of the input admittance. A device which is often employed with parallel-wire and coaxial lines consists of a short section of transmission line connected in parallel with the transmission line and terminated in a short circuit. This impedance-transforming device is called a single-stub transformer.

The problem illustrated in Fig. 5·3 is now considered as an admittance (shunt) problem. The normalized input impedance is converted to normalized input admittance by $\lambda/4$ rotation of the diagram. This corresponds to location of the reciprocal of the normalized input impedance, as previously demonstrated, at a point diametrically opposite the point A. The normalized input admittance is located at point M, and the normalized value of input admittance y_i is read as $0.4 - j0.2$. For a match, $G/Y_0 \pm jB/Y_0$ must have the value $1 \pm j0$. The line from the center of the chart through M intersects the wavelengths toward generator scale at 0.4630. The distance measured on the wavelength scale by traveling clockwise from the load admittance point M to the point P, where $B/Y_0 = 1$, is 0.1990λ. The normalized input admittance at this point is $1 + j1$ (a capacitive susceptance). An inductive susceptance connected *across* the line at this point results in the input admittance point moving to the center of the chart, where $G/Y_0 = 1$.

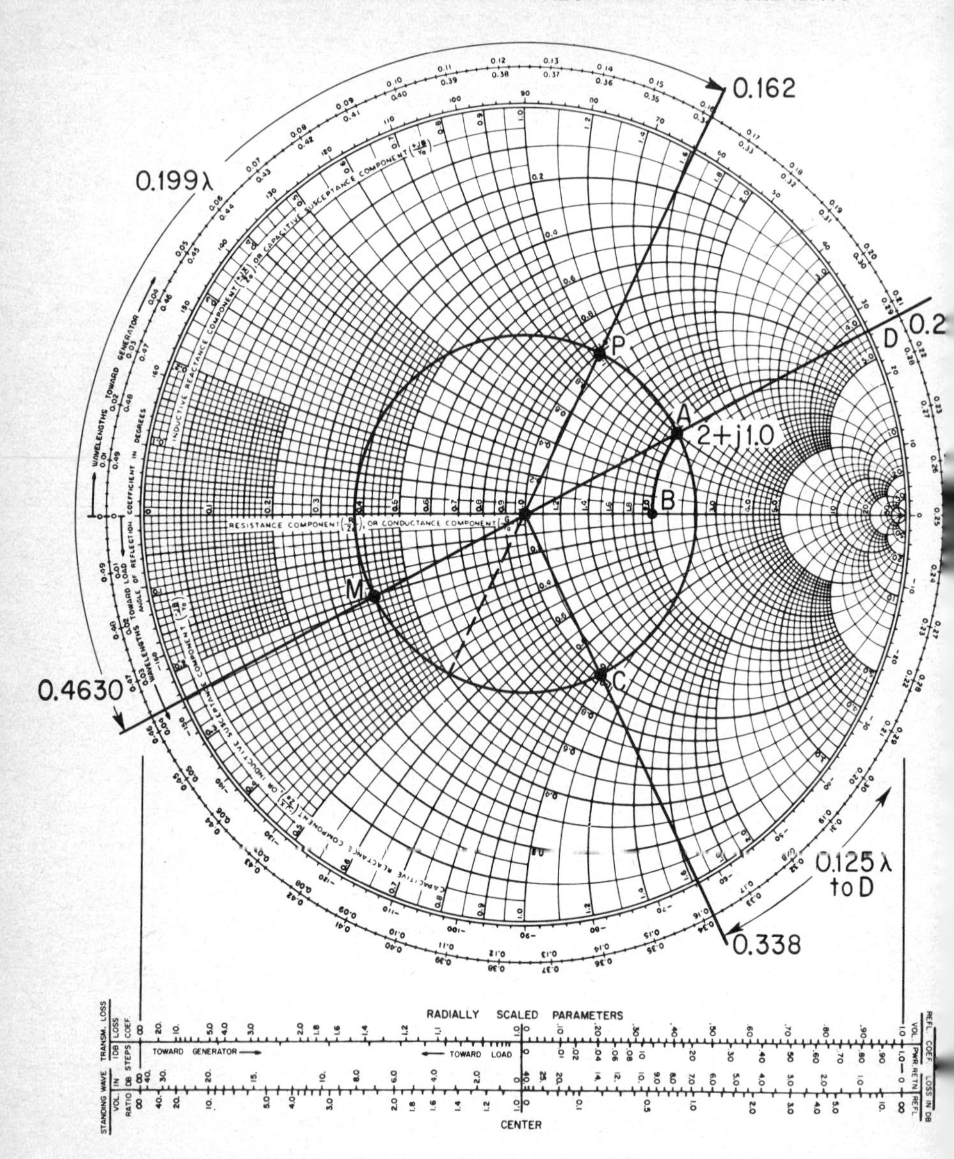

Fig. 5·3 Smith chart. ***(From Irving L. Kosow (ed.), "Microwave Theory and Measurements," by the Engineering Staff of the Microwave Division, Hewlett-Packard Company, © 1962, by permission of Prentice-Hall, Inc., Englewood Cliffs, N.J.)***

If the frequency of the source in the above problem is considered to be 600 Mc, the distance along the line at which the series reactance or shunt susceptance must be placed can be calculated from $\lambda = v/f = 50$ cm. The series inductance used to match the line at C would be placed $0.125(50) = 6.25$ cm from the load. The capacitive reactance to be inserted at a point on the line corresponding to point P on the diagram is $0.449(50) = 22.45$ cm from the load. The shunt susceptance (inductive) required to match the line is placed across the line at a point corresponding to P located $0.199(5) = 9.95$ cm from the load position. As in the case of impedance matching, two points exist where the transmission line can be matched. The other point on the line where the shunt susceptance (capacitive) is placed for a match is point C and is located $(0.375)(50) = 18.75$ cm from the load.

5·4 Double-stub matching

In the previous section the matching procedure required that a point be located on the line where the reactance or susceptance element is placed either in series or in shunt with the line. The single-stub method of matching a line is not suitable for most transmission line structures because of the difficulty of building a single stub adjustable in position. Therefore, the double-stub transformer is used. It consists of two stubs separated by any distance less than one-half wavelength, and each stub is an adjustable short circuit. A diagram of the double-stub unit is shown in Fig. 5·4. The separation of the stubs is usually ⅜ wavelength, as indicated on the diagram. The load admittance which is to be matched is the impedance as seen at the load side of stub 1, even though the actual load may be any distance from the stub. The separation between the load and tuner should not be great in terms of wavelengths, especially if the source exhibits frequency instability. If the tuner is many wavelengths from the load, a small change in frequency causes large changes in admittance (or impedance), as seen at the tuner. This is referred to as the *long line* effect. Even though a given spacing will not accommodate all impedances, the spacing can be chosen so that the range of impedances is sufficiently wide to allow for any slight variations in load or frequency that may be encountered.

The normalized input admittance y_i shown in Fig. 5·4*b* must be $1 + j0$ for a matched condition. Since the stubs can change susceptance only, the desired susceptance component is obtained by adjustment of stub 1. Stub 1 is used to add the proper amount of susceptance to Y_L so that the resulting admittance may be transformed by the ⅜ wavelength line to a point on the $G/Y_0 = 1$ circle. This is illustrated by calculating the admittance values and lengths of stubs required to match the load shown in Fig. 5·4*b*.

The normalized (per unit) load admittance is $1.6 + j1.0$ (point A). The point representing the per unit admittance of stub 1 plus the load admittance $(y_L + y_1)$ must lie somewhere on the $G/Y_0 = 1.6$ circle because stub 1 only adds susceptance to the load admittance. Each point of the $G/Y_0 = 1.6$ circle is rotated through an angle corresponding to the distance between stubs 1 and 2, which is ⅜ wavelength in this problem, and all possible values of the input admittance y_3 are located on this circle. This rotation of each point is obtained simply by rotating the G_L/Y_0 circle clockwise around the chart a

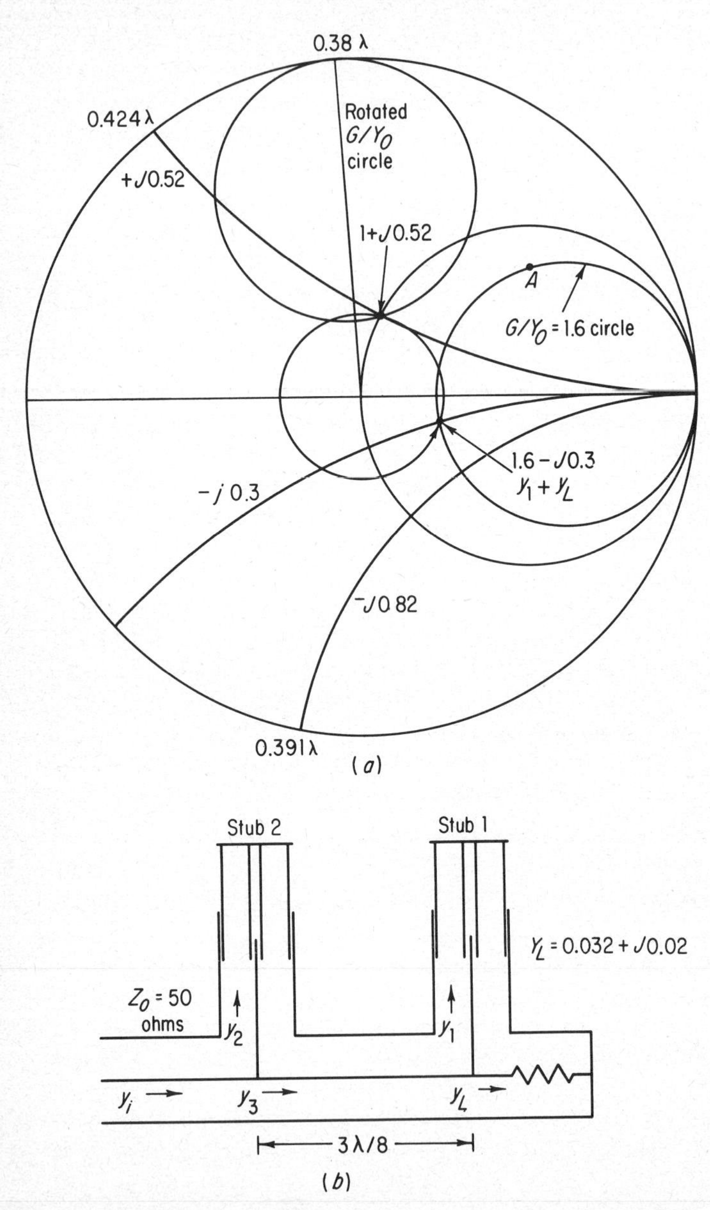

Fig. 5·4 Example illustrating the use of the double-stub transformer.

distance corresponding to the distance between the stubs. Since the normalized admittance $y_i = y_2 + y_3$ must be $1.0 + j0$ and since stub 2 can add susceptance only, it follows that the per unit admittance y_3 must equal $1 + jB$ and for the stated problem $y_3 = 1 + j0.52$. Stub 2 must be $-j0.52$, and the corresponding length is $0.424\lambda - 0.25\lambda = 0.174\lambda$.

The value of the admittance of stub 1 is determined by rotating the y_3 point ⅜ wavelength CCW along the constant VSWR circle to $g_L = 1.6$ circle. This point represents the total admittance $(y_1 + y_L) = (1.6 - j0.3)$, and if y_L is subtracted from this value, the value of y_1 is found to be

$$y_1 = (y_1 + y_L) - y_L = (1.6 - j0.3) - (1.6 + j1.0) = -j1.3$$

The length of the stub 1 is determined as

$$0.355\lambda - 0.25\lambda = 0.105\lambda$$

The following examples illustrate the limitations of the double-stub transformer:

1. If a circle corresponding to a normalized conductance $g_L = 2$ is drawn at the ⅜ wavelength distance, it is noted that this circle is tangent to the $G/Y_0 = 1$. Since the rotated g_L circle must intersect the $G/Y_0 = 1$ circle, it is obvious that if the normalized conductance value is greater than 2, these circles will not intersect and the matched condition cannot be obtained.
2. If the distance between the stubs is $\lambda/4$ and the normalized conductance g_L is 1.0, it is noted that the $G/Y_0 = 1$ circle and the g_L circle are tangent at the center of the chart. This signifies that any normalized conductance greater than 1 cannot be tuned out.

The limitation of the stub is a function of the distance between the stubs. The range of operation is least if the stub separation is one-quarter wavelength and increases if the separation distance is increased toward one-half wavelength or is decreased toward zero.

The operation of the three-stub tuner is even more difficult to describe using this step-by-step method.

5·5 Radial scaled parameters

The various parameters which are uniquely related to one another are shown on the Smith chart diagrams of Figs. 4·5 and 5·5 plotted radially from the center of the chart and labeled *radially scaled parameters.* A circular transmission line calculator with separately rotatable wavelength scale around the rim and with a transparent arm engraved with all the radially scaled parameters can be constructed. An adjustable cross-hair index along the radial arm permits reading any or all of the several scales at the intersection of the slider index. The relationship of these parameters may be

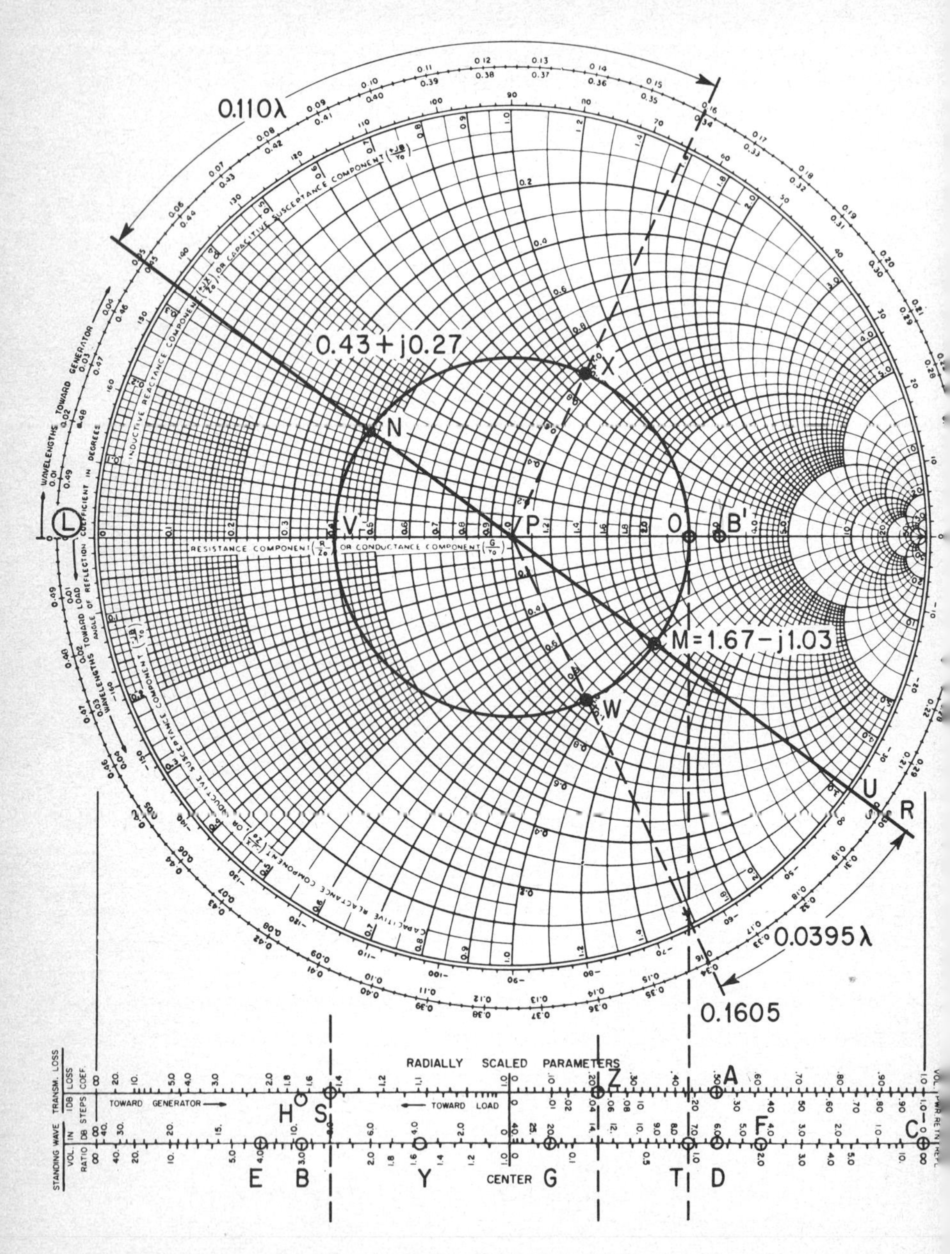

Fig. 5·5 Smith chart. (*From Irving L. Kosow (ed.), "Microwave Theory and Measurements," by the Engineering Staff of the Microwave Division, Hewlett-Packard Company,* © *1962, by permission of Prentice-Hall, Inc., Englewood Cliffs, N.J.*)

evaluated by using dividers or any other method for measuring the related radial distances on the chart and the radial scales. For convenience of explanation, the following discussions will consider the use of dividers.

Reflection Coefficient. The reflection coefficient is read on the right-hand side of the radial scale at the bottom of Fig. 5·5. There are two reflection coefficient scales labeled *voltage* and *power*. The voltage reflection coefficient or scale is above the line and is a plot of E_r/E_i and varies from *zero* at the center of the chart (Z_0 point) to *one* at the outside rim of the chart, which is the zero resistance circle. The adjacent scale underneath the same line is the power reflection coefficient scale and is designated Γ^2. An examination of the values shows that each value on this scale is precisely the square of the value of Γ noted above the line.

The angle of the reflection coefficient is zero on the Z_{max} resistance line where the incident and reflected waves are in phase. The angle of Γ from the zero point is linearly related to the distance traveled, and the sign of the angle is indicated on the outside rim of the impedance coordinate system. The relationship of various parameters to the reflection coefficient can be illustrated on the chart. Use dividers to measure $\Gamma = 0.5$ on the radial scale (point A). Comparing the measured distance to the VSWR scale reading on the right-hand axis of the chart shows a VSWR of 3.0, as indicated by B and B' in Fig. 5·5. The power reflection coefficient Γ^2 at A is noted to be 0.25, and if this value is multiplied by 100, the per cent of reflected power is obtained. The reflected power is 25 per cent of the incident value. When performing measurements on the radial scales and the chart proper, the center of the radial scales corresponds to the center of the chart and the end of the radial scales corresponds to the outside rim of the chart (zero resistance). Therefore, dividers can be used to compare measured values either from the center out, or from the outside rim toward the center, as long as one is consistent as to the chosen reference. The relationships of the remaining parameters to reflection coefficient can be obtained by measuring the corresponding distances on the scales of interest.

Return Loss. The *return loss* is the ratio of the incident power to the reflected power at a point on the transmission line and is expressed in decibels. The reflected power from a discontinuity is expressed as a certain number of decibels below the incident power upon that discontinuity.

$$R_{\text{loss}}(\text{db}) = 10 \log \frac{P_i}{P_r} = 20 \log \frac{E_i}{E_r} = 20 \log \frac{1}{\Gamma} \qquad (5\text{·}1)$$

The return loss scale is located on the right side of the radial scaled parameters on the line headed "Loss in db." The return loss values are indicated above the line. When the reflection coefficient is 1, the return loss is zero, as indicated at the extreme right which is on the outside rim of the impedance coordinate system (point C). This return loss value indicates that no signal was lost and that all of the signal incident upon the discontinuity was returned toward the source. As the reflection coefficient approaches zero, the return loss approaches infinity. That is, the more perfect the load, the less the reflection from that load. Practical applications of this parameter are all important to the microwave engineer or technician.

Assume that the 3-db attenuator in Fig. 5·6 is perfectly matched (input and output VSWR = 1.0). The indicated input power of 100 mw is decreased to 50 mw at the output of the 3-db attenuator. This 50 mw is reflected from the short circuit back through the attenuator in the reverse direction, and one-half of this reflected power is lost in the 3-db attenuator. The reflected power

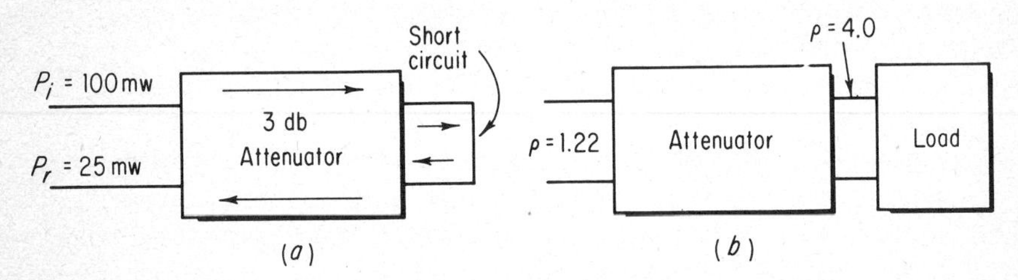

Fig. 5·6 Schematic representation of practical applications of return loss theory.

at the input is 25 mw. The definition of return loss signifies that return loss is the total *round-trip loss* of the signal, and in this case the round-trip loss is 6 db. Use a set of dividers and measure the distance from zero on the return loss scale to the 6-db line on the return loss scale. This point is labeled *D* on the return loss scale. Use this setting to mark the distance from the right-hand rim of the impedance coordinate system toward the center of the chart. The VSWR as read from the chart is 3.0 at *B'*. *This example is of special significance since it shows that the VSWR is decreased when attenuation exists on the line and also that a high VSWR can be decreased by placing attenuation in the line.* In many cases, it is not desirable to place attenuation in the line in order to reduce VSWR because of the loss of power in the forward direction. Special components of a unidirectional nature are required to provide the necessary attenuation and to dissipate minimum power in the desired direction of power flow. Using the same setting on the dividers, measure the distance from 1.0 on the voltage and power reflection coefficient scales to obtain the voltage reflection coefficient of 0.5 and power reflection coefficient of 0.25. This verifies that the reflected power is 25 per cent of the incident power. The voltage reflection coefficient is squared and multiplied by 100 in order to obtain the reflected power in per cent.

The second example is illustrated in Fig. 5·6*b*. In order to find the proper value of attenuator required to reduce the VSWR from 4.0 to 1.22, use the Smith chart of Fig. 5·5 and measure the distance on the VSWR scale from the left end of the scale (infinity) to the point on the scale which indicates a VSWR of 4.0 (point E). This distance can also be measured on the chart from the outside rim of the impedance coordinate system to the normalized resistance circle of 4.0. This same distance on the return loss scale indicates a return loss of 4.4 db (point F). Measure the scale length for a VSWR of 1.22 and mark off the same distance on the return loss scale. The return loss is found to be 20 db (point G). Therefore, 20 db − 4.4 db = 15.6 db, which is the value of return loss to be added. The attenuator value required to obtain this value of return loss is 7.8 db, since the round-trip loss must be 15.6 db.

In the above examples the distances were measured from the outside rim of the impedance coordinate system on the chart and the right end of the radial scales. The measurements can also be made from the center of the chart and the center line of the radial scales; one need only be consistent as to the reference point chosen.

Transmission Loss Scale. The transmission loss in 1-db steps is presented on the left-hand side of the radial scaled parameters. It is the equivalent *one-way* loss in the signal on the transmission line. By measuring the distance to D on the return loss scale and measuring off the same distance to H on the transmission loss scale, it is noted that the attenuation reading is one-half the reading on the return loss scale.

Standing-wave Ratio in Decibels. In many practical applications the standing-wave ratio is measured in decibels, and the corresponding VSWR and other parameters are obtained from this measurement. The standing-wave ratio in decibels is expressed as

$$\text{SWR (db)} = 20 \log \text{VSWR} \tag{5·2}$$

A radial scale plot of the standing-wave ratio in decibels is shown in Fig. 5·5, and values of SWR (db) can be compared to the various parameters by measuring equidistant points on the desired scales as demonstrated in previous examples.

Mismatch Loss (Reflected Loss). The mismatch loss is a measure of the loss caused by reflection. It is the ratio of incident power to the difference between incident and reflected power and is expressed in decibels as follows:

$$\text{Mismatch loss (db)} = 10 \log \frac{P_i}{P_i - P_r} = 10 \log \frac{E_i^2}{E_i^2 - E_r^2}$$

$$= 10 \log \frac{1}{1 - \Gamma^2} = 10 \log \frac{(1 + \rho)^2}{4\rho} \tag{5·3}$$

As an example, a VSWR of 3.0 represents a mismatch loss of 1.25 db. This point is shown, opposite the return loss of 6 db, at B in Fig. 5·5.

Example: A signal frequency of 3 Gc is applied to a transmission line which has a characteristic impedance of 50 ohms. The voltage-standing-wave ratio is 2.5:1, and the first voltage minimum is located 0.2 wavelength from the load. It is desired to find the value of load impedance, load admittance, reflection coefficient, angle of reflection coefficient, the maximum and minimum impedance on the line, mismatch loss, return loss, and the distance to the first voltage minimum in centimeters. Also, it is desired to find the point where a *series* reactance and the point where a *shunt* susceptance would be added in order to match the line.

A matched 3-db attenuator is inserted in the line in front of the load; it is desired to know the new values of VSWR, reflection coefficient, and return loss.

The problem is solved using the Smith chart of Fig. 5·5.

$$\text{Wavelength } \lambda = \frac{v}{f} = 10 \text{ cm}$$

The VSWR circle is drawn with a radius of 2.5. The load impedance is found by traveling from the point L on the fractional wavelength scale at the left side of the chart CCW toward the load 0.2λ to the point R. Draw a line from R to the point P at the center of the chart. Record the normalized load impedance at M $(1.67 - j1.03)$. Multiply this value by the characteristic impedance of 50 ohms to obtain the load impedance of (83.5 − 51.5) ohms.

The load admittance is found by extending the line RP to intersect the VSWR circle at N and reading the normalized admittance value $(0.43 + j0.27)$. $Y_L = (0.43 + j0.27)/50 = (0.0086 + j0.0054)$ mhos.

Use dividers to measure the distance PO, the 2.5 VSWR point on the Z_{max} axis. Use this value to obtain the mismatch loss of 0.88 db on right side lower radial scale. Also, the return loss is found at point T and is 7.3 db.

The angle of reflection coefficient is $-36°$ at point U.

The maximum impedance at O is $(\text{VSWR})(Z_0) = 125$ ohms, and the minimum impedance at V is $Z_0/\text{VSWR} = 20$ ohms.

The distance from the load to the first voltage minimum is 2 cm since the wavelength is 10 cm.

The series reactance required to match the line is located by traveling around the VSWR circle from the load at M to the point W. The impedance at this point is $1 - j0.95$. A series reactance of $+j0.95$ is required at this point to cancel the $-j0.95$ value. This reactance is located a distance of $0.20\lambda - 0.1605\lambda = 0.0395\lambda$, as indicated on the fractional wavelength scale. Since $\lambda = 10$ cm, the series reactance is located $0.0395(10) = 0.395$ cm from the load.

The point where the shunt susceptance is added is located by traveling around the constant VSWR circle to the load admittance value of $1 + j0.95$ at X. A shunt susceptance of $-j0.95$ is placed at this point to obtain a match. The distance from the load is obtained by noting the fractional wavelength change indicated at the intersection of the lines PN and PX extended to the outside rim of the chart. The values of $0.45 - 0.34 = 0.110\lambda$ from the load. The distance in centimeters is therefore $(0.11)(10) = 1.1$ cm.

The VSWR of 2.5 resulted in a return loss of 7.3 db, as indicated at point T. Insertion of the 3-db matched attenuator corresponds to the addition of a return loss of 6 db. The total return loss is now 13.3 db. The corresponding VSWR is 1.55 at Y, and the reflection coefficient is 0.215 at Z on the radial scale.

PROBLEMS

Use the Smith chart to work the following problems.

5·1 The load impedance of a 100-ohm transmission line operating at a frequency of 1.5 Gc is located 0.16λ from the first voltage minimum. The VSWR is 3.6:1.

a. Find Z_L, Y_L, Z_{min}, Z_{max}, the angle and magnitude of the reflection coefficient, and the distance in centimeters to the first voltage maximum.

b. What are the values of return loss and equivalent attenuation?

c. What is the standing-wave ratio in decibels?

d. If a 6-db attenuator (three wavelengths long) is placed in front of the load, what are the new values of the parameters listed in (*a*)?

5·2 An antenna is located 4.6 m from the source. The characteristic impedance of the lossless line is 50 ohms, the VSWR is 3:1, and the antenna presents a pure resistance load greater than the characteristic impedance. If the operating frequency is 50 Mc:

a. Find Z_L, Z_{min}, Z_{max}, the angle and magnitude of the reflection coefficient, Y_L, and the per cent of reflected power.

b. Find the values of return loss, mismatch loss, equivalent attenuation, and SWR in decibels.

c. Find the distance in centimeters to the first E_{max} and E_{min} points.

d. Find the input impedance and the input admittance at the generator.

e. Locate the point on the line and find the value of series reactance required to match the antenna.

f. Find the point on the line and the value of shunt susceptance required to match the antenna.

g. If the line had an attenuation loss of 0.2 db per meter, calculate the input impedance, VSWR, input admittance, and the return loss presented to the generator.

5·3 The first voltage minimum is located 0.23λ in front of the load, and the VSWR is 6:1 on a lossless transmission line which has a characteristic impedance of 100 ohms. The operating frequency is 100 Mc.

a. Find Z_L, Z_{max}, Z_{min}, return loss, and per cent of reflected power.

b. A 6-db attenuator (five wavelengths long) is placed in the line. What are the new values of the parameters listed in (*a*)?

5·4 What is the value of attenuation which is required to reduce a VSWR of 8:1 to 1.08:1? How much power is lost because of the attenuator?

CHAPTER

6

COAXIAL TRANSMISSION LINES AND MEASURING EQUIPMENT

Introduction. This chapter deals with coaxial transmission lines and measuring instruments used at ultrahigh frequencies. The study is concerned with the same general principles previously applied to unconfined fields. An introduction to basic measurement equipment and measurement techniques is presented in order to acquaint the student with practical applications of microwave techniques.

6·1 The coaxial line

The coaxial line is a special form of confined space composed of two concentric conductors separated by an insulating material. The coaxial line illustrated in Fig. 6·1 can be constructed in both rigid and flexible forms.

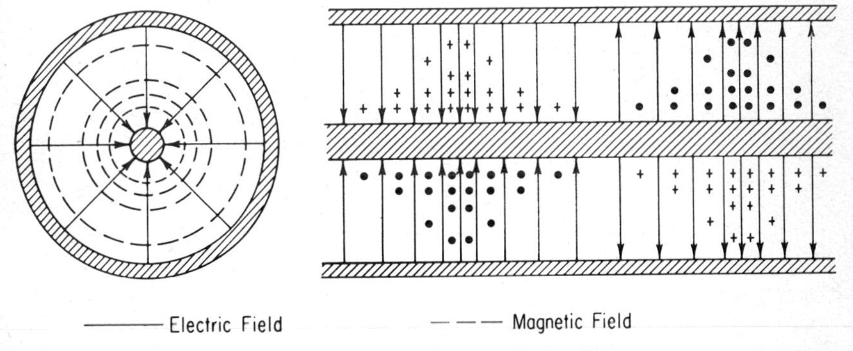

Fig. 6·1 The coaxial transmission line.

The dielectric in the rigid coaxial line is usually air, and the center conductor is concentrically located within the outer conductor by means of dielectric insulating supports called *beads.* In the case of the flexible cable, the center conductor is surrounded throughout its length by a flexible dielectric material such as polyethylene. The characteristic impedance of the coaxial line is in the order of 30 to 100 ohms as given by

$$Z_0 = \frac{138}{\sqrt{\epsilon'}} \log \frac{b}{a} \tag{6·1}$$

where a is the outer diameter of the inner conductor and b is the inner diameter of the outer conductor, or a and b can be the corresponding radii.

The electromagnetic field is restricted to the region between the inner and outer conductors, and this results in practically perfect shielding between fields inside and outside of the line. The dominant mode in coaxial lines is the TEM (transverse electromagnetic) in which the electric lines of force are radial and the magnetic lines of force are concentric circles as shown in the diagram. The strength of the electric field is proportional to the voltage difference between conductors and is inversely proportional to the distance between the axis of the line and the point in question. The magnetic field strength is proportional to the current flow and varies inversely with the distance from the axis of the line.

Both inner and outer conductors can supply charge and current distributions so that any rate of variation of the electric and magnetic fields can take place along the line. That is, the currents adjust themselves so that the field is transmitted for any applied frequency. Therefore, the coaxial line is not frequency sensitive, and it is a broad band device having no cutoff frequency. The electric and magnetic vectors for the dominant TEM mode are exclusively perpendicular to the direction of motion so that the *velocity of propagation is the same as for a wave in the same insulating medium without any conductors.*

Higher order modes of propagation may be present unless the conductor dimensions are chosen properly when operating in the microwave range. The higher order modes which are most likely to be encountered are the TE_{11} and TM_{01} waveguide-type modes, which are discussed in Chap. 7, Sec. 7·9. The cutoff wavelength for TE modes, which have variations around the circumference, is given by the approximate relation

$$\lambda_c \cong \frac{2\pi}{n} \frac{b + a}{2} \qquad n = 1, 2, 3, \ldots \tag{6·2}$$

where a and b are the inner and outer radii, respectively.

The approximate cutoff for TM modes is given by

$$\lambda_c = \frac{2}{p}(b - a) \qquad p = 1, 2, 3, \ldots \tag{6.3}$$

where a and b have been defined and p is the half wavelength spacing between conductors.

6·2 The slotted line

The slotted line is one of the important measuring instruments used at microwave frequencies. It is designed to measure the standing-wave pattern of the electric field intensity which is a function of the longitudinal position in the guiding structure. A probe is mounted on a carriage which slides along the outside of the section of coaxial line or waveguide which has a longitudinal slot. The probe extends into the slot and is provided with an adjustment for varying the probe penetration into the slot and with a tuning adjustment (usually a stub) used to cancel the reactive component of probe impedance. The probe is connected to a barretter or crystal detector which detects the r-f voltage. This voltage is amplified and applied to the appropriate indicating meter.

A slotted section used over the frequency range from about 300 to 5,000 Mc is shown in Fig. 6·2*a*. The slotted section in Fig. 6·2*b* is usable over the frequency range of 500 to 4,000 Mc.

The standing-wave ratio is measured by sliding the probe along the line for a maximum and minimum indication on the output meter. The standing-wave ratio is calculated from the above data or read directly from the indicating device if the indicator is calibrated in SWR.

The wavelength of the signal frequency can be measured by obtaining the distance between minima, since it was previously shown that the distance between successive minima or maxima is one-half wavelength.

6·3 Errors in slotted-line techniques

The possible sources of error associated with slotted-line measurements must be carefully evaluated and proper operating techniques must be applied in order to eliminate or minimize these errors. The accuracy of standing-wave ratio measurements is limited by the connectors on the coaxial slotted section since the slotted section probe responds to the combined reflections from the connector and the load beyond the connector. The uncertainty in the standing-wave ratio measurements can be evaluated when the inherent SWR of the slotted section is known.

Variation of the maxima and minima at different points on the line is referred to as the *slope* error. It can be caused by variation of the probe depth as the probe carriage is varied and also by energy leakage through the slot. This error can be adjusted to a minimum value in some slotted sections.

Probe Tuning Errors. One of the major sources of error in standing-wave measurements is excessive probe penetration. The presence of the probe affects the VSWR because it is essentially an admittance shunting the line.

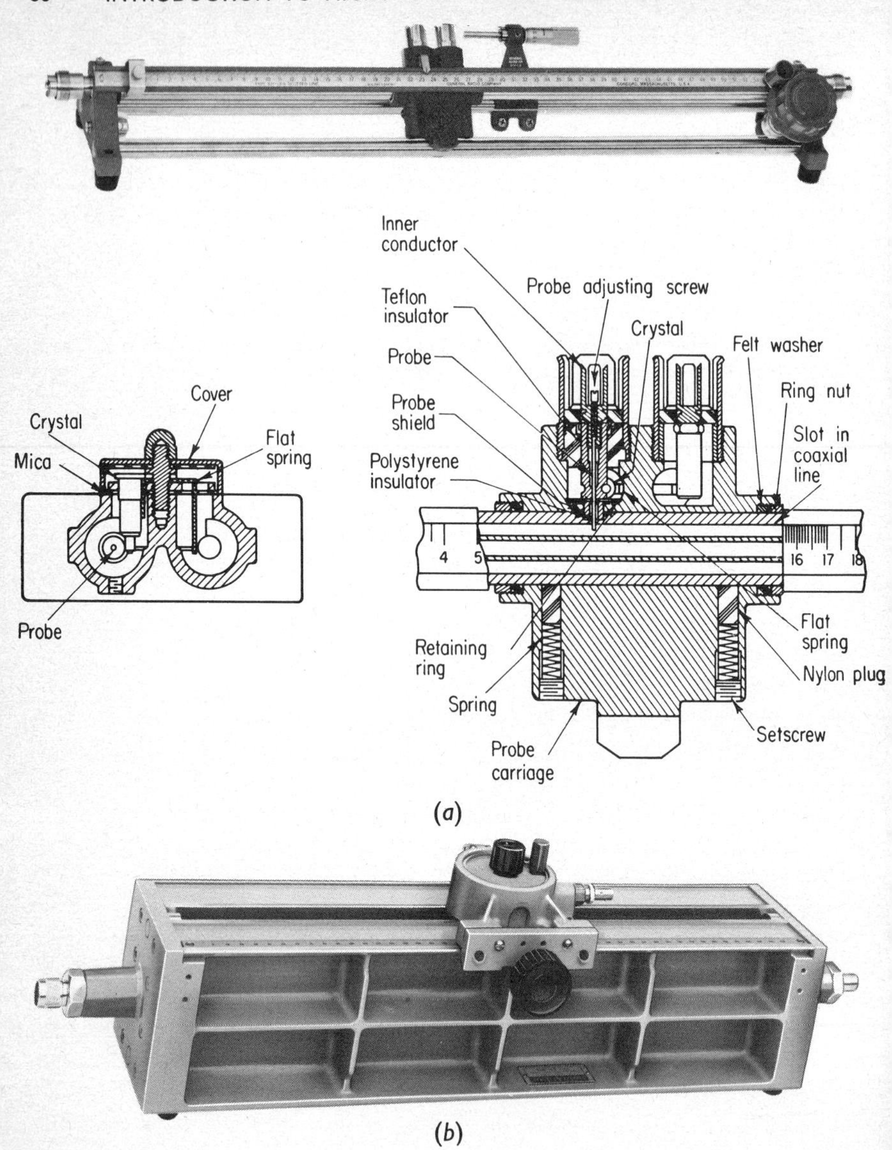

Fig. 6·2 Coaxial slotted sections. (a) General Radio Type 874 LBA slotted line. (*General Radio Company*.) (b) Model 805. (*Hewlett-Packard Company*.)

Excessive coupling to the line causes a shift in the maxima and minima and also causes the measured VSWR to be lower than the true VSWR. In addition to the distortion of the field pattern, reflections from the probe vary when the probe is moved. Errors in measurement of low VSWR arise when these reflections are re-reflected from a mismatched source. Therefore, the probe coupling should be kept as small as possible except in cases where it is only desired to examine the minimum point on the standing-wave pattern. Excessive probe penetration can be minimized by using a high sensitivity detector, assuming that there is adequate signal source power available.

Harmonics and Spurious Signals. Harmonics are usually present in signal sources that have coaxial outputs. Errors are possible if the probe is tuned to a harmonic of the fundamental. The frequency to which the probe is tuned can be easily checked by measuring the half-wavelength distance between two voltage minima on the line. Harmonics are usually reduced to a negligible value in coaxial systems by the use of low-pass filters.

Spurious signals usually arise from improper adjustment of modulating voltages used to square-wave-modulate a signal source such as a klystron which has several modes of oscillation. An illustration of improper modulation of a klystron is shown in Fig. 12·4.

Frequency Modulation. Variations of the instantaneous signal source frequency are referred to as frequency modulation (f-m). The minima of a standing-wave pattern are obscured in the presence of frequency modulation since the minima of the standing-wave patterns at the different frequencies do not appear at the same position on the line. If the f-m becomes excessive, it is possible that other portions of the standing-wave pattern can be distorted. Poor regulation of potentials applied to the oscillator is often responsible for f-m problems. In order to prevent frequency modulation of modulated signal sources, square-wave modulation is used. The presence of f-m on modulated sources is usually displayed on a scope using a frequency meter (frequency discriminating device). The f-m can also be detected by investigating the minimum of the standing-wave pattern when the slotted section is terminated with a short circuit.

Detector Characteristics. The characteristics of barretters and crystals determine the power levels in a measurement system. These detector elements are usually used with equipment which is calibrated in terms of the square-law response. In order to prevent a departure from the square-law response, the barretter should be operated at power levels less than 200 μw and the crystal should be operated at power levels less than 10 μw (Sec. 8·10). In either case, the departure from square law can be checked by noting the detector response on the standing-wave indicator for different levels of input power. This check is performed by measuring a fixed mismatch at different power inputs to the detecting element.

6·4 Standing-wave measurements

A typical setup for performing measurements with the slotted line in the frequency range below 5 Gc using an unmodulated source and a superheterodyne detector or receiver is shown in Fig. 6·3. It consists of a signal source, local oscillator, linear mixer, 30-Mc intermediate frequency (i-f)

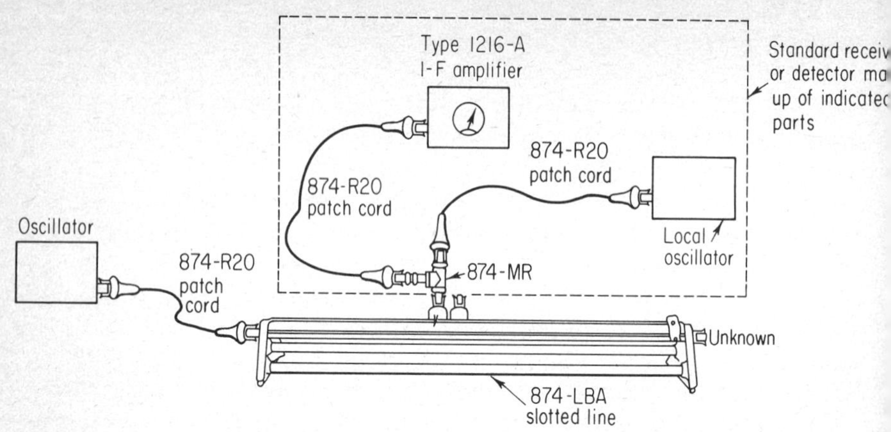

Fig. 6·3 A typical setup for measurements with the Type 874 LBA slotted line using an unmodulated source and superheterodyne detector or receiver. (*General Radio Company.*)

amplifier, and the slotted section. The continuous-wave (c-w) source frequency is sampled by the probe and mixed with the local oscillator frequency in the linear mixer. The resulting 30-Mc i-f signal is applied to the 30-Mc amplifier which incorporates a calibrated attenuator and output meter. The indicating meter has a voltage scale and decibel scale. The VSWR is calculated from the ratio of the maximum to minimum voltage as measured using the voltage scale, or it can be calculated by taking the antilog of the decibel variation since the SWR in decibels is given by

$$\text{SWR (db)} = 20 \log \text{VSWR} \tag{6·4}$$

When using this system, care must be taken to tune the local oscillator to beat with the desired signal and not with one of its harmonics. An amplifier output is produced by any two signals which beat together to produce a 30-Mc difference frequency. As an example, the local oscillator must be tuned to 470 or 530 Mc in order to produce the 30-Mc difference frequency when the source frequency is 500 Mc. If the local oscillator is tuned to 485 or

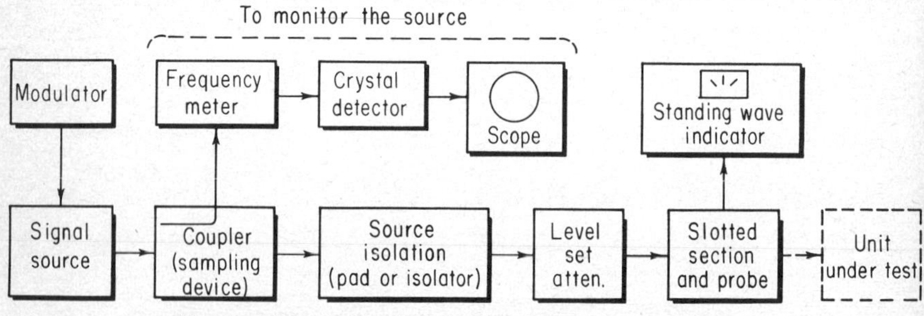

Fig. 6·4 Basic standing-wave measurement system.

515 Mc, it is noted that the second harmonics are 970 and 1,030 Mc, respectively. Either of these frequencies can mix with the second harmonic of 500 Mc to produce a 30-Mc difference frequency. In this case, the system would be operating at 1,000 Mc instead of the indicated 500 Mc.

A block diagram of the most commonly used standing-wave ratio measurement system is shown in Fig. 6·4. This basic system is used in all frequency ranges where the necessary system components are available. The source frequency is modulated at an audio frequency, usually 1,000 cycles per second. The standing-wave indicator is a selective tuned amplifier (tuned to the modulation frequency) which has a calibrated range switch and an output meter which is calibrated in terms of the square-law response of the detector. The standing-wave indicator also provides the necessary bias for barretters. The meter face of a standing-wave indicator is shown in Fig. 6·5. The VSWR measurement is made by moving the slotted section probe along the line to

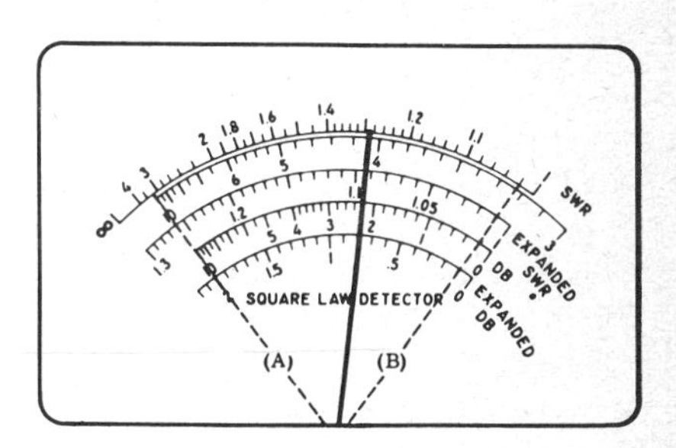

Fig. 6·5 Detail of meter face which is calibrated in terms of square-law detection. (*Hewlett-Packard Company.*)

obtain maximum deflection as indicated on the output meter. The amplifier gain is adjusted to place the meter pointer to one (1) on the SWR scale. The VSWR is read from the scale when the probe is adjusted for minimum deflection on the output meter. Standing-wave ratios up to 10:1 can be measured by switching downrange and reading VSWR values greater than 3:1 on the second scale. The expanded scale can be used to measure ratios less than 1.3:1. The SWR in decibels can also be obtained using this indicating meter.

Measurement of VSWR values greater than 10 : 1 requires special techniques. Accurate measurements can be made using a method referred to as the width-of-minimum or twice-minimum-power method. In using this method, it is necessary to establish the electrical distance Δl between points on the line at which the r-f voltage is $\sqrt{2}$ times the voltage at the minimum, as illustrated in Fig. 6·6. The validity of this method is based upon the assumption that the standing-wave pattern approximates a parabola in the vicinity of the minimum. If the output meter is calibrated for use with a square-wave detector, the ratio of 1.4:1 corresponds to 3 db, as shown on the meter dial in Fig. 6·5. The VSWR is related to the spacing (Δl in centimeters) and the waveguide wavelength (λ_g in centimeters) by the expression

$$\text{VSWR} \cong \frac{\lambda_g}{\pi \Delta l} \tag{6·5}$$

Measurements of high standing-wave ratios can be performed by measuring the standing-wave ratio in decibels. A precision calibrated attenuator is placed in the line with the level set attenuator (Fig. 6·4). With the precision

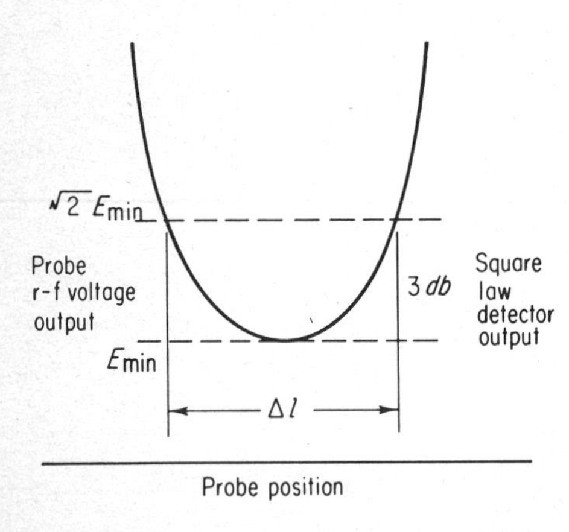

Fig. 6·6 Width of voltage minimum for determination of high VSWR.

attenuator set to zero db, the slotted-line probe is moved to a minimum of the standing-wave pattern, and a convenient reference is set on the standing-wave indicator meter. The slotted-line probe is then moved to the maximum of the standing-wave pattern, and the precision attenuator is used to decrease the standing-wave indicator reading to the original reference. The decibel change in attenuation is used to calculate the VSWR using Eq. (6·4).

6·5 Impedance measurements

A very substantial part of microwave techniques is centered around impedance concepts which explain the nature of transmission lines. Therefore, the measurement of impedance is important in many microwave applications. It has been shown that the input impedance varies with position along a transmission line which has standing waves. Therefore, the impedance measurement must be referred to some reference plane.

The load impedance can be determined on a lossless transmission line from the VSWR and the position of a voltage minimum with a load connected and with the line shorted. The VSWR of the load is measured, and the position of the minimum of the standing-wave pattern is obtained. The load is replaced with the short circuit and the shift in the minimum (less than $\lambda/4$) is obtained. The impedance is a function of the VSWR, the shift in the minimum, and the direction in which the minimum shifts. The reference plane is the plane of the short circuit. The voltage variations with a load connected and with the short connected are illustrated in Fig. 6·7. The load impedance is determined as follows:

1. Connect the load to the slotted section, measure the VSWR, and record the position of a voltage minimum as read on the slotted section scale.
2. Remove the load and connect a short circuit in its place. Record the shift in the minimum of the standing wave. The shift cannot be greater than a quarter of a wavelength. Note whether the shift in minimum is *toward the load* or *toward the generator.*

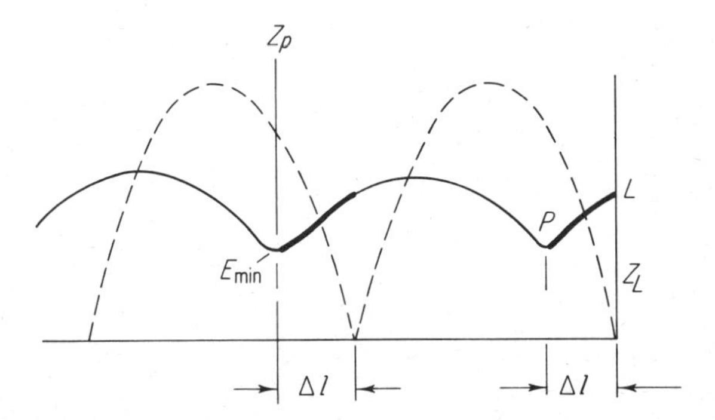

Fig. 6·7 Voltage variation along a transmission line with a load connected (solid lines) and with a short circuit connected at the load position (dotted line).

3. The normalized impedance can be computed from the equation

$$Z_L = Z_0 \frac{Z_p - jZ_0 \tan \beta(\Delta l)}{Z_0 - jZ_p \tan \beta(\Delta l)}$$

$$z_L = \frac{1 - j(\text{VSWR}) \tan \beta(\Delta l)}{\text{VSWR} - j \tan \beta(\Delta l)} \tag{6·6}$$

since $Z_p = Z_{\min}$ in the measurement. $\beta(\Delta l) = \dfrac{2\pi}{\lambda}(\pm\Delta l)$ in which case (Δl) is *positive* when the minimum shifts *toward the load* and is *negative* when the minimum shifts *toward the generator.*

The calculation of impedance transformation using the equations above can be time-consuming. The Smith chart can be used to solve the same problem as illustrated by the following example which is plotted on the Smith chart of Fig. 6·8.

1. The load VSWR is 2.4 and the VSWR circle is drawn through this point (A) on the chart.

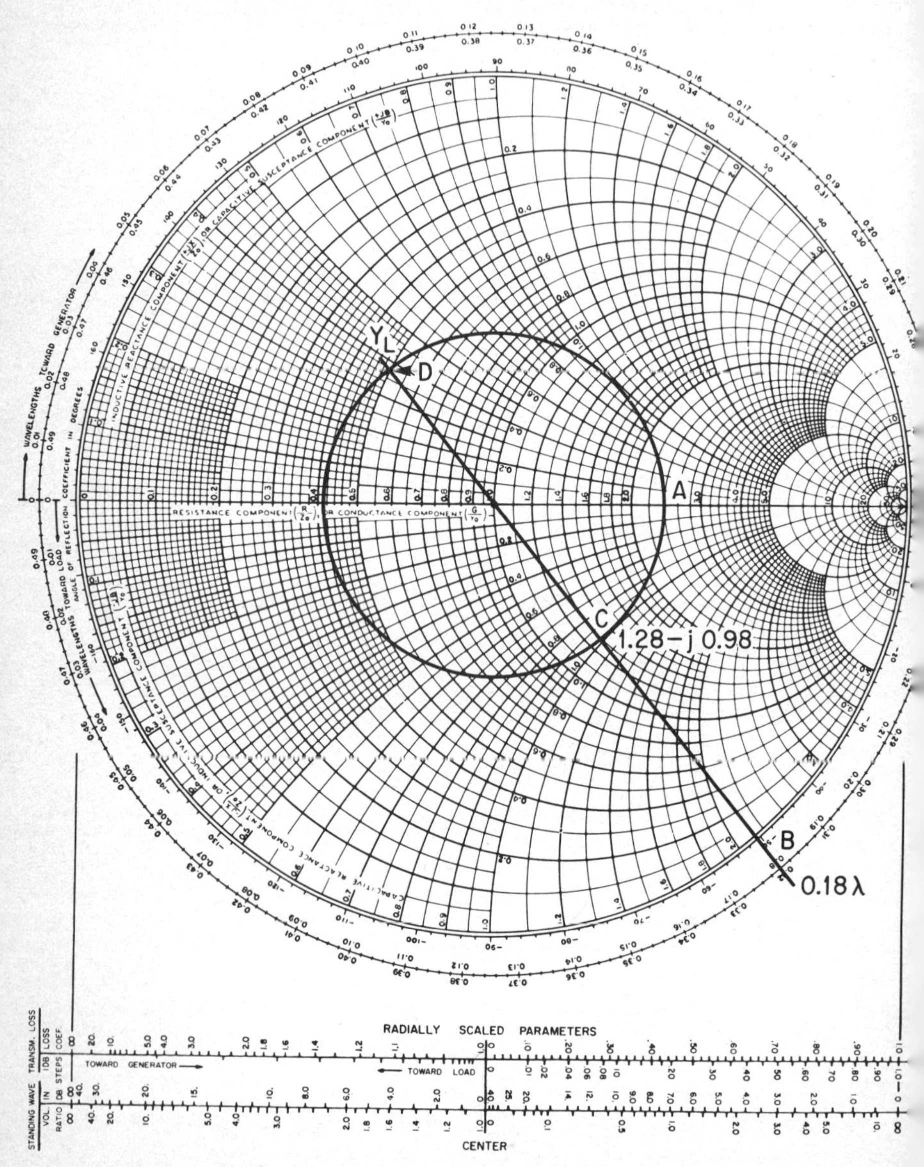

Fig. 6·8 Smith chart. (From *Irving L. Kosow (ed.), "Microwave Theory and Measurements," by the Engineering Staff of the Microwave Division, Hewlett-Packard Company, © 1962, by permission of Prentice-Hall, Inc., Englewood Cliffs, N.J.*)

2. The centimeter scale reading on the slotted section at the voltage minimum was 26 cm.
3. The minimum shifted to 15 cm when the short was connected to the line, and the shift in minimum was *toward the load.*
4. $\Delta l = 26 - 15 = 9$ cm.
5. The wavelength was found to be 50 cm, therefore $\Delta l/\lambda = 0.18$ wavelengths.
6. Proceed from the short-circuit impedance point on the Smith chart *toward the load* (CCW) to the 0.18 wavelength on the inside wavelength scale and draw a line from this point (B) to the center of the chart and extend the line to intersect the VSWR circle on the opposite side of the chart.
7. The normalized load impedance is located at (C) and is approximately $1.28 - j0.98$. The normalized load admittance is located diametrically opposite point C at D.

It can be seen in Fig. 6·7 that the shift of Δl represents the actual distance of E_{min} from the load, and, as one proceeds around the VSWR circle from E_{min} to point C on the Smith chart, the same distance is represented by the points P and L in Fig. 6·7.

If the minimum shifts *toward the load* when the short circuit is connected, the load is *capacitive*. If the shift in minimum is *toward the generator*, the load is *inductive*.

6·6 Residual VSWR of a slotted section

The VSWR caused by reflection from one or more discontinuities in the slotted section system is called the *residual* VSWR. The reflections which are predominant in causing the residual VSWR are fixed phase reflections. Therefore, the discontinuities can be considered as a *single lumped discontinuity*. This residual VSWR must be known in order to determine the possible effects of its interaction with the VSWR being measured.

In the ideal case, the residual VSWR could be determined by terminating the slotted section with a load having no reflection. The "perfect load" does not exist because all loads have some reflection. The effect of a perfect load can be obtained by using a measurement technique in which the slotted section is terminated with a low-loss sliding load. The reflections from the sliding load can be distinguished from those caused by the discontinuities of the slotted section. It should be noted that *any discontinuity in the connector of the load is combined with the residual of the slotted section and that only the VSWR of the load can be isolated from the rest of the system.*

The technique for determining the residual VSWR is illustrated by the relationships of the incident and reflected voltages in Fig. 6·9. The phasor combination of the two voltages is also discussed in Chap. 12.

The sliding load is connected to the slotted section in place of the usual load, and the vector addition of the voltage reflected from the load E_L and the voltage reflected from the slotted section discontinuity E_D add at some random phase, as indicated in Fig. 6·9*a*. The position of the slotted section probe and the position of the sliding load are varied for maximum output on the standing-wave indicator connected to the slotted section probe. This is the in-phase addition of E_L and E_D as shown in Fig. 6·9*b*, which represents the highest obtainable VSWR. The adjustments require some care because they are interdependent.

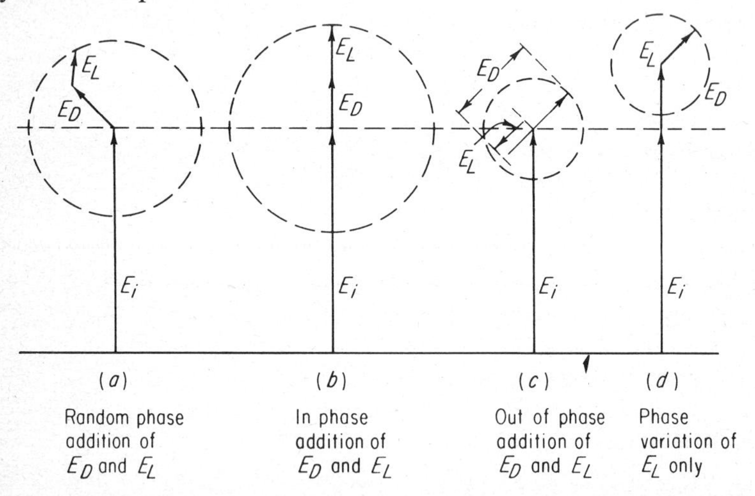

Fig. 6·9 Vector relations of incident and reflected voltages associated with the residual VSWR measurement.

The VSWR is measured with the slotted section, and the value obtained is

$$\rho_{\max} = \frac{E_i + (E_D + E_L)}{E_i - (E_D + E_L)} \tag{6·7}$$

which is represented in Fig. 6·9*b*.

The slotted section probe is returned to the maximum of the standing-wave pattern, and the sliding load is adjusted to obtain a minimum indication on the output at the standing-wave indicator. E_L and E_D are out of phase, and the result is indicated in Fig. 6·9*c*. The VSWR is measured again, and the new value obtained is

$$\rho_{\min} = \frac{E_i + |E_D - E_L|}{E_i - |E_D - E_L|} \tag{6·8}$$

The maximum and minimum reflection coefficients are calculated from

$$\Gamma_{\max} = \frac{\rho_{\max} - 1}{\rho_{\max} + 1} \tag{6·9}$$

$$\Gamma_{\min} = \frac{\rho_{\min} - 1}{\rho_{\min} + 1} \tag{6·10}$$

The reflection coefficient of the discontinuity Γ_D combines with the load reflection coefficient Γ_L to obtain

$$\Gamma_{max} = \Gamma_L + \Gamma_D \tag{6·11}$$

$$\Gamma_{min} = \Gamma_L - \Gamma_D \quad \text{or} \quad \Gamma_D - \Gamma_L \tag{6·12}$$

The load and discontinuity reflection coefficients are determined from the above equations.

$$\Gamma_L \quad \text{or} \quad \Gamma_D = \frac{\Gamma_{max} + \Gamma_{min}}{2} \tag{6·13}$$

$$\Gamma_L \quad \text{or} \quad \Gamma_D = \frac{\Gamma_{max} - \Gamma_{min}}{2} \tag{6·14}$$

The load and discontinuities can be evaluated, and it is only necessary to determine Γ_L or Γ_D. The equivalent VSWR values are obtained and the load can be varied in order to establish its VSWR when the slotted section probe is kept fixed in position. This variation of the load is shown in Fig. 6·9*d*. If the two VSWR values are almost the same value, it may be necessary to measure the residual with another load, or to deliberately introduce an additional reflection from the load (such as placing a small piece of metal foil on the load) and then perform the measurements again.

Another set of equations will be considered in order to avoid the previous time-consuming calculations. The interaction of two mismatches results in maximum and minimum mismatches given by the approximations

$$\rho_{max} = \rho_1 \rho_2 \tag{6·15}$$

$$\rho_{min} = \frac{\rho_1}{\rho_2} \tag{6·16}$$

assuming that ρ_1 is greater than ρ_2.

Solving for ρ_1 and ρ_2 it is found that

$$\rho_1 = \sqrt{\rho_{max}\rho_{min}} \tag{6·17}$$

$$\rho_2 = \sqrt{\frac{\rho_{max}}{\rho_{min}}} \tag{6·18}$$

It is only necessary to measure the maximum and minimum VSWR values and calculate ρ_L and ρ_D from Eqs. (6·17) and (6·18). ρ_L and ρ_D are uniquely determined as related previously.

6·7 Admittance meter

The General Radio Type 1602-B uhf admittance meter is used to measure the admittance and impedance of coaxial circuits in the frequency range from 40 to 1,500 Mc.

The schematic diagram of the circuit is shown in Fig. 6·10. Three adjustable loops couple to the magnetic fields in the coaxial lines. The input voltage is the same for each line since all lines are fed from a common source at a common junction point. The device balances in the same manner as a bridge, and zero output is obtained when the loops are properly oriented.

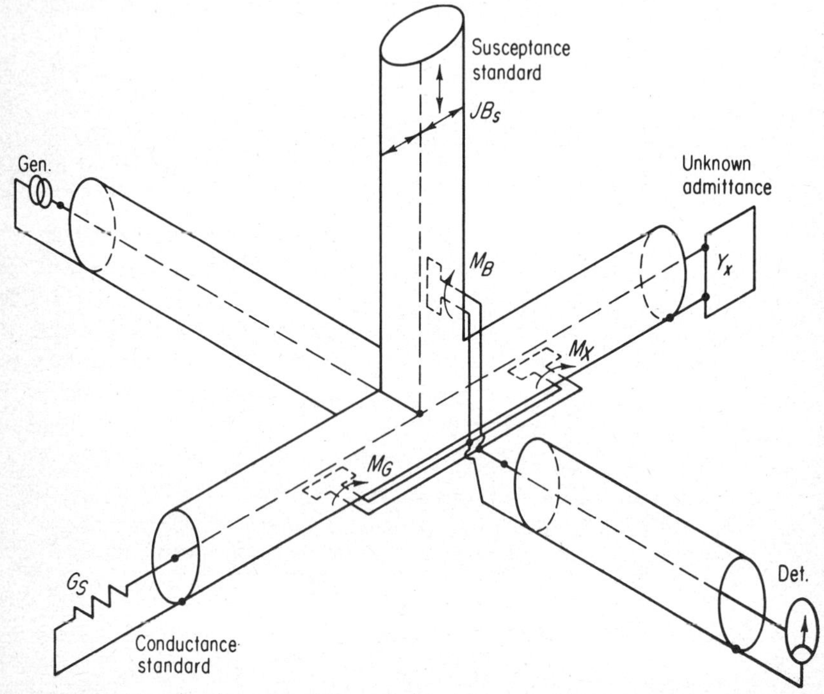

Fig. 6·10 Schematic diagram of the admittance meter. (*General Radio Company.*)

The admittance meter actually measures the admittance at a point inside the junction block directly under the center of the loop coupling to the unknown line. Therefore, a line-length correction is necessary in order to obtain the admittance or impedance at the point of connection to the line. The line-length correction can be eliminated if a low-loss adjustable constant-impedance line is used to obtain an integer multiple of a half wavelength between the admittance meter and the unknown, so that the unknown admittance is the same as the measured admittance. The meter balance is obtained by adjusting the line length when the adjustable constant impedance line is terminated with an open circuit. The unknown is then connected in place of the open circuit and causes the unbalance at a multiple of one-half wavelength from the load. When the meter is rebalanced, the unknown admittance is indicated by the position of the conductance and susceptance indicator arms.

The impedance measurement is performed in a similar way except that the adjustable constant impedance line is terminated with a short circuit prior to initial balance. The short circuit is removed, the load is connected, and the necessary adjustments are performed to obtain a balance or null. The conductance and susceptance scale readings are multiplied by the appropriate multiplying factors in order to obtain the impedance value.

Fig. 6·11 The admittance meter. (*General Radio Company.*)

6·8 Strip transmission lines

Strip transmission lines are essentially modifications of the two-wire line and coaxial lines. The form of the line called *microstrip*, which corresponds to the two-wire line, is shown in Fig. 6·12*a*. It is composed of a flat strip of transmission line, which may be photoetched, and a ground plane using only one copper-clad dielectric sheet. This type of line is more economical to manufacture compared to other types, but since it is incompletely shielded, it exhibits substantial radiation loss and stray coupling effects. Also, the phase velocity and characteristic impedance are unpredictable due to changes in thickness and dielectric constant of the copper-clad dielectric material.

The flat symmetrical-plane strip line of Fig. 6·12*b* can be compared to the coaxial line. The magnetic field lines circle the center conductor in conformance with its geometric shape. All electric field lines are in the transverse plane and terminate at the ground planes in the region of the center strip. The symmetrical strip line is also an open line but not to the same degree as

Fig. 6·12 Strip transmission lines. (*a*) Microstrip. (*b*) Symmetrical-strip line.

microstrip. There are several types of strip transmission line structures in addition to the examples in Fig. 6·12.

Strip lines possess the wide band characteristics of coaxial lines and have dominant modes with zero cutoff frequencies. They are well suited for use in complete microwave circuits where economy, light weight, and compactness are of foremost importance. Strip lines have been used extensively in many types of couplers, attenuators, and filters.

PROBLEMS

6·1 An antenna is connected to a slotted section. The VSWR is 1.8 and the operating frequency is 600 Mc. The antenna is replaced with a short circuit and the minimum of the standing-wave pattern shifts 6.2 cm toward the generator. What is the impedance of the antenna if Z_0 is 50 ohms?

6·2 The sliding load technique is used to measure the residual VSWR of the transmission system. $\rho_{\max}$ is 1.68 and $\rho_{\min}$ is 1.04. If the sliding load has the lowest VSWR, what is the residual VSWR of the system?

6·3 Calculate Z_0 of a coaxial cable if b and a dimensions are 0.402 and 0.175 in., respectively.

6·4 Repeat Prob. 6·3 with $b = 0.58$ in.

CHAPTER

7

WAVEGUIDES

Introduction. The definition of a transmission line which was given in Chap. 2 did not distinguish between the various forms of guiding structures. Any type of transmission line is a waveguide since it is designed primarily to "guide" or conduct energy from one point to another. The term "waveguide" as used in this text refers to guiding structures which enclose the electric and magnetic fields.

The most commonly used waveguide takes the form of a rectangular hollow metal pipe. Therefore, the waveguide theory developed in this chapter will be primarily centered around the rectangular waveguide, and a less detailed examination of transmission through circular pipes and dielectric rods will follow.

7·1 Review of basic concepts

It might be well to review some of the concepts of the electromagnetic wave and the electromagnetic properties of different media before going further into this subject. We do not know the exact nature of the electromagnetic field, as with gravity, but laws that it obeys have been devised by observation of its effects and by hypothesis. Some of the more important concepts are the following:

1. The source of electromagnetic waves is time-varying currents.
2. The electromagnetic wave may be pictured as a vector field consisting of two interrelated components, the electric vector and the magnetic vector.
3. Propagation of the fields is radial from the source in wave fashion except as affected by a change in media of propagation.
4. The properties of media affecting transmission are permittivity, permeability, and conductivity.
5. At the boundary between media of different electrical properties, the

electromagnetic wave will, in the general case, be partially transmitted and partially reflected.

a. The tangential components of the electric and magnetic fields are equal on the two sides of any boundary between physically real media.

b. The tangential component of electric field vanishes at the surface of a perfect conductor.

c. The normal component of a time-varying magnetic field vanishes at the surface of a perfect conductor.

6. Electric field lines may begin and end on charges. If an electric field ends on a conductor, it must represent a charge induced on the conductor.
7. Magnetic field lines can never end, since magnetic charges are not known physically. Magnetic fields must always form continuous closed loops, surrounding either a conduction current or a changing electric field (displacement current).
8. Electric field lines may form continuous closed paths, surrounding a changing magnetic field.
9. Electromagnetic waves are classified according to their equiphase surfaces or wavefronts, and the polarization is defined by the direction of the electric field.

Rays Lines indicating the direction of propagation.

Plane waves Wavefronts of the electric and magnetic fields are parallel to the plane of the source.

Transverse waves Electric and/or magnetic fields lie in a plane perpendicular to the direction of propagation.

1. TEM (transverse-electromagnetic)—Both the electric and magnetic fields are transverse to the direction of propagation.
2. TE (transverse-electric)—The electric field is transverse to the direction of propagation.
3. TM (transverse-magnetic)—The magnetic field is transverse to the direction of propagation.

Linear polarized waves The components of the electric and magnetic fields in the transverse plane do not change in direction from instant to instant or point to point.

Vertical polarization The electric field is in the vertical direction.

Horizontal polarization The electric field lies in the horizontal plane.

Circular polarization Resultant electric field when equal amplitude horizontal and vertical polarized waves 90° out of phase are combined. The direction of this wave varies with time and distance. The wave is elliptically polarized if the horizontal and vertical waves are not equal in amplitude.

7·2 Engineering aspects of guiding systems

The exact solution of the fields and currents of a guiding system involves several vector quantities varying in both magnitude and direction as a function of time and three-dimensional space. As would be expected, such a solution is at least a long and arduous process. Fortunately, however, it is rarely necessary to obtain an exact solution for field and current distribution. The factors of most interest in the transfer and distribution of electromagnetic energy in waveguides are the efficiency (losses), reflections (impedance), power-handling capacity, band width, network and special purpose application, physical aspects, and cost. To analyze these factors, or any other engineering problems, it is first wise to reduce the problem to the most simple form for solution. This is especially true in waveguide work due to the complexity of field theory. Actually, a large percentage of waveguide work concerns the application of transmission line theory which may be applied to waveguides almost without reservations. A reduction to simple circuit theory is possible in a number of instances. Although it is seldom necessary to actually derive field equations in waveguide work, it is quite important to have knowledge of the possible field and current distributions in the guiding system of interest in order to fully understand the problems that arise.

7·3 Elemental concept of the waveguide

Since it is difficult to attach physical significance to the mathematical procedure involved in determining the possible field and current distributions along the waveguide structure, we will attempt to justify the resulting field in

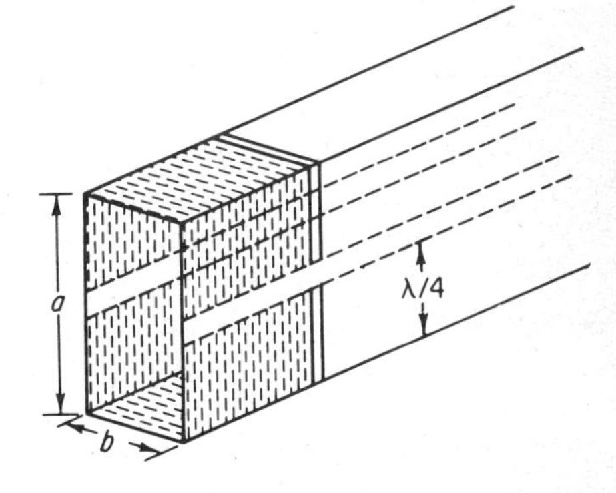

Fig. 7·1 Section of waveguide developed from an infinite number of elementary quarter-wave sections.

other ways. One method commonly used to derive the rectangular waveguide from the two-wire transmission line is illustrated in Fig. 7·1. The transmission line is supported by two quarter-wavelength sections, and since the input impedance of each section is theoretically infinite, they have no effect on the transmission of power. If the number of stubs is increased to infinity, the rectangular waveguide is formed as illustrated. It can be seen that the a dimension of the waveguide cannot be less than one-half wavelength. In fact, it must be slightly more than one-half wavelength in order to completely accommodate the transmission line function and at the same time preserve the insulating properties of the quarter-wave sections. Any frequency lower than that which makes the a dimension less than one-half wavelength will cause the circuit to become an inductive shunt and there is no propagation. The frequency at which the a dimension is one-half wavelength (free-space wavelength) is called the *cutoff* frequency and is designated f_c. The free-space wavelength associated with this cutoff frequency is the cutoff wavelength designated λ_c; ($\lambda_c = 2a$).

7·4 Advantages of hollow waveguides

There is no power loss by radiation from metal pipes of any type, including coaxial lines, if the ends are closed. Hollow waveguides are superior to coaxial lines in their ability to handle large concentrations of charges since these concentrations can be kept farther apart so that the electric field is less intense.

The construction of the waveguide is more simple than that of the coaxial cable since the inner conductor and its supports are eliminated. It is also more rugged and less susceptible to vibration and shock. Elimination of the insulating supports also results in a decrease of attenuation. Waveguides are usually air-filled, and for practical purposes they are considered to have no dielectric loss.

The current-carrying capacity is greater since, in practice, the waveguide is likely to have much greater conducting surface than a coaxial line. The overall power lost as heat in the walls of the waveguide is lower than the heat dissipated in the conductors of conventional size coaxial lines.

7·5 Reflections from a metal surface

The uniform plane wave is a highly idealized field pattern in which the electric and magnetic field wavefronts are parallel to the source. This idealized wavefront is never quite attained in practice but is assumed here in order to simplify the explanations. A great variety of complex field patterns is possible and may be classified in accordance with the shape of their wavefronts or equiphase surfaces. The locus of points at which the fields are in the same phase of variation is called the equiphase surface, and the direction of propagation is perpendicular to this surface. Consequently, the field varies from point to point except in the case of the idealized uniform plane wave.

Assume a plane wave incident upon a perfect conducting surface placed at an oblique angle as shown in Fig. 7·2*a*. The line which indicates the direction of propagation is referred to as the *incident ray* and is perpendicular to the equiphase surface. It intersects the conducting surface at the point P and makes an angle θ with the perpendicular at P as shown. Assume that the particular wavefront used for illustration represents a maximum equiphase surface of the wave, in which case, the lines of electric field intensity are pointing out of the page. The *ray* of the reflected wave lies in the plane of the incident ray and the normal to the boundary. The corresponding angles of reflection and incidence are equal. These equal angles θ are indicated in Fig. 7·2*a*. The ray of the incident and reflected waves is the free-space velocity of the electric and magnetic fields and is labeled v on the diagram. The angle between the incident wave and the conducting surface is also θ. The direction of the magnetic fields is shown, and all vectors of v, **E**, and **H** are properly directed to agree with the right-hand rule (Poynting's theorem).

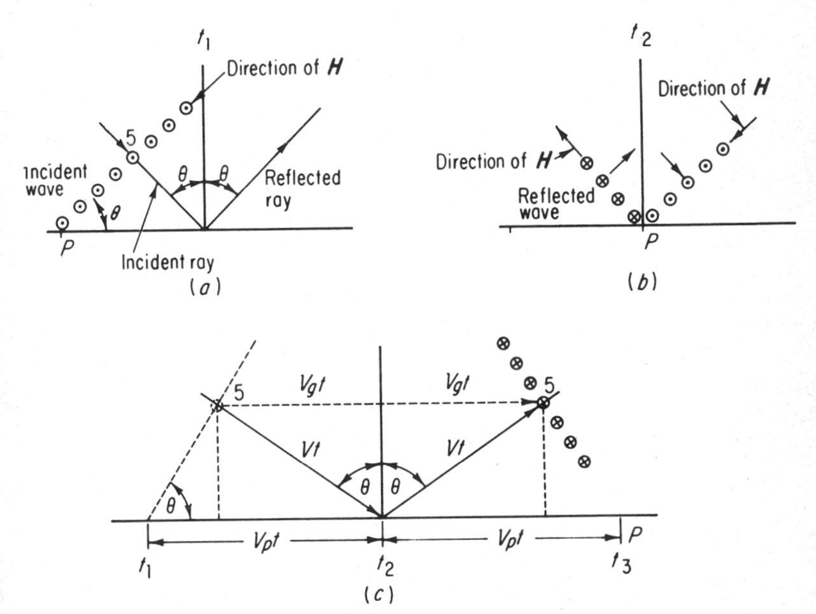

Fig. 7·2 Reflections from a conducting surface.

In order to satisfy the boundary conditions at the interface of a perfect conductor, the electric field is reflected without change in amplitude but with a reversal of phase. The tangential component of the magnetic field is reflected without change in amplitude or phase. This results in an electric force of zero everywhere along the interface and a tangential component of magnetic force equal to 2**H** at the interface.

The properties of this wave can be considered from two points of view. Assume that a nearsighted observer is located at *P* in Fig. 7·2*a* and that he wishes to single out the line of electric force labeled 5. This observer cannot see far beyond the point *P*, therefore he is unable to distinguish one line of force from another until the line of force is in the vicinity of some point *P* near the interface. The observer would have to travel parallel to the interface from the point *P* at t_1 in Fig. 7·2*a* to the point *P* at t_2 in Fig. 7·2*b* in order to single out line 5. The distance traveled is labeled v_pt in Fig. 7·2*c*. The velocity parallel to the interface is called the *phase velocity* and is given by the relation

$$v_p = \frac{v}{\sin \theta} \tag{7·1}$$

This is the velocity at which the point of incidence of the wavefront moves along the surface. It is an *apparent velocity* since the actual velocity of the electric line of force parallel to the interface is less than the free-space velocity.

If the angle θ becomes smaller, this apparent velocity increases, and, when the wave approaches the surface at perpendicular incidence, the phase velocity approaches infinity. The most common analogy is that of a long ocean wave which approaches the shore at an angle θ with the shore. The point at which the wave strikes the shore moves parallel to the shore at a velocity considerably greater than the velocity with which the wave actually advances perpendicular to the wavefront. If the wavefront approaches the shore at a perpendicular angle, θ is zero and the velocity of the point of wave contact is infinite.

If a second observer is capable of singling out particular lines of force, a different view is obtained. If he observes line 5 from time t_1 in Fig. 7·2*a* to time t_2 in Fig. 7·2*b*, he notes that the line of force travels a distance v_t (Fig. 7·2*c*) and that the actual progress parallel to the interface is $v_g t$ as shown in Fig. 7·2*c*. v_g is the effective velocity with which energy is propagated parallel to the metal surface and is called *group velocity*.

$$v_g = v \sin \theta \tag{7·2}$$

This velocity approaches zero when the wavefront approaches the surface at perpendicular incidence.

7·6 Field patterns obtained by oblique reflections

The field patterns in front of the plane conductor placed at an oblique angle to the direction of wave propagation may be determined through use of principles set forth in the previous section. Figure 7·3*a* shows the structure of the incident and reflected waves in terms of the electric field. The (+) and (−) lines are at right angles to the direction of power flow and represent wavefronts or regions of constant electric field strength and, for the purpose of explanation, are assumed to be straight and parallel. The velocity of the wavefront is the free-space velocity. The dark solid lines represent zero voltage points of the incident and reflected waves. The incident wave is reflected from the reflecting surface with the same amplitude but with a reversal of phase. The total electric field is zero over the entire surface, thus satisfying the boundary conditions.

Along the $E_{\max}$ line the (+) incident and reflected waves are in phase and the (−) incident and reflected waves are in phase. These waves combine and produce a resultant wave which is *longer* than the applied wave and which travels parallel to the surface of the conductor; along the $E_{\min}$ line the (+) and (—) incident and reflected waves are out of phase, and zero electric field exists at all points along the line. This process continues as indicated by the next $E_{\max}$ line, and it can be seen that a series of zero and maximum lines exists parallel to the metal surface.

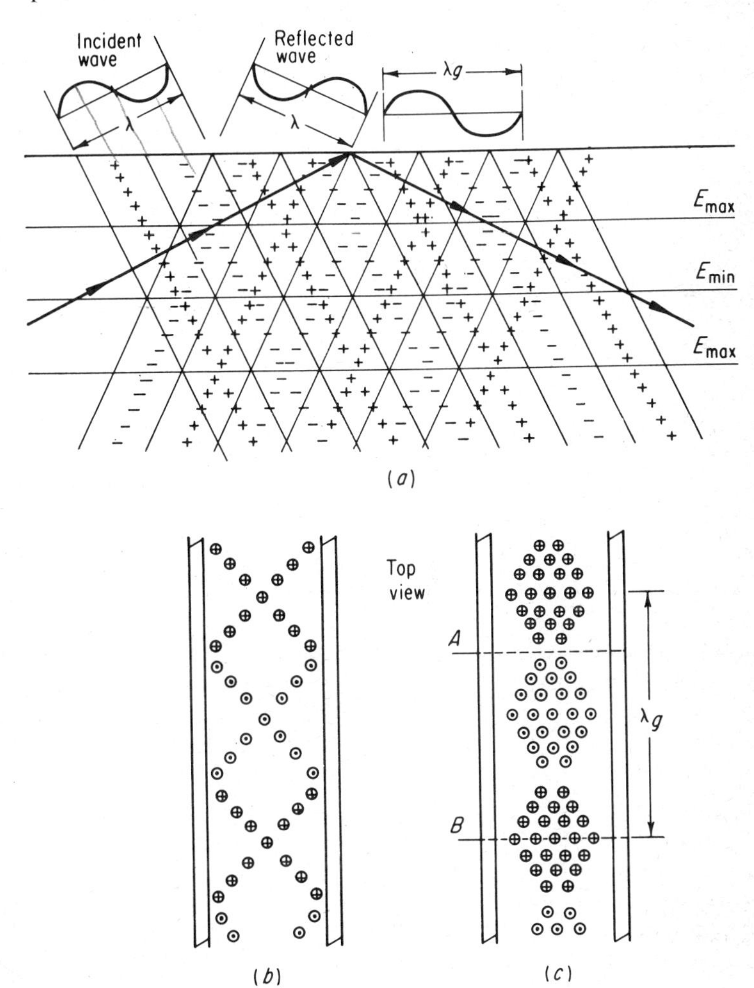

Fig. 7·3 (*a*) Field patterns obtained by oblique reflections from a plane conductor. (*b*) and (*c*) indicate the electric field distributions inside a waveguide.

Since a metal surface represents the zero voltage condition, it follows that wherever a zero voltage line occurs, a second metal-reflecting surface could be placed parallel to the first with no distortion of the wavefront pattern. It is apparent that many possible locations are available for the second metal surface, and each different location would result in a different field configuration. Each particular set of field configurations is referred to as a *mode*.

7·7 Waveguide transmission

Figure 7·3*b* represents the incident and reflected electric fields in a waveguide when viewed from the top. This is the same distribution between the surface and the first $E_{\min}$ as previously pointed out. The lines indicate the electric field distribution for the fundamental mode of operation. Each of the various sets of configurations can be considered as the resultant of a series of plane waves traveling with a velocity characteristic of the medium inside the guide, and all multiply reflected between opposite walls.

Figure 7·3*c* represents the electric field distribution in the waveguide. If we pass laterally across the guide at *A*, the instantaneous value of the electric field is everywhere zero. If we cross the guide at *B*, the electric field varies sinusoidally beginning at zero at either wall and reaching a maximum in the middle of the waveguide. If we travel along the *Z* axis of the guide, the electric field at any instant varies sinusoidally with distance. Thus, a *standing wave* is produced across the guide, and as time goes on, this standing wave moves down the guide. The peak value of this traveling wave is a function of the normal distance x from the reflecting surface and is given by

$$E = 2E_{im} \sin [(\beta \cos \theta)x] \qquad (7\cdot3)$$

$(\beta \cos \theta)$ is the phase constant of the standing wave in the x direction (across the waveguide). The peak value therefore varies between zero and $2E_{im}$ in accordance with variations of $\sin (\beta \cos \theta)x$.

The nature of the field patterns and the relationship of the free-space wavelength λ to the waveguide wavelength λ_g are illustrated in Fig. 7·4*a*, *b*, and *c*. The distance between the sides of the waveguide is labeled *a*. In Fig. 7·4*a* the free-space wavelength λ is long (low in frequency) and approaches the *a* dimension of the guide. The incident and reflected wavefronts make contact with the sides of the waveguide at a small angle θ. The rays (normal to the wavefronts) indicate that the wavefronts bounce back and forth across the waveguide many times in traveling down the guide. Therefore, power progresses down the guide at a velocity far below the free-space velocity with which the waves travel zigzag across the guide. From a previous discussion of a wavefront reflected from a metal surface, it was noted that the phase velocity of the wave became great as the angle θ became smaller. One-half of a waveguide wavelength $\lambda_g/2$ is indicated on the diagram by the intersections of the incident and reflected wavefronts and can be compared to $\lambda_g/2$ on the diagram of Fig. 7·3.

If the angle θ approaches zero, it can be seen that the phase velocity v_p approaches infinity and the waveguide wavelength λ_g approaches infinity. At the same time, the rays are normal to the two surfaces and the wavefronts are bouncing back and forth without making any progress down the waveguide (the group velocity v_g becomes zero). The free-space wavelength λ is equal to $2a$ for this particular case. The frequency at which this condition exists is called the *cutoff frequency*, and the corresponding free-space wavelength is called the *critical* or *cutoff wavelength*. At wavelengths greater than cutoff no appreciable amount of power is propagated through the guide.

Figure 7·4*b* and *c* illustrates the configurations when the frequency is increased. The wavefronts approach the sides of the guide at a greater angle,

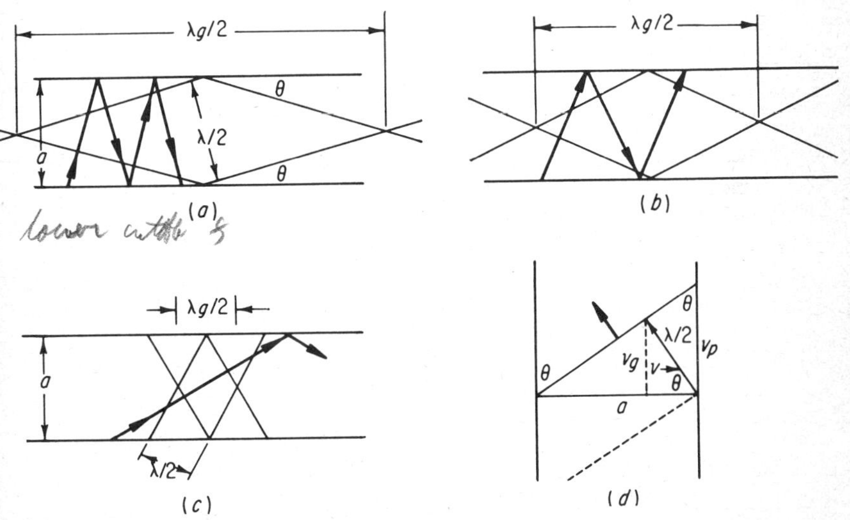

Fig. 7·4 Relationship of free-space wavelength and waveguide wavelength. Low frequency and the corresponding long wavelength is represented at (*a*). At (*b*), the frequency is higher than at (*a*) and lower than at (c). (*d*) Velocity and wavelength relationships.

the phase velocity decreases, λ_g decreases, the group velocity v_g increases, and there are less reflections from the sides of the guide. As the frequency is continually increased, the group velocity v_g and the phase velocity v_p approach the speed of light. The corresponding wavelengths approach the free-space wavelength.

The mathematical relationships related to waveguide transmission can be obtained from the geometrical relation between the free-space wavelength and the width (*a* dimension) of the waveguide as shown in Fig. 7·4*d*.

A right triangle is constructed with *a* and $\lambda/2$ as the sides. This illustration shows that

$$\cos\theta = \frac{\lambda}{2a} \tag{7·4}$$

The velocity at which any point of incidence of the wavefront moves along the guide is given by the relation

$$v_p = \frac{v}{\sin\theta} \tag{7·5}$$

and since

$$\sin\theta = \sqrt{1 - \cos^2\theta}$$

$$\sin\theta = \sqrt{1 - \left(\frac{\lambda}{2a}\right)^2} \tag{7·6}$$

By substitution into Eq. (7.5) the phase velocity is

$$v_p = \frac{v}{\sqrt{1 - (\lambda/2a)^2}} \tag{7·7}$$

Also

$$v_p = \frac{v}{\sqrt{1 - (\lambda/\lambda_c)^2}} = \frac{v}{\sqrt{1 - (f_c/f)^2}} \tag{7·8}$$

The waveguide-wavelength λ_g equation is obtained from the triangle of Fig, 7.4*d* as follows

$$\sin\theta = \frac{v}{v_p} = \frac{\lambda}{\lambda_g}$$

and

$$\lambda_g = \frac{\lambda}{\sin\theta} = \frac{\lambda}{\sqrt{1 - (\lambda/\lambda_c)^2}} = \frac{\lambda}{\sqrt{1 - (f_c/f)^2}} \tag{7·9}$$

In the particular configuration just described, only the electric field was considered and this field was *everywhere transverse*. A complete account of transmission must also include a consideration of the lines of magnetic force. The magnetic component may be either longitudinal or transverse, depending on the point in the guide at which the observations are made.

Along the line *A* in Fig. 7·3*c*, the magnetic vector is a maximum near each wall and decreases cosinusoidally to zero at the middle of the guide. In this type of wave, the magnetic lines of force form closed loops, and the electric field extends from the upper to the lower walls of the guide. Along the line *B* in Fig. 7·3*c*, the magnetic force is zero at the walls and increases to a maximum at the center of the guide. The field configurations are shown in Fig. 7·5.

A wave of this type is of great importance, and it is convenient to think of it as a new type of wave rather than a combination of two plane transverse waves. This wave can take one of two forms, designated as TE or TM modes.

In the TE (transverse-electric) mode, the electric field is transverse to the direction of propagation. In the TM (transverse-magnetic) mode, the magnetic field is transverse to the direction of propagation.

In addition, subscripts are used to describe the electric and magnetic field configurations. The general symbol will be TE_{mn} or TM_{mn} where the subscript *m* indicates the number of half-wave variations of the electric field intensity along the *a* (wide) dimension of the guide. The second subscript *n* indicates the number of half-wave variations of the electric field in the *b* (narrow) dimension of the guide. The field distributions for the TE_{10} mode are shown in Fig. 7·5. The TE_{10} mode has the *longest operating wavelength* and is designated as the *dominant mode*. It is the mode for the lowest frequency that can be propagated in a waveguide. When the *a* dimension is less than one-half wavelength, there is no propagation down the guide, therefore the

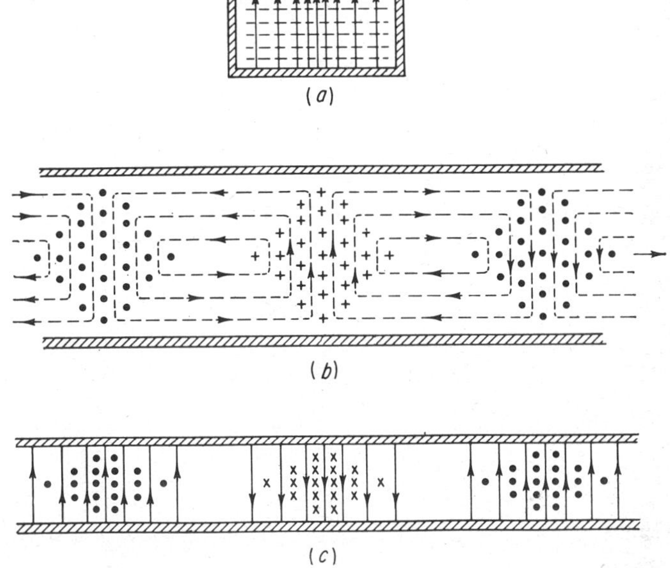

Fig. 7·5 Field configurations of the TE_{10} mode in the rectangular waveguide. (*a*) End view. (*b*) Top view. (*c*) Side view.

waveguide acts as a *high-pass filter* in that it passes all frequencies above a critical or cutoff frequency. For the standard rectangular waveguide the cutoff frequency is given as

$$f_c = \frac{2.998 \times 10^{10}}{2\sqrt{\mu'\epsilon'}}\sqrt{\left(\frac{m}{a}\right)^2 + \left(\frac{n}{b}\right)^2} \tag{7·10}$$

$$\lambda_c = \frac{2}{\sqrt{(m/a)^2 + (n/b)^2}} \tag{7·11}$$

where *a* and *b* are measured in centimeters.

The magnetic field has a definite relationship to the currents in the waveguide walls. If the walls are perfect conductors, the current is confined to sheets of near zero thickness at the inside surface of the walls. The current lines are everywhere perpendicular to the magnetic lines at the conductor surface. The magnetic field at the surface of the waveguide is longitudinal, as indicated in previous considerations, and the flow of current is parallel to the *b* dimension as shown in Fig. 7·6.

This current pattern propagates as a wave in the direction of propagation. The components of current in the top and bottom of the guide are related to the longitudinal components of the magnetic field and can be accounted for by the fact that the charge concentration that must be present to allow the beginning and termination of the electric field lines must travel along the guide. The distribution of the wall currents is important since it is often necessary to cut small slots in the walls to sample the fields inside the guide.

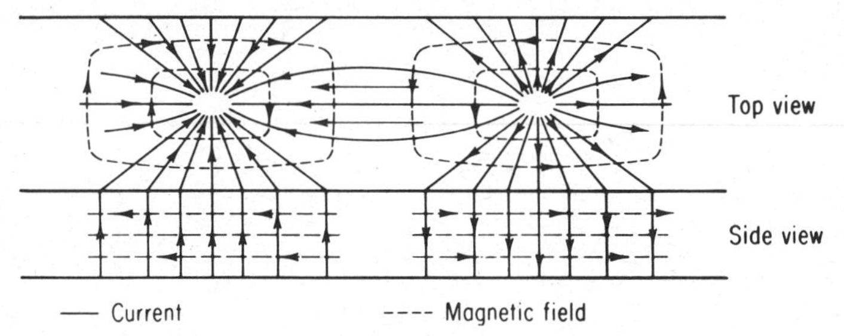

Fig. 7·6 Current distribution in the rectangular waveguide. The *dominant* (TE_{10}) *mode*.

These slots must be small in width and must be parallel to the direction of current flow in order to avoid disturbing the field inside the waveguide. The current path is completed by displacement current from top to bottom of the guide in the central regions shown in the illustration. The displacement current in the transverse plane is maximum where the electric field intensity is zero.

7·8 Higher order modes

A double infinity of higher order modes is possible in the rectangular pipe, as indicated by the integral number of half waves that may exist between the *a* and *b* dimensions. Practical use of waveguides is centered around the dominant mode, and these higher order modes are mainly of academic interest. In general, these waves are more highly attenuated and more difficult to recover from the waveguide system than the corresponding dominant wave.

If the source frequency is increased so that the *a* dimension is greater than one wavelength, the TE_{20} mode can exist. Also, half-wave distributions can be obtained in the *b* dimension resulting in, for example, the TE_{11} mode. The wave tends to remain in the dominant mode so that even though the source frequency may be high enough for higher modes to exist, they do not necessarily exist unless through deliberate or accidental distortion of the fields.

Higher order modes are present in the vicinity of a discontinuity where the fields are distorted. Since the waveguide dimensions, for the particular operating frequency, are such that the higher order mode cannot be propagated, these waves exist as fringing effects which become negligible at relatively short distances from their source.

Higher order modes are frequently given relative cutoff frequency values. These numbers are obtained by normalizing the cutoff frequency of the mode in question to the cutoff frequency of the TE_{10} mode in a rectangular waveguide. As an example, for the TE_{11} mode in a circular waveguide which has a diameter equal to the a dimension of the rectangular waveguide.

$$\text{Relative cutoff frequency} = \frac{f_c(\mathrm{TE}_{11})}{f_c(\mathrm{TE}_{10})} = \frac{7{,}700}{6{,}557} = 1.17$$

Several of the higher order modes are shown in Fig. 7·8.

7·9 Practical operating range

The dominant mode propagation without the presence of higher order modes is usually desirable since, for a given excitation frequency, it has

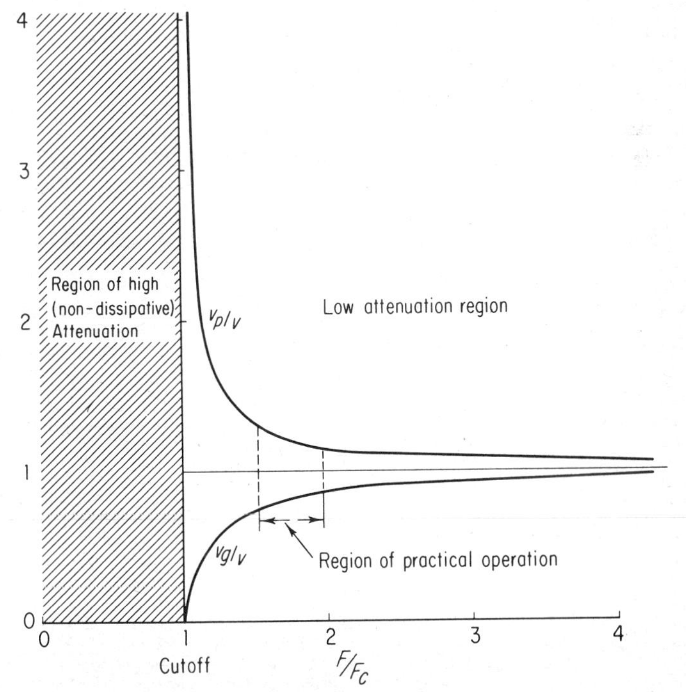

Fig. 7·7 Variation of phase velocity and group velocity with frequency.

lower power dissipation, requires smaller, lighter, and cheaper guiding structures, and requires simpler associated components.

The practical operating range of the TE_{10} mode in a rectangular waveguide where $b/a = 0.5$ is from 62 to 95 per cent of cutoff of the TE_{20} mode. Thus, the a dimension must be *greater* than $\lambda/2$ and *less* than λ.

The region of practical operation is shown in Fig. 7·7, which shows the relative phase velocity and group velocity for various conditions of waveguide operation.

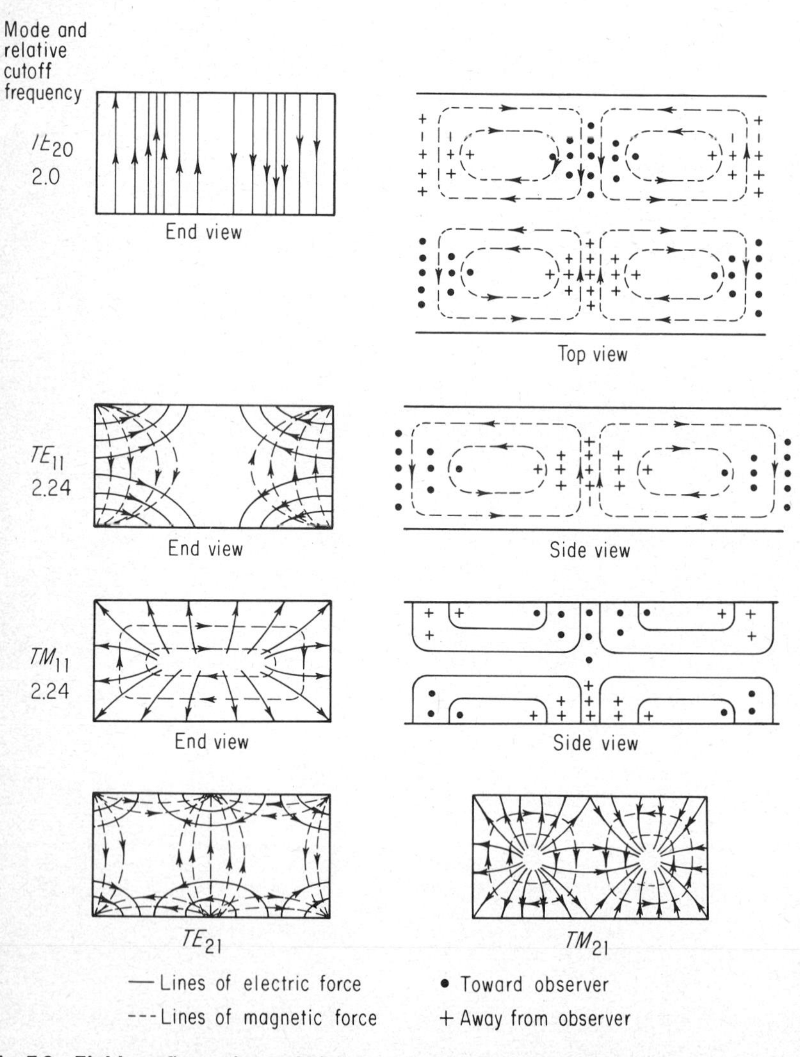

Fig. 7·8 Field configurations of higher order modes in the rectangular waveguide.

Reviewing again the previous analysis, we find that below cutoff the waveguide acts as a *nondissipative attenuator*, the group velocity is zero, and the phase velocity is infinite. As the operating frequency is increased far above the cutoff frequency, the phase and group velocities approach the free-space velocity.

7·10 Waveguide dimensions

Rectangular waveguides are usually chosen so that only the dominant mode exists over a certain frequency range. The above consideration determines the a dimension. The b dimension is important because of the following considerations:

a. The attenuation loss is greater as the b dimension is made smaller.

b. The b dimension determines the voltage breakdown characteristics and therefore determines the maximum power capacity.

The power transmitted by a waveguide has been determined from the Poynting vector evaluated at points on a cross section of guide. The total power transmitted is **P**.

$$\mathbf{P} = \tfrac{1}{4}\mathbf{E}_m\mathbf{H}_{xm}ab$$

and

$$\mathbf{H}_{xm} = \frac{\mathbf{E}_m}{Z_i}\frac{\lambda}{\lambda_g} = \frac{\mathbf{E}_m}{Z_i}\sqrt{1 - \left(\frac{\lambda}{\lambda_c}\right)^2}$$

therefore

$$\mathbf{P} = \tfrac{1}{4}\frac{\mathbf{E}_m{}^2}{Z_i}\sqrt{1 - \left(\frac{\lambda}{\lambda_c}\right)^2}\,ab \tag{7.12}$$

The power capacity depends on the maximum electric field intensity, a, and b. Therefore, a and b should be made as large as possible for higher power capacity. In practice, the dimension b is usually chosen to be about one-half of a. The ratio of b/a is ½. Standard waveguide dimensions are shown in Table 7·1. The accepted waveguide designations are given in Col. 1. The MDL letter designations are given in Col. 2. The WR-90 waveguide is usually referred to as *X band* as shown. However, the letter designations have not been standardized among the various manufacturers of microwave waveguide components.

7·11 Waveguide wave impedance

The *characteristic wave impedance* is analogous to the characteristic impedance of the two-wire and coaxial lines. The wave impedance represents the ratio of the electric to the magnetic fields, in which case, the electric field is the analogue of voltage and the magnetic field is the analogue of current.

The actual wave impedance is of little use and is seldom determined. There is no advantage in extending calculations beyond the determination of the

$$Z_0 = \frac{120\pi}{\sqrt{1-\left(\frac{f_c}{f}\right)^2}}$$

Table 7·1 Reference table of rigid rec-

				Waveguide						
EIA designation WR( )	MDL designation ( ) band	JAN designation RG( )/U	Material alloy	Dimensions (in.)					Recommended operating range for TE_{10} mode	
				Inside	Tol.	Outside	Tol.	Wall thickness nominal	Frequency (kmc/sec)	Wavelength (cm)
2300	2300	. . .	Alum.	23.000–11.500	±0.020	23.250–11.750	±0.020	0.125	0.32–0.49	93.68–61.18
2100	2100	. . .	Alum.	21.000–10.500	±0.020	21.250–10.750	±0.020	0.125	0.35–0.53	85.65–56.56
1800	1800	201	Alum.	18.000–9.000	±0.020	18.250–9.250	±0.020	0.125	0.41–0.625	73.11–47.96
1500	1500	202	Alum.	15.000–7.500	±0.015	15.250–7.750	±0.015	0.125	0.49–0.75	61.18–39.97
1150	1150	203	Alum.	11.500–5.750	±0.015	11.750–6.000	±0.015	0.125	0.64–0.96	46.84–31.23
975	975	204	Alum.	9.750–4.875	±0.010	10.000–5.125	±0.010	0.125	0.75–1.12	39.95–26.76
770	770	205	Alum.	7.700–3.850	±0.005	7.950–4.100	±0.005	0.125	0.96–1.45	31.23–20.67
650	*L*	69 103	Copper Aluminum	6.500–3.250	±0.005	6.660–3.410	±0.005	0.080	1.12–1.70	26.76–17.63
510	510	. . .	. . .	5.100–2.550	±0.005	5.260–2.710	±0.005	0.080	1.45–2.20	20.67–13.62
430	*W*	104 105	Copper Aluminum	4.300–2.150	±0.005	4.460–2.310	±0.005	0.080	1.70–2.60	17.63–11.53
340	340	112 113	Copper Aluminum	3.400–1.700	±0.005	3.560–1.860	±0.005	0.080	2.20–3.30	13.63–9.08
284	*S*	48 75	Copper Aluminum	2.840–1.340	±0.005	3.000–1.500	±0.005	0.080	2.60–3.95	11.53–7.59
229	229	. . .	. . .	2.290–1.145	±0.005	2.418–1.273	±0.005	0.064	3.30–4.90	9.08–6.12
187	*C*	49 95	Copper Aluminum	1.872–0.872	±0.005	2.000–1.000	±0.005	0.064	3.95–5.85	7.59–5.12
159	159	. . .	. . .	1.590–0.795	±0.004	1.718–0.923	±0.004	0.064	4.90–7.05	6.12–4.25
137	X_B	50 106	Copper Aluminum	1.372–0.622	±0.004	1.500–0.750	±0.004	0.064	5.85–8.20	5.12–3.66
112	X_L	51 68	Copper Aluminum	1.122–0.497	±0.004	1.250–0.625	±0.004	0.064	7.05–10.00	4.25–2.99
90	*X*	52 67	Copper Aluminum	0.900–0.400	±0.003	1.000–0.500	±0.003	0.050	8.20–12.40	3.66–2.42
75	75	. . .	. . .	0.750–0.375	±0.003	0.850–0.475	±0.003	0.050	10.00–15.00	2.99–2.00
62	K_U	91 — 107	Copper Aluminum Silver	0.622–0.311	±0.0025	0.702–0.391	±0.003	0.040	12.4–18.00	2.42–1.66
51	51	. . .	. . .	0.510–0.255	±0.0025	0.590–0.335	±0.003	0.040	15.00–22.00	2.00–1.36
42	*K*	53 121 66	Copper Aluminum Silver	0.420–0.170	±0.0020	0.500–0.250	±0.003	0.040	18.00–26.50	1.66–1.13
34	34	. . .	. . .	0.340–0.170	±0.0020	0.420–0.250	±0.003	0.040	22.00–33.00	1.36–0.91
28	K_A	— — 96	Copper Aluminum Silver	0.280–0.140	±0.0015	0.360–0.220	±0.002	0.040	26.50–40.00	1.13–0.75
22	*Q*	— 97	Copper Silver	0.224–0.112	±0.0010	0.304–0.192	±0.002	0.040	33.00–50.00	0.91–0.60
19	19	. . .	. . .	0.188–0.094	±0.0010	0.268–0.174	±0.002	0.040	40.00–60.00	0.75–0.50
15	*V*	— 98	Copper Silver	0.148–0.074	±0.0010	0.228–0.154	±0.002	0.040	50.00–75.00	0.60–0.40
12	12	— 99	Copper Silver	0.122–0.061	±0.0005	0.202–0.141	±0.002	0.040	60.00–90.00	0.50–0.33
10	10	. . .	. . .	0.100–0.050	±0.0005	0.180–0.130	±0.002	0.040	75.00–110.00	0.40–0.27

* This is an MDL flange number.
SOURCE: Microwave Development Laboratories, Inc.

tangular waveguide data and fittings

Cutoff for TE_{10} mode		Range in $\frac{2\lambda}{\lambda c}$	Range in $\frac{\lambda g}{\lambda}$	Theoretical attenuation lowest to highest frequency (db/100 ft)	Theoretical c-w power rating lowest to highest frequency (mw)	Fittings		EIA designation WR ( )
						Flange		
Frequency (kmc/sec)	Wavelength (cm)					Choke UG( )/U	Cover UG( )/U	
0.256	116.84	1.60–1.05	1.68–1.17	0.051–0.031	153.0–212.0	...		2300
0.281	106.68	1.62–1.06	1.68–1.18	0.054–0.034	120.0–173.0	...	FA168A*	2100
0.328	91.44	1.60–1.05	1.67–1.18	0.056–0.038	93.4–131.9	...		1800
0.393	76.20	1.61–1.05	1.62–1.17	0.069–0.050	67.6–93.3	...		1500
0.513	58.42	1.60–1.07	1.82–1.18	0.128–0.075	35.0–53.8	...		1150
0.605	49.53	1.61–1.08	1.70–1.19	0.137–0.095	27.0–38.5	...		975
0.766	39.12	1.60–1.06	1.66–1.18	0.201–0.136	17.2–24.1	...		770
0.908	33.02	1.62–1.07	1.70–1.18	0.317–0.212 0.269–0.178	11.9–17.2	... ...	417A 418A	650
1.157	25.91	1.60–1.05	1.67–1.18	...	...	...		510
1.372	21.84	1.61–1.06	1.70–1.18	0.588–0.385 0.501–0.330	5.2–7.5	... ...	435A 437A	430
1.736	17.27	1.58–1.05	1.78–1.22	0.877–0.572 0.751–0.492	3.1–4.5	... ...	553 554	340
2.078	14.43	1.60–1.05	1.67–1.17	1.102–0.752 0.940–0.641	2.2–3.2	54A 585	53 584	284
2.577	11.63	1.56–1.05	1.62–1.17	...	...	...		229
3.152	9.510	1.60–1.08	1.67–1.19	2.08–1.44 1.77–1.12	1.4–2.0	148B 406A	149A 407	187
3.711	8.078	1.51–1.05	1.52–1.19	...	...	...		159
4.301	6.970	1.47–1.05	1.48–1.17	2.87–2.30 2.45–1.94	0.56–0.71	343A 440A	344 441	137
5.259	5.700	1.49–1.05	1.51–1.17	4.12–3.21 3.50–2.74	0.35–0.46	52A 137A	51 138	112
6.557	4.572	1.60–1.06	1.68–1.18	6.45–4.48 5.49–3.83	0.20–0.29	40A 136A	39 135	90
7.868	3.810	1.57–1.05	1.64–1.17	...	...	...		75
9.486	3.160	1.53–1.05	1.55–1.18	9.51–8.31 ... 6.14–5.36	0.12–0.16	541 FA190A* ...	419 FA191A* ...	62
11.574	2.590	1.54–1.05	1.58–1.18	...	...	...		51
14.047	2.134	1.56–1.06	1.60–1.18	20.7–14.8 17.6–12.6 13.3–9.5	0.043–0.058	596 598 ...	595 597 ...	42
17.328	1.730	1.57–1.05	1.62–1.18	...	...	...		34
21.081	1.422	1.59–1.05	1.65–1.17	... ... 21.9–15.0	0.022–0.031	600 FA1241A* ...	599 FA1242A* ...	28
26.342	1.138	1.60–1.05	1.67–1.17	... 31.0–20.9	0.014–0.020	... ...	383 ...	22
31.357	0.956	1.57–1.05	1.63–1.16	...	...	...		19
39.863	0.752	1.60–1.06	1.67–1.17	... 52.9–39.1	0.0063–0.0090	... ...	385 ...	15
48.350	0.620	1.61–1.06	1.68–1.18	... 93.3–52.2	0.0042–0.060	... ...	387 ...	12
59.010	0.508	1.57–1.06	1.61–1.18	...	...	...		10

per-unit impedance since the actual measurement of impedance in a waveguide results in a value of normalized impedance. The wave impedance equations for TE and TM modes are

$$Z_{\mathrm{TE}} = \frac{\eta}{\sqrt{1 - (\lambda/\lambda_c)^2}} = \frac{\eta}{\sqrt{1 - (f_c/f)^2}} \tag{7·13}$$

$$Z_{\mathrm{TM}} = \eta\sqrt{1 - \left(\frac{\lambda}{\lambda_c}\right)^2} = \eta\sqrt{1 - \left(\frac{f_c}{f}\right)^2} \tag{7·14}$$

where η is the intrinsic impedance of the medium (377 ohms for free space).

7·12 Comparison and summary of modes of propagation

1. The characteristic impedance and phase velocity are essentially independent of frequency in the TEM mode of propagation. Lossless transmission lines carrying TEM waves are nondispersive.
2. Propagation occurs in a waveguide only above a critical frequency called the *cutoff frequency*. Thus, a waveguide carrying a TE or TM wave may be considered to be a high pass filter as far as the particular mode is concerned. At frequencies far above cutoff, the propagation characteristics approach that of free space, i.e., the waveguide wavelength approaches the free unbounded wavelength in the medium contained inside the waveguide.
3. In the waveguide, the phase and group velocities and the wave impedance of TE and TM waves are functions of frequency. Therefore, the waveguide is a *dispersive* medium in that it tends to spread out waves of different frequencies which may be applied simultaneously at a common point. This dispersion phenomenon would cause distortion in waves which have modulation components extending over a wide frequency range.
4. TEM waves cannot exist in closed waveguides without inner conductors, but TE and TM waves may exist as higher order modes on two-conductor or coaxial lines if the operating frequency is sufficiently high.

7·13 Ridged waveguide

A ridged waveguide may be constructed by adding a longitudinal metal strip to the top and/or bottom of a standard rectangular guide as shown in Fig. 7·9*a*. The ridge acts as a uniform distributed loading and reduces the characteristic impedance of the waveguide and lowers the phase velocity. The reduction in phase velocity is accompanied by a lowering of the cutoff frequency of the TE_{10} mode by a factor that may be more than 5 to 1. At the same time, the cutoff frequency of the TE_{20} and TE_{30} modes is higher, depending on the ridge width. In this case, the ridge should be less than half the total a dimension of the guide for the TE_{20} mode and between one-third and two-thirds of the total dimension. This increase in bandwidth is accompanied by an increase in the losses in the boundary walls and a decrease in power-handling capacity. By suitable variations of ridge dimensions, it is possible to vary the characteristic impedance of the guide by a factor of 25 or more and increase the attenuation by several hundred. The ridge waveguide is useful in certain coupling and matching requirements since the

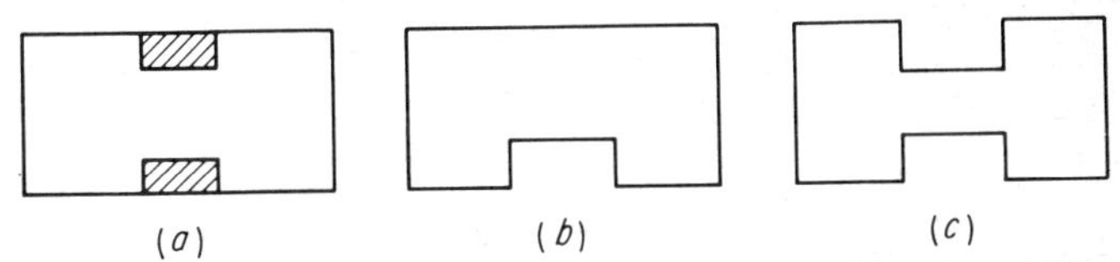

Fig. 7·9 Single-ridge (*b*) and double-ridge waveguide (*a*) and (*c*).

characteristic impedance can be changed easily by gradually tapering the ridge.

7·14 Dielectric rod or slab guides

If the dielectric constant of a material is substantially higher than the surrounding space, the interface between the two media acts as the guiding discontinuity, and waves very similar to those propagated through metal pipes may be propagated through the rod or slab of dielectric material. This follows from the concept of total reflection by which a wave traveling in a dense dielectric can strike a boundary of less dense dielectric at an angle of incidence greater than a certain critical angle where all energy is reflected. At frequencies lower than this critical frequency the dielectric does not act as a perfect waveguide. For large diameters, the velocities in dielectric wires and metal pipes are substantially the same. There is no definite cutoff and power is propagated even when the diameter of the guide is relatively small. The necessary guiding discontinuity is not, for most materials, as definite as that provided by a metal surface, and, as a result, the fields associated with the transmitted power reside partly inside and partly outside the guiding medium. One of the most important practical uses of dielectric rod guides has been for radiation where the leakage along the rod is permitted in order to form an end-fire antenna array.

7·15 Circular waveguides

The circular waveguide is used in many special applications in microwave techniques. The circular guide has the advantage of greater power-handling capacity and lower attenuation for a given cutoff wavelength, but it has the disadvantage of somewhat greater size and weight. Also, the polarization of the transmitted wave can be altered due to minor irregularities of the wall surface of the circular guide, whereas the rectangular cross section definitely fixes the polarization.

The wave of lowest frequency or the *dominant mode* in the circular waveguide is the TE_{11} mode. The subscripts which describe the modes in the circular waveguide are different than for the rectangular waveguide. For the circular waveguide, *the first subscript m indicates the number of full-wave*

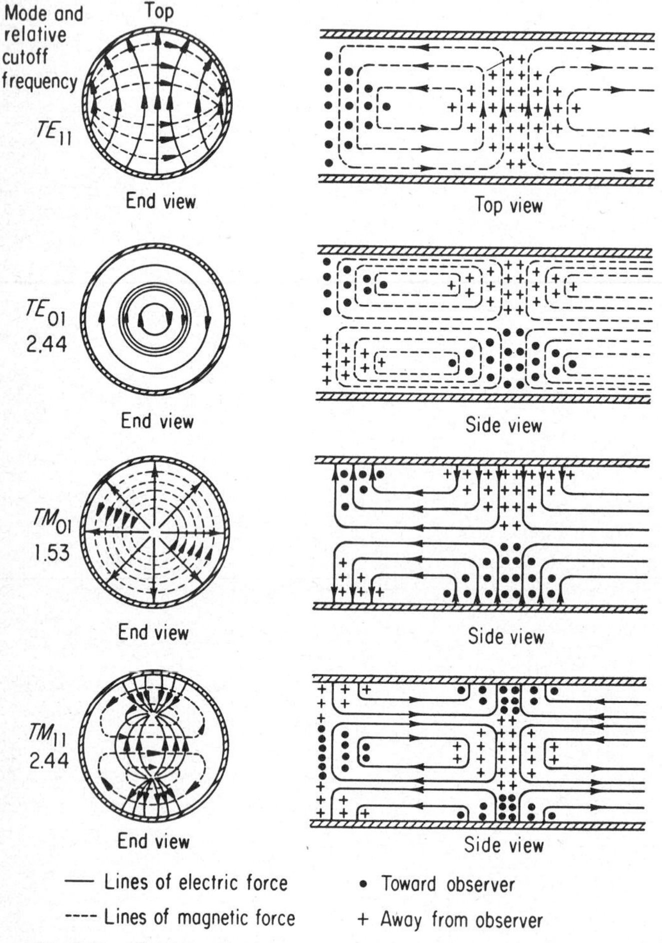

Fig. 7·10 Field configurations of modes in the circular waveguide.

variations of the radial component of the electric field around the circumference of the guide. The second subscript n indicates the number of half-wave variations across a diameter. Also, the first subscript indicates the number of diameters that can be drawn perpendicular to all electric field lines. The *second* subscript is the number of conducting cylinders (plus one for the guide) that can be inserted to divide the pattern, and in the case of TE_{0n} waves, it indicates the half-wave variation of the electric field across a radius of the guide.

The higher order modes which have properties of importance are the TM_{01} and TE_{01} modes. The TM_{01} mode has a circular symmetry which makes it adaptable for use in rotating joints. The TE_{01} mode has the property of decreasing attenuation as the frequency of the source is increased relative to cutoff. This mode is used for high-Q frequency meters.

The electromagnetic equations for circular waveguides usually appear as Bessel's functions. The theory of Bessel's functions is beyond the scope of this book but the functions may be used in our present requirement without any knowledge of the functions themselves. The cutoff wavelength and also the conditions for higher order modes are related to the waveguide radius a

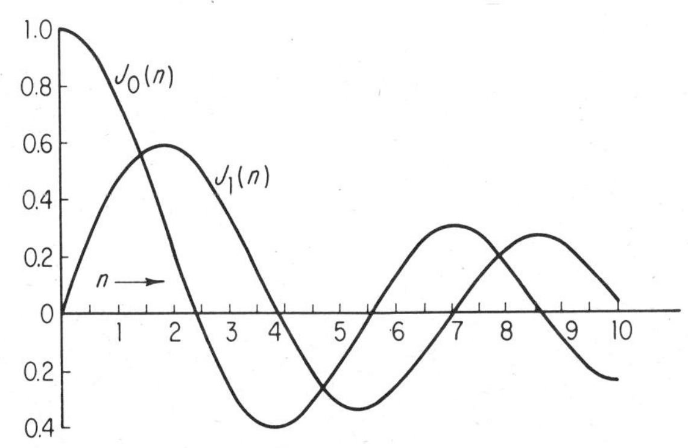

Fig. 7·11 The Bessel functions.

by way of the zeros of the particular Bessel function as designated by the subscripts of the mode being considered.

The approximate curves of the Bessel functions $J_0(n)$ and $J_1(n)$ are shown in Fig. 7·11. The TM_{mn} waves or modes may be supported when

$$\frac{2\pi a}{\lambda} = p_{mn}$$

where $J_m(n) = 0$ and p_{mn} is the nth root of $J_m(n) = 0$.

For TM_{0n} modes (the subscript m is zero)

$$\frac{2\pi a}{\lambda} = 2.405,\ 5.52,\ 8.65,\ \text{etc.}$$

These values correspond respectively to the subscripts $n = 1, 2, 3$, etc., as previously defined in terms of the field configurations inside the guide. As an example, consider the TM_{02} mode in which case p_{mn} equals p_{02} equals the second root of $J_0(2) = 0$. This means that the Bessel function under consideration is $J_0(n)$ and that the value of $2\pi a/\lambda$ is found where this curve crosses the x axis the *second* time. The value is 5.52, which can be closely read from the curve.

For TM_{1n} waves (the subscript $m = 1$)

$$\frac{2\pi a}{\lambda} = p_{mn} \qquad \text{where} \qquad J_1(n) = 0$$

$$\frac{2\pi a}{\lambda} = 3.83,\ 7.02,\ 10.17,\ \ldots$$

$$n = 1,\ 2,\ 3,\ \text{etc.}$$

For TE_{0n} modes or waves

$$\frac{2\pi a}{\lambda} = p'_{mn} \qquad \text{where} \qquad J_0'(n) = 0$$

$J_0'(n)$ is the first derivative of $J_0(n)$ and $J_0'(n) = J_1(n)$. Therefore, the values of $2\pi a/\lambda$ are the same for TE_{0n} and TM_{1n} modes. For TE_{1n} modes or waves

$$\frac{2\pi a}{\lambda} = p_1'(n) = 0$$

$$\frac{2\pi a}{\lambda} = 1.84,\ 5.33,\ 8.54,\ \ldots$$

and in Fig. 7·11 the approximate values are read where the slope of the $J_1(n)$ curve is zero.

The cutoff frequency and cutoff wavelength for circular waveguide modes are given by

$$\lambda_c(TM_{mn}) = \frac{2\pi a}{p_{mn}} \tag{7·15}$$

$$f_c(TM_{mn}) = \frac{p_{mn}}{2\pi a\sqrt{\mu\epsilon}} \tag{7·16}$$

$$\lambda_c(TE_{mn}) = \frac{2\pi a}{p'_{mn}} \tag{7·17}$$

$$f_c(TE_{mn}) = \frac{p'_{mn}}{2\pi a\sqrt{\mu\epsilon}} \tag{7·18}$$

7·16 Coupling to transmission lines

Several methods which can be used to couple energy from one transmission structure to another are shown in Fig. 7·12.

Magnetic coupling into a waveguide by means of a loop is shown at *a*. The loop may be placed at any position as long as it links with the magnetic lines of force. The degree of coupling can be altered by rotating the loop relative to the direction of the magnetic lines. The probe can sometimes be positioned where the magnetic field is weaker.

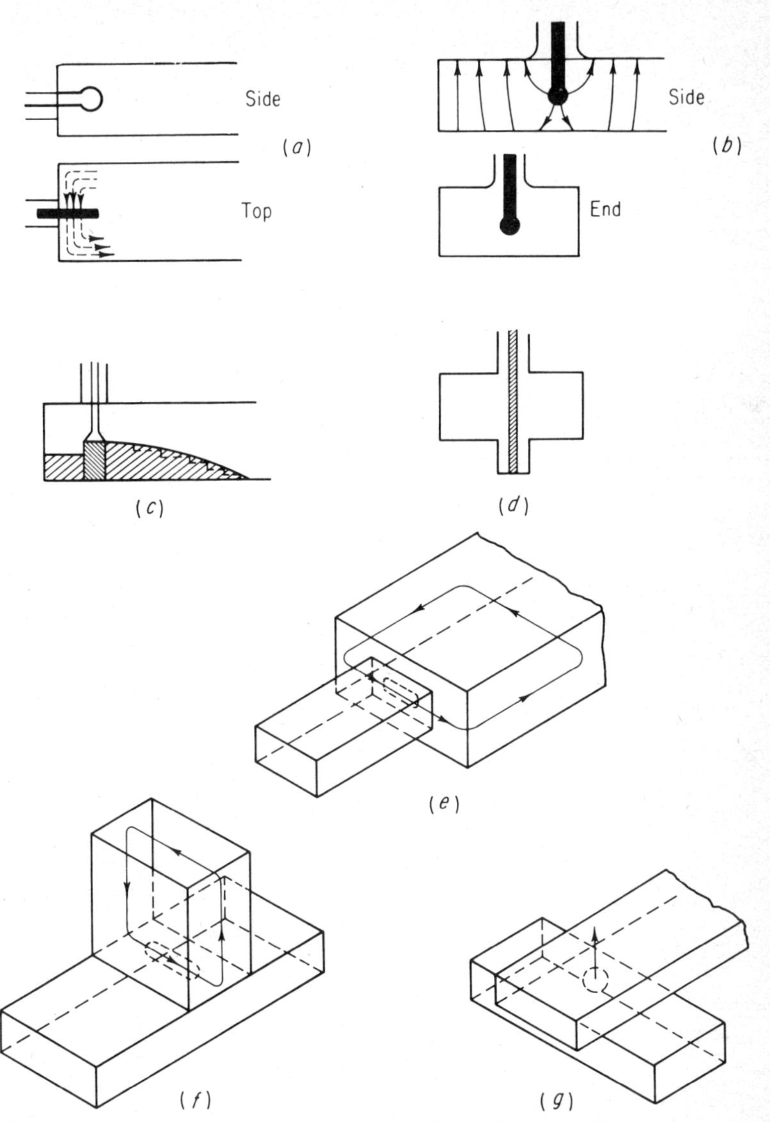

Fig. 7·12 Examples of coupling between transmission lines. (*a*) Side and top views. (*b*) Side and end views. (*c*) Tapered waveguide to coaxial transition. (*d*) Coaxial to waveguide transition. (e) Slot coupling. (*f*) Series slot coupling. (*g*) Transverse electric coupling.

The coaxial line may be coupled to the waveguide by extending the center conductor of the coaxial line into the waveguide parallel to the electric field at or near a point where the electric field has its maximum value. The conductor is located in the center of the guide as shown in Fig. 7·12*b*. The outer conductor terminates at the wall of the guide. The waveguide is usually closed at one end, and the probe is located one-quarter wavelength from the closed end so that the waves from the probe are reinforced by waves reflected from the closed end of the waveguide.

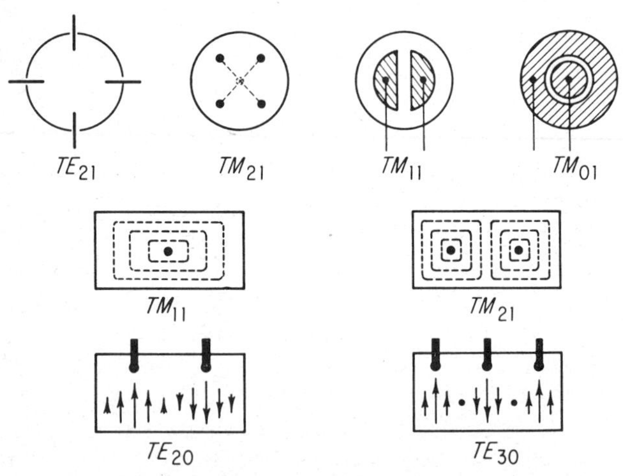

Fig. 7·13 Excitation of higher order modes in waveguides.

The coupling probe can be extended across the waveguide as shown at *d*. An adjustable short circuit can be connected to the center conductor beyond the junction so that the transition can be adjusted for best match over an appreciable frequency band.

The tapered transition at *c* is used when it is necessary to lower the waveguide impedance appreciably. The taper can be either a continuous or step type, as indicated by the dotted lines. This type transition is therefore useful for broadband applications.

There will be electric coupling through a round hole when there are components of electric field normal to the common surface as shown at *g*. There can be magnetic coupling if the magnetic field has a component parallel to the magnetic field at the adjoining surface. These conditions are sometimes determined by field distributions set up by certain types of restricted enclosures connected to the waveguide. Resonant cavities fall in this category.

Magnetic coupling occurs through narrow slots. The electric coupling is usually small, but magnetic coupling is appreciable.

There are many variations and combinations of the different types of coupling structures. Other types of coupling will be encountered in subsequent discussions.

Several methods[1] which have been used to excite higher order modes are shown in Fig. 7·13.

PROBLEMS

7·1 The a dimension of a rectangular waveguide is 0·9 in. and the b dimension is 0.4 in.

a. Calculate the cutoff frequency and cutoff wavelength of the waveguide for the dominant mode.

b. Calculate the wave impedance, phase velocity, and group velocity at 8.2, 9.0, and 12.4 Gc.

7·2 Plot curves of the wave impedance for TE and TM modes. Use $Z_{TE}/377$ and $Z_{TM}/377$ as the ordinate (wave impedance normalized to 377 ohms) and plot f/f_c as the abscissa. Compare the curves with Fig. 7·7.

7·3 Calculate the relative cutoff frequency for the TE_{21} and TM_{21} modes in rectangular waveguides. (Use WR-90 waveguide dimensions.)

7·4 Calculate the relative cutoff frequency for TE_{02} and TE_{12} modes in circular waveguides (relative to WR-90 waveguide).

7·5 Draw illustrations of TE_{01} and TE_{32} modes in a rectangular waveguide.

7·6 Draw illustrations of TE_{22} and TE_{02} modes in a circular waveguide.

7·7 Illustrate loop coupling and the associated magnetic fields in a rectangular waveguide with the loop located in the wide dimension. Repeat for a loop located in the narrow dimension. Show top and side views.

7·8 Make a complete comparison chart of the significant characteristics of TEM, TE, and TM modes. Six columns of the chart are as follows: Type of mode, cutoff frequency, type of line, velocity of propagation (v_p and v_g), attenuation (for a line of finite losses), and variation of characteristic impedance with frequency for lossless lines.

7·9 A K_u-band (WR-62) waveguide is filled with a material which has a dielectric constant of 2.4. Calculate the cutoff frequency and cutoff wavelength. Calculate the wave impedance at 12.4, 15.0, and 18.0 Gc.

7·10 Calculate the relative cutoff wavelength for the following circular waveguide modes in terms of a, where a is the radius of the guide.

a. TE_{02}, TE_{12}, TE_{11}, TM_{11}, TE_{03}, TE_{01}, TM_{01}, and TM_{12}.

b. Arrange the modes in order of increasing cutoff *frequency*.

Note: The relative cutoff frequencies of circular waveguide modes can be calculated from the cutoff wavelength values obtained in this problem. However, since relative cutoff wavelength and frequency require that the diameter of the circular waveguide be equal to the broad dimension of the rectangular guide (relative cutoff to TE_{10} mode in a rectangular guide), it is necessary to divide each λ_c calculation by 2 in order to obtain an equation in terms of the diameter.

Example:

$$\frac{\lambda_c(\text{TE}_{10})\text{ rectangular}}{\lambda_c(\text{TE}_{11})\text{ circular}} = \frac{2a}{3.412a/2} = 1.17$$

REFERENCES

1. G. C. Southworth, Some Fundamental Experiments with Waveguides, *Proc. IRE*, vol. 25, no. 7, July, 1937.

L. J. Chu, and W. L. Barrow, Electromagnetic Waves in Hollow Metal Tubes of Rectangular Cross Section, *Proc. IRE*, vol. 26, December, 1938.

CHAPTER

8

WAVEGUIDE ELEMENTS AND WAVEGUIDE COMPONENTS

8·1 Waveguide impedance-matching elements

Impedance-changing devices are introduced into the waveguide near the sources of reflected waves in order to eliminate standing waves. If diaphragms of good conductivity and of thickness small compared to wavelength are extended into the waveguide, the necessary susceptance can be introduced to reduce the standing waves to nearly zero as desired. Such diaphragms therefore play an important part in microwave techniques. The element illustrated in Fig. 8·1*a* adds inductive susceptance across the waveguide and is called an *inductive window* or *inductive iris*. The *capacitive window* or *iris* is shown at *b* and can be explained by the fact that the field present in the vicinity of the diaphragm is mainly electric. The inductive iris has an advantage over the capacitive iris in high-power waveguide installations. If the combined effects of the inductive and capacitive windows are employed, the composite type illustrated in *c* and *d* is obtained and is referred to as the *resonant window* or *iris*. The circular iris is usually preferred in the circular waveguide. The circular iris, which provides inductive susceptance, is shown at *e*.

The tuning screw shown in *f* is capacitive, but if the screw is long enough and is inserted on into the guide, it becomes series resonant when the depth of penetration equals one-quarter of a wavelength and becomes inductive if the depth is made greater. When the screw is inserted across the guide, the inductive post in *g* is obtained. If the diameter of the post or screw is decreased, or if they are moved away from the center of the waveguide, the susceptance of the equivalent inductor or capacitor decreases. A considerable range of susceptance can be obtained by variation of the probe insertion and distance along the line by the structure shown in *h*. The illustrations in Fig. 8·1 are only representative of the many possible configurations and combinations of these elements.

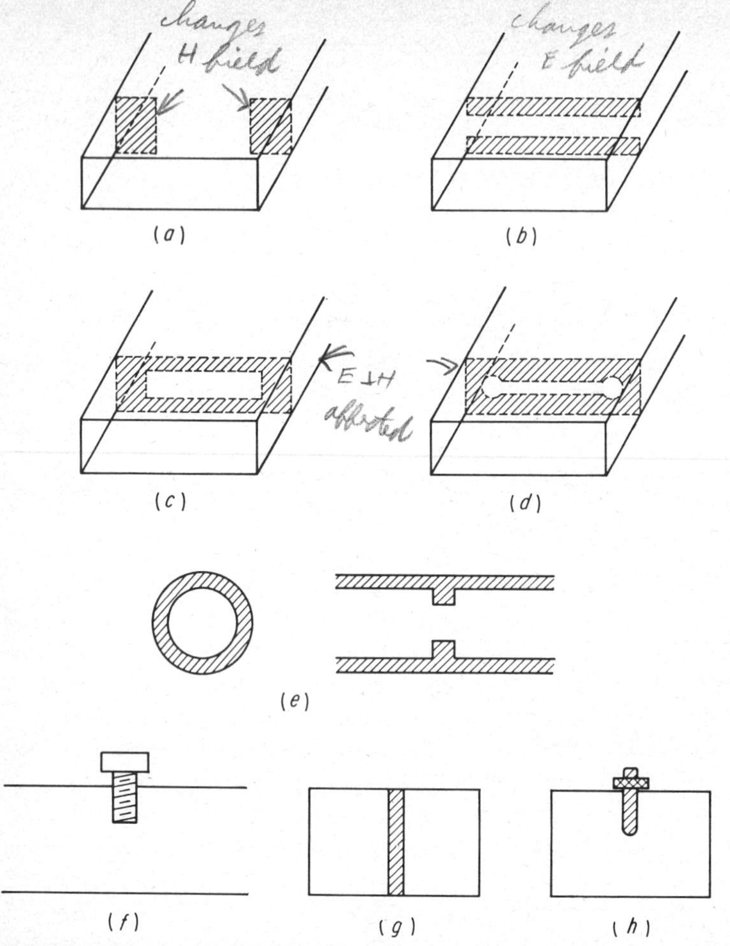

Fig. 8·1 Waveguide impedance-changing devices. (*a*) Symmetrical inductive iris. (*b*) Symmetrical capacitive iris. (*c*) and (*d*) Resonant windows. (*e*) Circular iris. (*f*) Capacitive tuning screw. (*g*) Inductive post. (*h*) Probe adjustable along the line with variable probe insertion.

8·2 Waveguide short circuit

One of the most useful waveguide elements is the waveguide closed at one end. It may be fixed closed at the end or an adjustable short-circuiting plunger may be provided. The adjustable short is more convenient and finds many applications in microwave techniques. A metal piston with contacting fingers riding against the sides of the waveguide is used in many applications, but a design which finds wide application at present is shown in Fig. 8·2 and is referred to as the *dumbbell* type.

8·3 Waveguide tees

Waveguide junctions play an important part in waveguide techniques; two of the common forms are shown in Fig. 8·3. The interesting properties

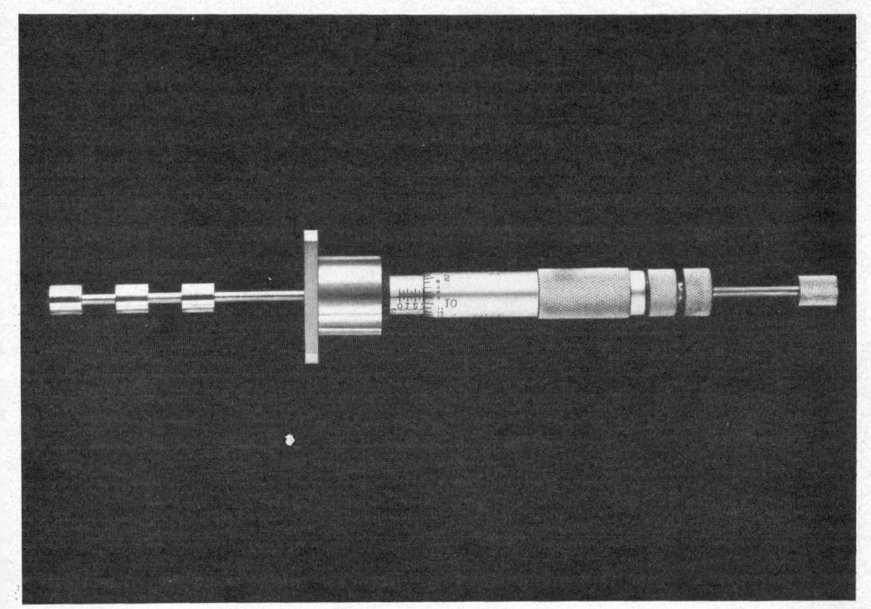

Fig. 8·2 Dumbbell type waveguide short circuit (quick-sliding and precisely adjustable). (*Hughes Aircraft Company, Microwave Standards Laboratory.*)

and characteristics of these junctions can be examined by considering the behavior of the electromagnetic fields in passing through the junctions. These junctions are referred to as waveguide *tees*, and they differ with regard to the plane in which the branch lies. The tee in Fig. 8·3*a* is called an **E** *plane* or *series tee* because the axis of the side arm is parallel to the E field of the main transmission line. The tee at b is called an **H** *plane* or *shunt tee* because the axis of the side arm is parallel to the planes of the magnetic field of the main transmission line.

The progress of the representative line of electric force of the wavefront is shown in the series tee. The line of electric force bends as it leaves the side

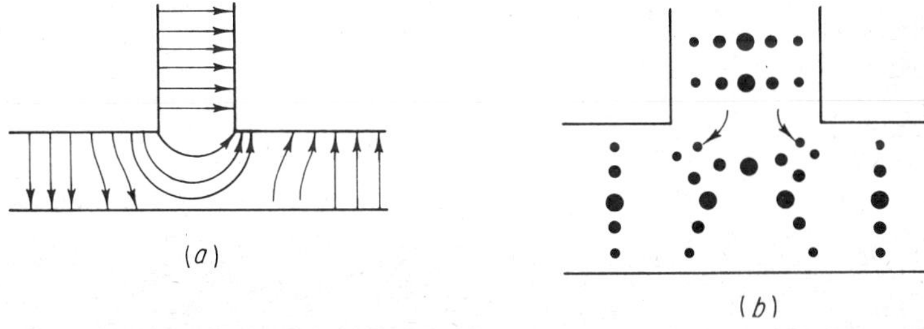

Fig. 8·3 Waveguide tee junctions. The E *plane* or *series* tee is shown at (*a*). (*b*) The H *plane* or *shunt* tee.

arm and causes the electric fields in the side arms to be of opposite polarity (out of phase). These waves set up in the side arms are equal in magnitude if the junction is completely symmetrical.

If the shunt tee is completely symmetrical, a similar analysis shows that a wave enters the **H** side arm and leaves the side arms equal in magnitude and in phase, as shown at *b* where a single cross section of electric lines of force is followed through the junction.

8·4 The magic tee

A combination of the **E**-plane tee and the **H**-plane tee forms a hybrid waveguide junction called the *magic tee*. The magic tee is illustrated in

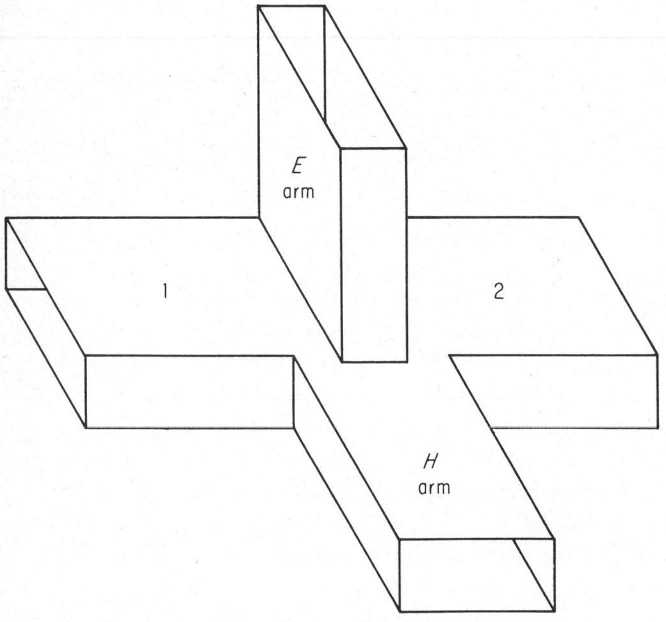

Fig. 8·4 The magic tee.

Fig. 8·4. The term magic tee is sometimes reserved for the hybrid junction in which matching structures have been introduced to improve the match of the **E** and **H** arms to the junction.

The characteristics of the hybrid in which the **E** and **H** arms are symmetrically placed are such that energy applied to either the **E** or **H** arm divides equally between arms 1 and 2, and no energy emerges from the opposite arm. If the fields entering arms 1 and 2 are equal in amplitude and of the same polarity, the net field in the **E** arm is zero, and the total energy emerges from the **H** arm.

At the junction each arm is effectively terminated by two other arms of equal impedance. Therefore, discontinuities will be inevitable unless special precautions are observed. Several forms of compensated junctions have been developed. One manner in which the magic tee can be matched is by the use of tuning rods. One rod is placed normal to the **E** field in the series arm, and one is placed normal to the **E** field in the shunt arm. The precise location and configuration of the various matching structures are usually determined empirically.

The magic tee can be used as a phase shifter when the series and shunt arms are terminated with adjustable short circuits. An extensive use of the magic tee has been in connection with microwave receivers where crystal detectors are placed in arms 1 and 2. The signal frequency enters the **H** arm, and the local oscillator signal is fed into the **E** arm. The tee provides isolation between sources of the two signals which are mixed in the crystals. The magic tee can be used in microwave phase-measuring systems and various microwave bridge circuits.

8·5 Phase shifters

The difference in phase shift between any two points is determined by the velocity of propagation and is therefore a function of the medium. There are several methods by which the effective velocity of propagation may be modified to introduce varying amounts of phase shift between two points. The phase may be changed by restricting the cross section of the guide, introducing inductive or capacitive irises into the guide, or by placing diametral rods across a circular guide. Our attention is mainly centered on commercial phase shifters of the variable type.

One method of obtaining a variable phase shift is to insert a dielectric vane into the waveguide. The vane may be attached to supporting rods and moved across the *a* dimension of the guide or the dielectric vane may be inserted through a slot in the top of the guide. The dielectric vane is properly tapered to give minimum reflection over the desired frequency range. This type of phase shifter usually has a dial gauge indicator and does not read phase shift directly.

A precision waveguide phase shifter which is direct reading in phase shift is the Hewlett-Packard Type 885A. It consists basically of three sections of round waveguide, each of which contains a plate of dielectric material. Rectangular-to-round transition sections provide the required waveguide input and output.

A functional drawing of the phase shifter is shown in Fig. 8·5*a*. The input and output differential phase sections are fixed in position at an angle of 45° in the waveguide. The center section is free to rotate. The input and output sections are called quarter-wave plates since an electric vector in the plane of the dielectric vane is delayed in phase by 90°. The central rotary vane is

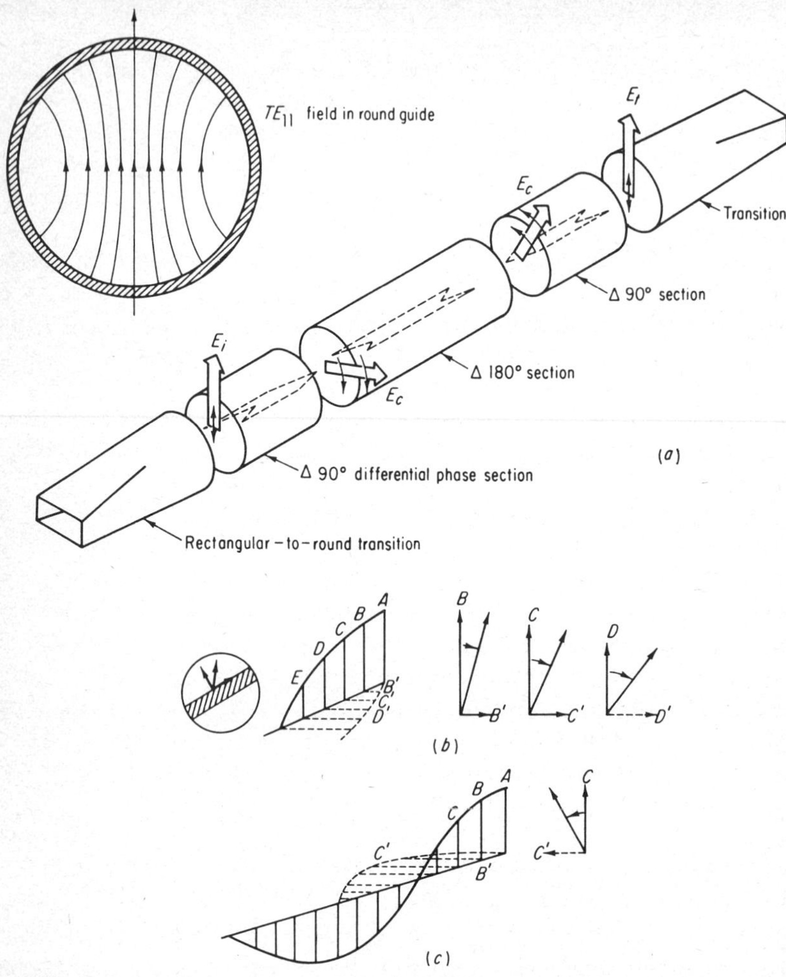

Fig. 8·5 Functional drawing of the Type 885A phase shifter. (*Hewlett-Packard Company*.) Clockwise rotation of the electric field due to the quarter-wave section is shown at (b). Counterclockwise rotation due to the half-wave section is illustrated at (c).

referred to as a half-wave plate since the electric vector in the plane of the dielectric vane is delayed by 180°.

The TE_{10} dominant mode in the rectangular waveguide is converted to the dominant TE_{11} mode in the circular waveguide. The linearly polarized electric vector is resolved into two mutually perpendicular vectors, one in the plane of the dielectric vane and the other perpendicular to the vane, as shown in Fig. 8·5*b*.

The representative illustration in Fig. 8·5*b* indicates the relative magnitudes of the two waves emerging from the end plane of the quarter-wave plate. *A* indicates maximum amplitude in the vertical plane. *B-B′* indicates the magnitudes of the two emergent waves an instant later. The orientation at the output plane of the quarter-wave plate indicated that the original linearly polarized wave is now a *circularly polarized* wave rotating in a clockwise direction.

Figure 8·5*c* indicates a vector which has traveled through a half-wave plate, in which case the electric vector in the plane of the vane has been delayed by 180°. Applying the previous method of analysis at the output plane of the half-wave section, it is noted that the circularly polarized wave is now rotating in a counterclockwise direction.

The output quarter-wave plate converts the counterclockwise rotating wave back to a linearly polarized wave.

Rotation of the 180° section through an angle results in an angular displacement of the electric vector at the output of 2θ. Therefore, 180° mechanical rotation results in 360 electrical degrees phase shift between the input and output electric vectors.

8·6 Attenuators

Attenuators are used to control the amount of power transferred between points on a transmission line by absorbing and/or reflecting some of the microwave power. Attenuators that operate on the reflection principle employ sections of waveguide-below-cutoff, and attenuators that operate on the power absorption principle use dissipative elements.

The attenuator placed in the transmission line should present a good impedance match at each terminal in order to decrease the dependence of the attenuator value upon the circuit in which the attenuator is used. Various techniques are used to obtain the desired impedance-match conditions. Other desirable properties have been obtained by selection of the materials used in the dissipative element and by the particular construction details of the dissipative element. One class of attenuators consists of thin metallic films coated on glass. A baked-on metallic film combining platinum and palladium and also an evaporated film of chromium or nichrome with a protective film of magnesium fluoride have been satisfactorily used. Glass was chosen as the base material because it does not react with the film, its surface is very smooth, it will not warp or change shape, and its melting point makes it usable for most processes.

Coaxial Attenuators. Two basic types of fixed coaxial resistive-film attenuators in common use are shown in Fig. 8·6*a*. The T-section attenuator is most useful in the frequency range below 4 Gc, and the distributed type is used for the upper frequencies. The center conductor is a glass tube which has two resistive metalized film sections on either side of a metalized circular disk. The distributed type attenuator consists of a higher resistive main

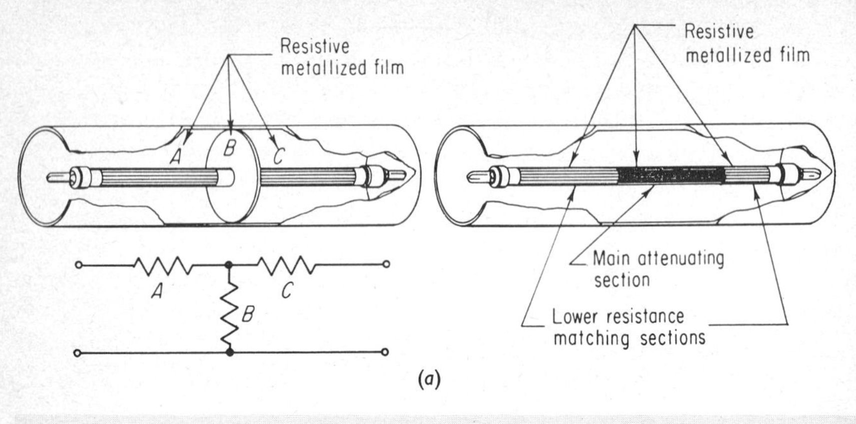

(a)

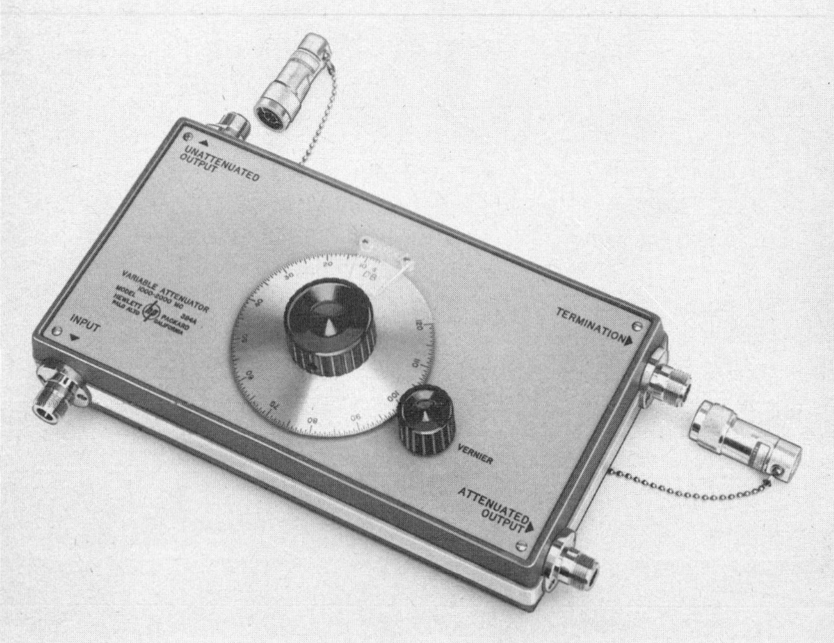

(b)

Fig. 8·6 Coaxial attenuators. (a) Basic types of resistive-film fixed attenuators. (*PRD Electronics*.) (b) Variable attenuator. (*Hewlett-Packard Company*.)

attenuation section in the center and lower resistance matching sections on each end. The resistance of the inner conductor is made high by maintaining the thickness of the film less than the skin depth of the film material. The attenuation value of this type of unit changes as a function of frequency. Therefore, if the precise value of the attenuator is required, the attenuator has to be calibrated at the particular operating frequency in question.

The variable coaxial attenuator in Fig. 8·6*b* has an attenuation range from 6 to 120 db. The instrument can be used as an attenuator or directional coupler or for mixing two signals. It is a type of waveguide-below-cutoff attenuator. The fields in a waveguide operating below cutoff attenuate exponentially with distance along the axis. The amount of power is varied by means of a movable element that is similar to the input coupling element. The plot of attenuation in decibels versus the displacement of the coupling

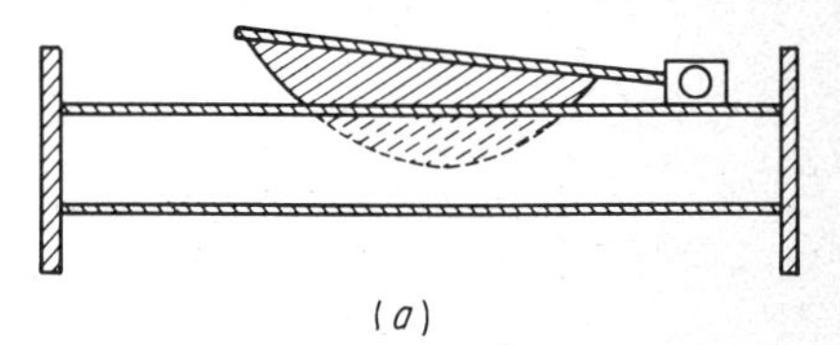

Fig. 8·7 Waveguide attenuators. (*a*) Flap attenuator. (*b*) Glass-vane attenuator.

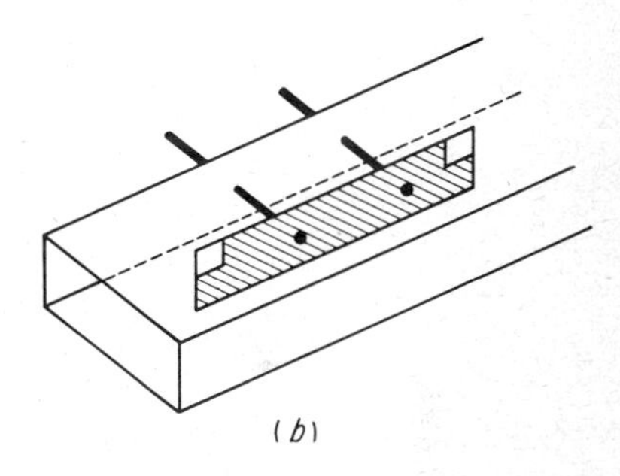

elements is, except for close spacing, a straight line. If a waveguide-below-cutoff attenuator is designed to have a cutoff well above the frequencies at which it is to be used, the variation of attenuation with frequency will be negligible.

Waveguide Attenuators. The waveguide attenuator consists of an attenuation plate supported within the guide by means of metal rods or by moving the plate in and out of the guide at a fixed location. As an example of the latter, a resistance card is inserted through a longitudinal slot in the top (center) of the waveguide. This attenuator is referred to as a *flap* attenuator. Absorbing material is necessary next to the slot in order to lower the leakage.

Cross sections of the two types of waveguide attenuators are shown in Fig. 8·7. A drive mechanism is attached to the struts in order to move the vane in and out of the electric field to obtain the variable attenuation characteristic. Also, a dial mechanism is attached to provide a means of accurately calibrating the attenuator. Various methods are used to match the waveguide attenuators. The resistive transformer shown in the diagram is only one of

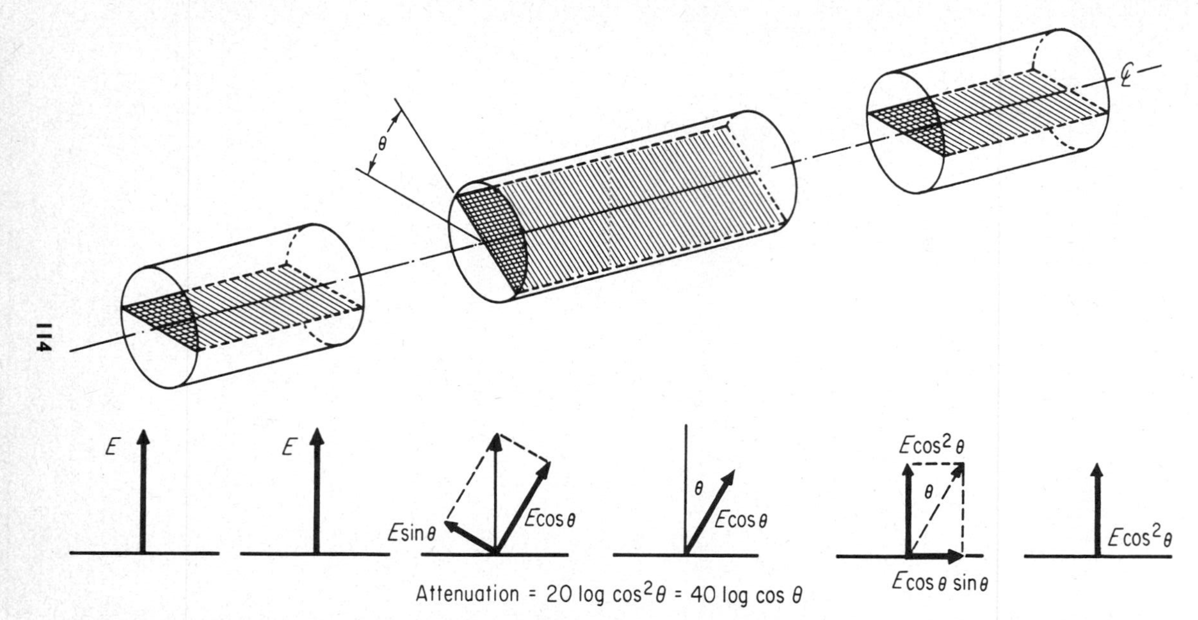

Fig. 8·8 Functional drawing including operating principle of the rotary-vane type attenuator. (*Hewlett-Packard Company*.)

several of its type, and, in addition to this method, a match can be obtained by tapering the vane at each end. A one-half-wavelength taper provides a good match. In addition to the frequency sensitivity of the variable glass-vane attenuator, considerable phase shift is encountered since the moving vane is a dielectric material. Typical curves for a complete line of glass-vane attenuators are found in the PRD catalog of microwave components.

Rotary-vane Attenuator. A direct-reading precision waveguide attenuator which obeys a mathematical law is the rotary-vane attenuator. The attenuator has a calibrated range of 50 db which is accurate within 2 per cent of the decibel reading at any frequency in the waveguide band. The phase-shift variation of the attenuator is less than one degree between 0- and 40-db variations. A functional drawing indicating the operating principle of this type attenuator is shown in Fig. 8·8.

Basically, the attenuator consists of three sections of waveguide in tandem. In each section, a resistive film is placed across the guide. The middle section is a short length of round guide which is free to rotate axially with respect to the two fixed end sections. The end sections are rectangular-to-round waveguide transitions in which the resistive films are normal to the field of the applied wave. The construction is symmetrical and the device is bidirectional.

When all films are aligned, the **E** field of the applied wave is normal to the films. No current then flows in the films, and no attenuation occurs. If the center film is now rotated to some angle θ, the **E** field can be considered to be split into two components: $\mathbf{E} \sin \theta$ in the plane of the film, and $\mathbf{E} \cos \theta$ at right angles to the film. The $\mathbf{E} \sin \theta$ component will be absorbed by the film, while the $\mathbf{E} \cos \theta$ component, oriented at an angle θ with respect to the original wave, will be passed unattenuated to the third section. When it encounters the third film, the $\mathbf{E} \cos \theta$ component will be split into two components. The $\mathbf{E} \cos \theta \sin \theta$ component will be absorbed, and the $\mathbf{E} \cos^2 \theta$ component will emerge at the same orientation as the original wave. The attenuation is thus ideally proportional only to the angle at which the center film is rotated and is completely independent of frequency. The attenuation, in decibels, is $40 \log \cos \theta$. The attenuation of the device is limited by the attenuation of the center rotating vane which normally has an attenuation value of 70 db or more.

8·7 Directional couplers

A useful form of hybrid known as a directional coupler is constructed by placing an auxiliary section of uniform waveguide along the narrow or wide dimension of the main guide with appropriately located apertures connecting the two.

The *Bethe hole* coupler consists of a single coupling aperture in the wide dimension of the guide. The electric field which leaks into the auxiliary guide is a combination of two fields, the fringing electric field through the hole

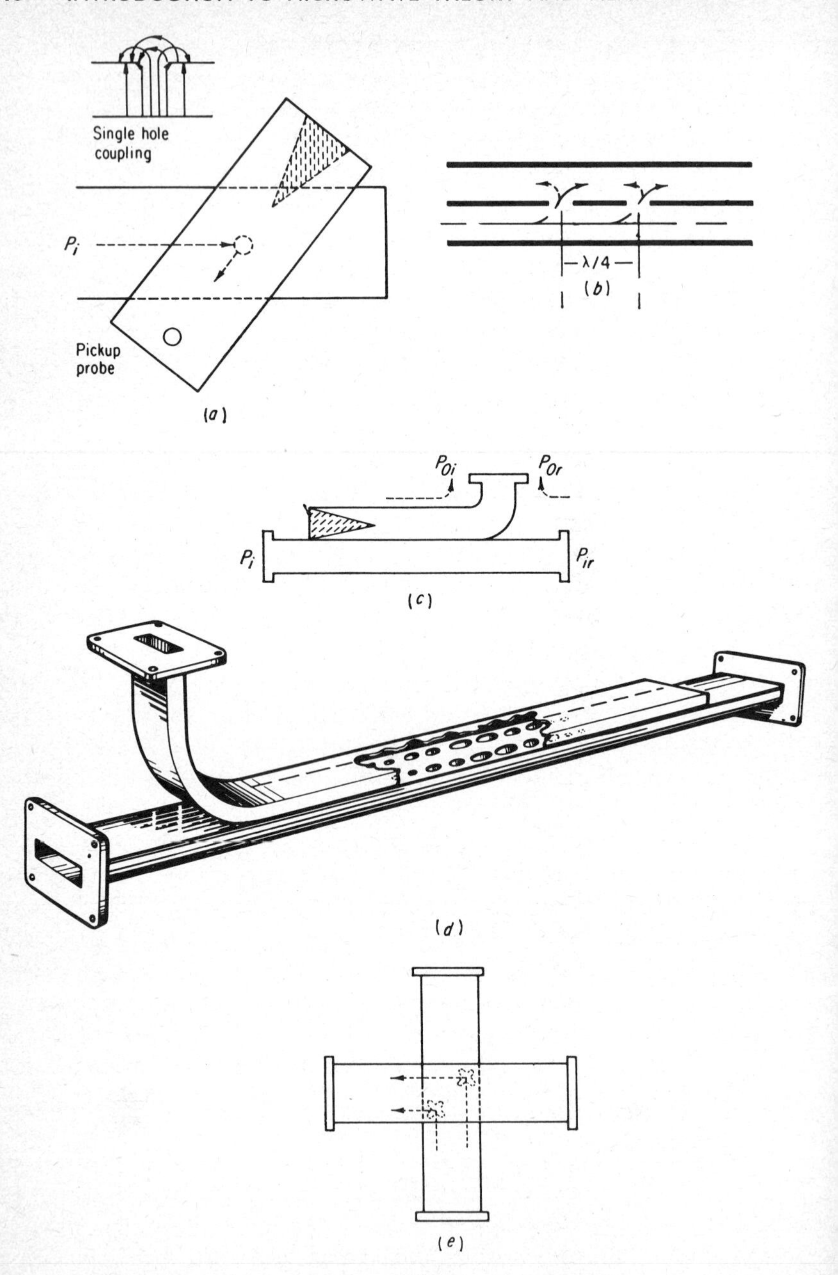

Fig. 8·9 Illustrations of waveguide couplers. (*a*) Bethe hole coupler. (*b*) Cross section of a two-hole coupler. (*Hewlett-Packard Company*.) (c) Side view of the multihole coupler. (*d*) Construction details of the multihole coupler. (*Hewlett-Packard Company*.) (e) Crossguide coupler.

and the electric field across the hole caused by the flow of charge out through the hole. This flow of charge is caused by the transverse component of the magnetic field of the wave. The directions of the component electric fields in the auxiliary guide are such that the fields tend to cancel in the forward direction from the aperture and reinforce in the opposite direction. The magnetic coupling is a function of the angle between the guide axis, whereas the electric coupling is essentially independent of the angle. Therefore, the maximum directional property can be obtained by an optimum setting of the angle between the auxiliary guide and the main guide, as shown in Fig. 8·9*a*. The coupling and directivity are sensitive to changes in frequency.

The two-hole directional coupler is illustrated in the cross-section diagram in Fig. 8·9*b*. The two auxiliary holes are placed one-quarter wavelength apart. The signal which travels back to the first hole from the second hole is 180° out of phase, and the two signals tend to cancel in the reverse direction. The signals in the forward direction are in phase and reinforce each other. The coupling and directional properties are impaired by off-frequency operation when the distance between the holes is no longer one-quarter wavelength.

The coupling factor is defined as the ratio, expressed in decibels, of the power entering the main line input to the power output of the auxiliary guide.

$$C = 10 \log \frac{P_i}{P_0}$$

From the definition and illustration of Fig. 8·9*c*, the directivity is given by

$$D = 10 \log \frac{P_{0i}}{P_{0r}}$$

Directivity is defined as the ratio, expressed in decibels, of the powers out of the auxiliary guide when a given amount of power is alternately applied in the forward and reverse directions in the main guide.

The leakage signal due to the intrinsic directivity of the coupling holes, the reflected signals from the internal termination, and any discontinuities at the output flange of the coupler determine the total directivity signal.

Multihole couplers operate on the same basic principle as the two-hole coupler. The coupling array is illustrated in Fig. 8·9*d*; it provides high directivity. The coupling versus frequency variations are ±0.5 db over the entire waveguide frequency range. Coupling is obtained through a series of graduated holes which have been accurately machined along the broad or narrow face of the waveguide. Couplers designed for high-power operation usually have the holes in the narrow dimension of the waveguide.

In applications where a multihole coupler is not required, the less expensive and more compact crossguide coupler shown in Fig. 8·9*e* is used.

8·8 Terminations (matched loads)

Matched loads or terminations are special devices designed to absorb incident energy without appreciable reflection. Terminations are constructed by mounting power-absorbing material in the space near the end of a closed section of waveguide. A list of low-power-absorbing materials includes powdered iron and a binder, carbon mixed with a binder and deposited on a strip of dielectric, and porcelain containing silicon carbide. At higher powers, graphite mixed with cement, and Aquadag-coated sand have been used.

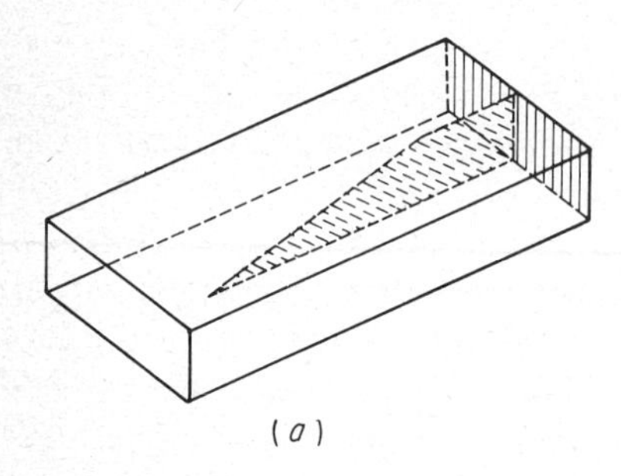

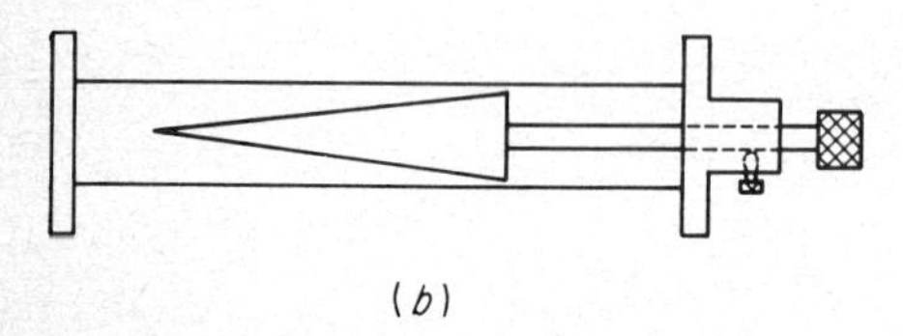

Fig. 8·10 Low-power waveguide terminations. (*a*) Tapered resistance strip. (*b*) Movable polyiron termination.

Radiating fins are added to the waveguide casing to aid in removal of heat. Noncirculating and circulating water loads have been used extensively. Space is not available for illustrations of the various configurations and material forms such as the symmetrical taper in circular waveguides, the tapers in the electric and magnetic plane in rectangular waveguides, etc. Figure 8·10 illustrates two of the most commonly used low-power loads, the tapered resistance strip and the polyiron pyramid. The polyiron load is shown on a sliding-load assembly which finds many applications in microwave measurement techniques. The sliding termination is variable over at least one-half waveguide wavelength at the lowest waveguide operating frequency.

Mismatched loads or *standard reflections* are precision loads used to set up exact reflections for standardizing standing-wave and reflection-coefficient measuring setups. The a dimension of the precision casing is the same as the standard waveguide, but the b dimension is reduced the proper distance to establish the required reflection at the junction. The precision-tapered load is movable so that reflections caused by the load can be isolated from the standard calibrated discontinuity.

8·9 Tuners

Tuners are used primarily for correcting discontinuities in a microwave system. The double-stub coaxial tuner was discussed in a previous chapter, and its limitations were pointed out. Tuning out mismatches with waveguide irises or tuning screws inserted into the guide was also discussed.

Three types of waveguide tuners are shown in Fig. 8·11. The slide-screw tuner consists of a slotted section of waveguide with a carriage on which is mounted an adjustable probe. The position of the probe along the line and the penetration into the waveguide can be adjusted to set up a reflection of the proper phase and amplitude to cancel out existing reflections in a system. VSWR values of 20:1 can be tuned out with this type tuner, and the insertion loss of a 20:1 mismatch is usually less than 2 db.

The **E-H** tuner is particularly useful where power leakage is undesirable. VSWR values of 20:1 result in insertion loss of usually less than 3 db. The series and shunt arms are terminated with adjustable short circuits which reflect the proper susceptance to the junction of the main guide.

The broadband five-stub tuner shown in Fig. 8·11 assures a high degree of measurement repeatability and does not have energy leakage.

8·10 Detectors (bolometers and crystals)

A *bolometer* is a temperature-sensitive, resistive element. The useful property of this device is that its resistance changes when it absorbs electromagnetic radiation (power), and the change in resistance is a function of the power dissipated. Bolometers are usually classed as *barretters* or *thermistors*. There is no strict adherence to the above classification since the most common usage of the term "bolometer" refers to the barretter class only. The barretter has a *positive* temperature coefficient (the resistance increases with an increase in temperature), whereas the thermistor has a *negative* temperature coefficient. The bolometer element is small compared to the wavelength of the microwave signal, and its resistance change is large enough to be measured accurately when small power changes occur. The change in barretter resistance is not the same for d-c and r-f heating because of lengthwise temperature distribution in the r-f case, but since the barretter and thermistor elements are small, the discrepancy between the r-f and d-c heating is negligible.

Barretters. The barretter consists of a thin Wollaston wire mounted in structures such as the Sperry Type 821 cartridge and the threaded Type 560 and 825 constructions shown in Fig. 8·12*a* and *b*. The coaxial type element holders are represented in Fig. 8·12*c*. The metal used in Wollaston wires is platinum. Barretters are normally biased to an operating resistance of from

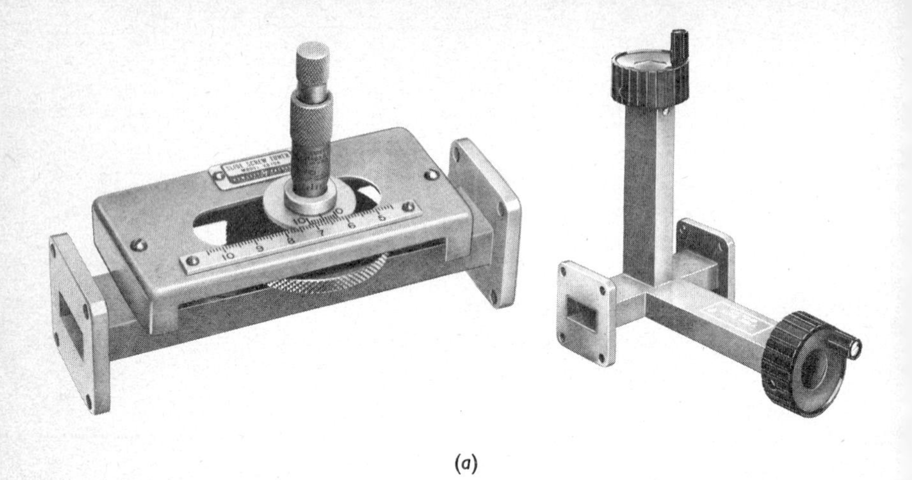

(*a*)

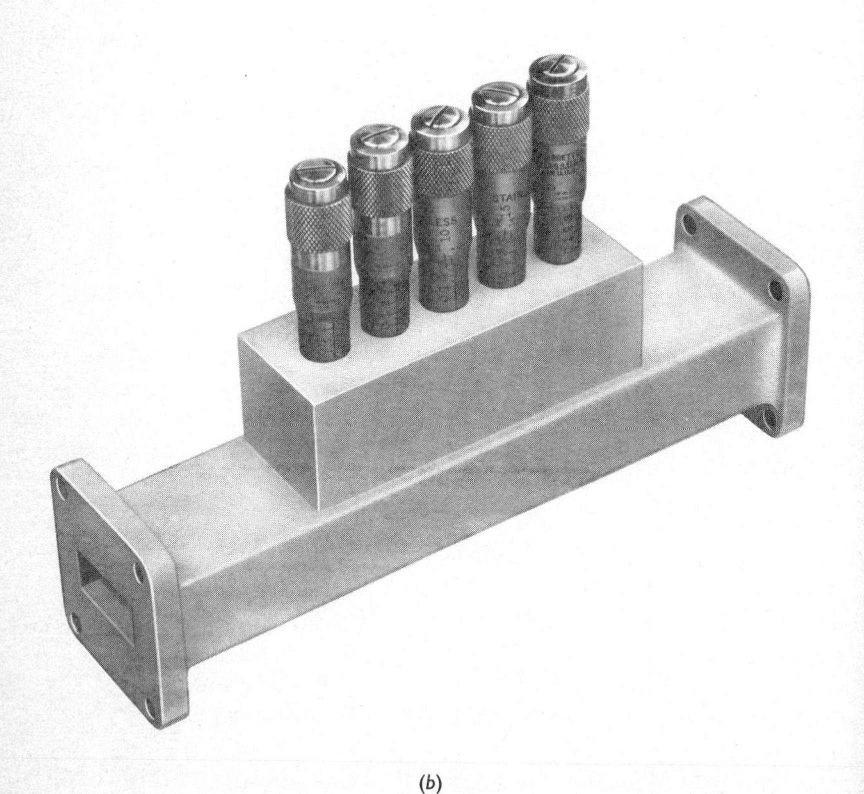

(*b*)

Fig. 8·11 Waveguide tuners. (*a*) Slide-screw tuner and E-H tuner. (*Hewlett-Packard Company.*) (*b*) Five-stub tuner. (*Hughes Aircraft Company, Primary Standards Laboratory.*)

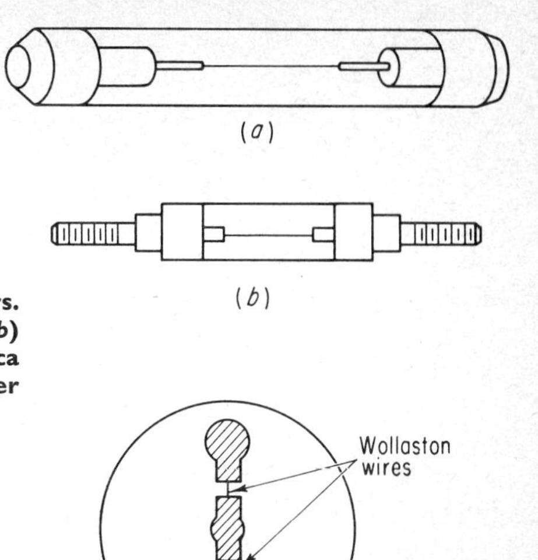

Fig. 8·12 Wollaston wire barretters. (*a*) Sperry Type 821 Barretter. (*b*) Sperry Type 560 and 825. (*c*) Mica disk (Series r-f bypass under center section).

50 to 400 ohms with 200 ohms operation being most common. A biasing circuit is shown in Fig. 8·13 where the barretter acts as a transfer device to convert r-f power change to audio-voltage changes when a constant bias current flows through the barretter. This is a common application in which power-ratio measurements are made with the r-f source 100 per cent modulated at an audio rate.

If the output voltage is proportional to the r-f power input, the detector is said to be *square law*. Since audio-voltage ratios can be measured with a high

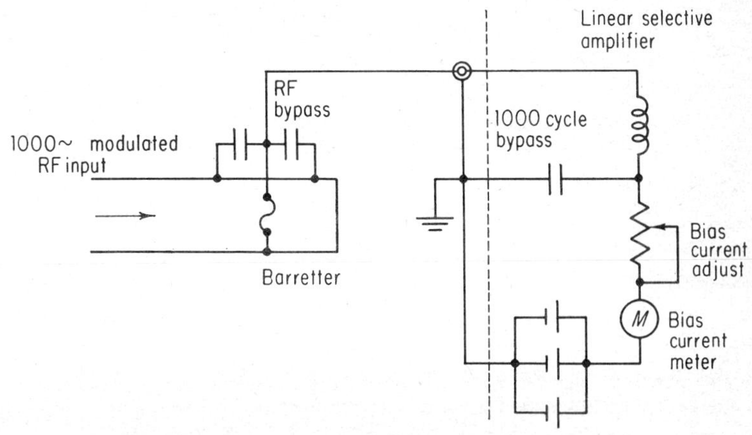

Fig. 8·13 Basic circuit of linear selective amplifier.

degree of accuracy, the accuracy of such a system depends mostly on the deviation of these devices from square law. If the bias-adjust resistance is very large compared to the 200 ohms of the barretter, the current through the barretter is relatively constant. Therefore, since $P = I^2R$, and in this case I is constant, the device is linear and the change in output voltage (due to change in barretter resistance with change in input power) is proportional to the change in the input r-f power.

The per cent deviation from square law is given by

$$\left(\frac{R_1 + \Delta R}{R_1} \times \frac{P_1}{P_1 + \Delta P} - 1\right)100$$

where R_1 is the barretter operating resistance, ΔR is the change in barretter resistance when microwave power is applied, P_1 is the d-c bias power, and ΔP is the microwave power.

The square-law device must be able to follow the r-f power change sufficiently fast in order to obtain a usable audio output. This ability is limited by the thermal time constant of the element. Typical time constants for Wollaston wires are from 80 to 300 μsec; this limits the modulation frequencies to values below 2,000 cps.

The range of r-f power ratio which can be accurately measured is limited at the upper end by the deviation of the detecting element from the square-law characteristic, and at the lower end by the noise which is produced by the detector itself and by the following amplifying system. By rule of thumb, the deviation of the barretter from square law is *one* per cent per 100 μw of applied microwave power. For accurate attenuation measurements, the ratio of minimum signal to noise should be at least 10:1, in which case the decibel range to 2 per cent deviation from square law is 46 db for the barretter when used with an amplifier which has a 25-cycle bandwidth.

Measurements on several types of barretters have indicated that the output voltage, across the element, ranges from 20 to 28μv per microwatt of r-f power.

The sensitivity of the Wollaston wire barretter is usually in the range of 4.5 ohms per milliwatt with the sensitivity of some types approaching 10 ohms per milliwatt of applied power.

The characteristics of the barretter do not change appreciably up to the burnout point. The burnout power for the Sperry 821 barretter is approximately 32 mw and includes 15 mw bias power. The lower current PRD 610 barretter has a power rating of 7 mw, which includes approximately 4.5 mw bias power.

Thermistors. The *thermistor* is a temperature-sensitive resistor with a large *negative* temperature coefficient. It consists of a small bead of semiconducting material supported between two wires, as illustrated in Fig. 8·14. Thermistors are made of complex metallic-oxide compounds using oxides of manganese, nickel, copper, cobalt, and sometimes other metals. The bead which is formed by these metallic-oxide mixtures is coated with a thin film of glass, which makes the assembly strong, heat resistant, and stable. The thermistor element is sometimes in the form of a thin film deposited on mica.

The thermistor is far more sensitive than the barretter and requires higher bias current. The thermistor is biased to 100 or 200 ohms. The sensitivity ranges from 40 to 140 ohms per milliwatt of applied power. The sensitivity decreases as the power dissipation is increased, and a large mismatch exists if the applied power is excessive. This high VSWR protects the thermistor when overloading occurs because a large portion of the applied power never

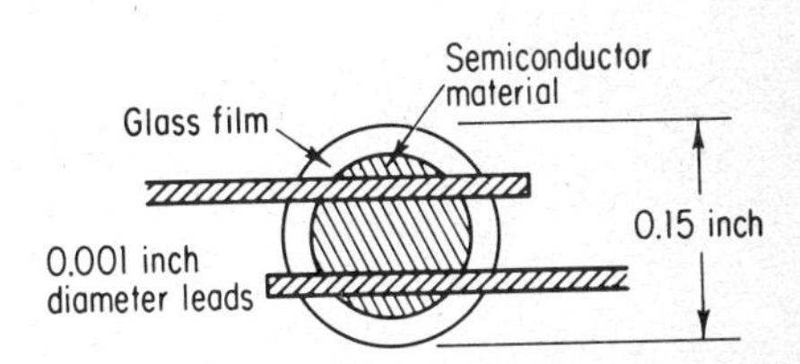

Fig. 8·14 Basic form of microwave thermistor bead.

reaches the thermistor element. The high sensitivity creates difficulties at low levels since the element responds to ambient temperature variations.

The thermal time constant of the thermistor is in the order of seconds. This can be an advantage or disadvantage depending upon the specific application.

Crystals. The crystal rectifier is the most sensitive and the simplest of all rectifying devices. It consists of a fine gold-plated tungsten wire (cat whisker) which is carefully pointed and brought into contact with a suitable semiconductor such as silicon or germanium. The volume surrounding the contact is usually filled with wax for mechanical stability and to prevent moisture from accumulating on the elements. The two types of physical structures shown in Fig. 8·15 illustrate the placement of the semiconductor and cat whisker between the base and contact prong. The cat whisker and semiconductor are reversed in position for reversed polarity types. The simplified equivalent circuit of the crystal is shown in Fig. 8·15*c*.

Microwave crystal diodes are designed as nonlinear circuit elements for frequency conversion, rectification, modulation, detection, and harmonic generation.

When crystals are used in mixer applications, the degree of crystal performance is defined by the noise figure, noise ratio, i-f impedance, r-f impedance, and conversion loss. The crystal noise is a combination of thermal noise, barrier noise caused by the diode rectification action, and noise known as fluctuation noise. The noise is high in the audio range and approaches thermal noise above 500 kc. The noise ratio of a crystal excited with 0.5 to 1 mw of c-w power may be in the range of 3 to 1 (referenced to the noise power generated by a 300-ohm resistor excited by the same amount of power).

The *i-f impedance* is the impedance of the diode and holder when looking from the output terminals while the diode is excited at a microwave frequency. The *r-f impedance* is the impedance looking into the r-f terminals of the mixer at the local oscillator frequency and power level. The i-f and r-f impedance are functions of the crystal geometry, the crystal holder, and the impedance

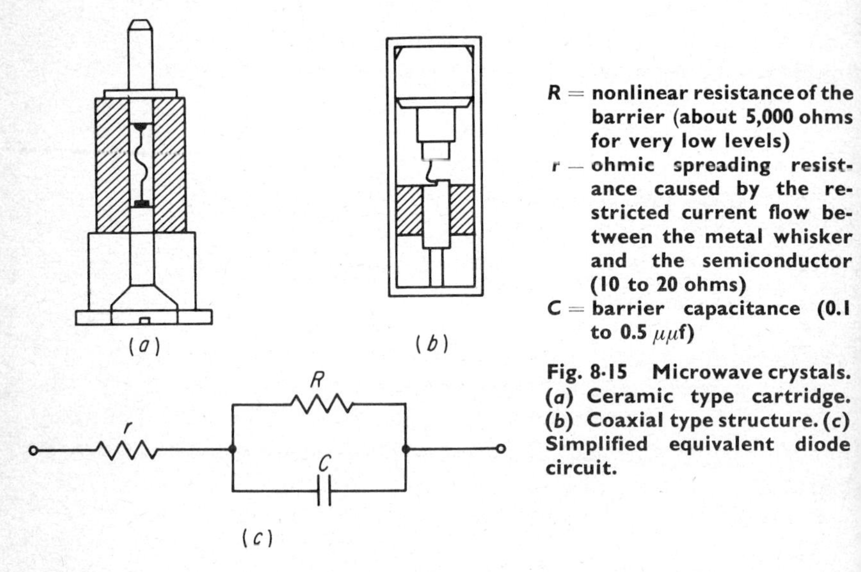

R = nonlinear resistance of the barrier (about 5,000 ohms for very low levels)
r = ohmic spreading resistance caused by the restricted current flow between the metal whisker and the semiconductor (10 to 20 ohms)
C = barrier capacitance (0.1 to 0.5 $\mu\mu$f)

Fig. 8·15 Microwave crystals. (*a*) Ceramic type cartridge. (*b*) Coaxial type structure. (*c*) Simplified equivalent diode circuit.

under excitation. Therefore, the impedance of a crystal rectifier is of significance only in terms of the circuit in which it is measured.

The conversion loss is defined as the ratio of available r-f power input to the measured i-f power output at the mixer. The conversion loss equation is

$$L = 10 \log \frac{(\Delta P)^2}{(\Delta I)^2 2PR_L}$$

where L = conversion loss, db
ΔP = change in power level, mw
ΔI = change in crystal current, μa
P = average power level, mw
R_L = load resistance

Crystal harmonic generators are useful for a variety of reasons. They are widely used in frequency-multiplying chains in the generation of standard frequency signals. The source power is introduced into the crystal mounted in a waveguide structure appropriate to the input frequency. The output waveguide is provided with tunable shorting plungers which provide accurate tuning of the desired harmonic and minimization of other harmonics which are simultaneously present in the output. The amplitude of the harmonic is much smaller than that of the fundamental and is a function of the fundamental power level and the increasing order of the harmonic used. The conversion loss increases rapidly with harmonic number.

One of the major applications of microwave crystals is video detection. The performance criteria used in determining the suitability of a detector are simplicity and response time, video impedance or video resistance, tangential signal sensitivity, and square-law response.

The following material is devoted to the characteristics of the crystal as a low-level detector since we are mainly concerned with the problems associated with the detection of microwave power.

The video impedance is the dynamic impedance of the crystal diode considered as a constant-current generator in the microwatt region and is determined by measuring the current flow when 5 mv is applied. If the video impedance is too high, the pulse shape is distorted because of splitting of the high-frequency components.

The sensitivity depends upon the nature of the semiconductor material and the contact area. Crystals are frequency sensitive, and the magnitude of this variation in output signal is further related to the applied power. The sensitivity decreases at high power levels. Typical video characteristic curves for type 1N31 and 1N32 crystals indicate that the open-circuit output is in the range of 4,000 to 5,000 μv output per microwatt input. The sensitivity decreases with increasing frequency and, in order to obtain the output level at a constant value in the range of 3,000 to 10,000 Mc, may result in input power changes as great as 100:1.

The d-c output resistance R of the crystal is not constant but decreases from a value of several thousand ohms at very low levels to a few hundred ohms at higher levels. Therefore, the matching of a crystal to a line is difficult to maintain. The crystal can suffer damage from overload, and the resulting changes in characteristics can cause inaccuracies of measurements.

The *minimum detectable signal* and *tangential sensitivity* are used as measurement levels for video diodes. The minimum detectable signal is the amount of signal power, below a 1-mw reference level, which exists when its presence is barely discernible in the noise. The tangential sensitivity is the amount of signal power, below a 1-mw reference level, required to produce an output pulse whose amplitude is sufficient to raise the noise fluctuation by an amount equal to the average noise level and is approximately 4 db above the minimum detectable signal.

The pulse response of a crystal video detector is a function of the crystal itself and the video output circuit. The rise time is controlled by the applied microwave power level, and the decay time is controlled by the external load resistance.

The burnout power is a complicated function of the particular crystal, its impedance match—whether c-w or pulsed, duty cycle, pulse width, and the particular load circuit. With the appropriate mixer circuit, pulse width, and duty cycle, some units can be subjected to as much as 25 watts peak power. For most video detection the c-w power level should be limited to 300 mw maximum.

The forward response of the crystal diode obeys the square law (rectified current is proportional to input power) in the region below approximately 10 μw. In some crystals the deviation from square law may take place at levels as low as 1 μw. The termination impedance is significant in determining the operational characteristics of the crystal. A certain amount of control may be exercised over the deviation from square law by proper selection of the load resistance. If the crystal is used with an amplifier which has a 25-cycle bandwidth, the decibel range from 10 db above noise level to 2 per cent deviation from square law is about 34 db.

For accuracies in the order of 0.2 db, the crystal diodes are capable of measuring r-f power ratios over a 38-db range with a maximum r-f power input of 0.1 μw average below approximately 5 Gc. This can be compared to the barretter which is capable of covering a range of 53 db with the same accuracy but requires a maximum input power of 1 mw average.

Crystals are not too well suited for high precision measurements. In general, high accuracy requirements call for alternate detection techniques. When a crystal is employed, it becomes of utmost importance that the crystal be operated in a closely controlled microwave and video environment in order to minimize errors. Also, operation must be checked as a part of the measurement.

8·11 Detector mounts

The usefulness of the microwave-detecting elements is intimately related to the design of the mounting structure. The mounting structure should be arranged to transform the r-f impedance of the element to equal the impedance of the reflections. For greatest usefulness the structure should maintain this condition (low VSWR) over as broad a frequency range as possible. The importance of obtaining a match is indicated by the fact that approximately one per cent of the input power is reflected from a VSWR of 1.22.

If the detecting element is connected directly across the open end of the waveguide, it turns out that there is a conductance component associated with radiation into open space and a susceptance component associated with reflections from the open end of the waveguide. The radiation is eliminated and the open circuit is obtained at the detecting element by attaching a closed quarter-wave section beyond the detecting element.

A broadband thermistor mount is illustrated in Fig. 8·16*a*. The transmission line impedance is matched to the resistive component of the

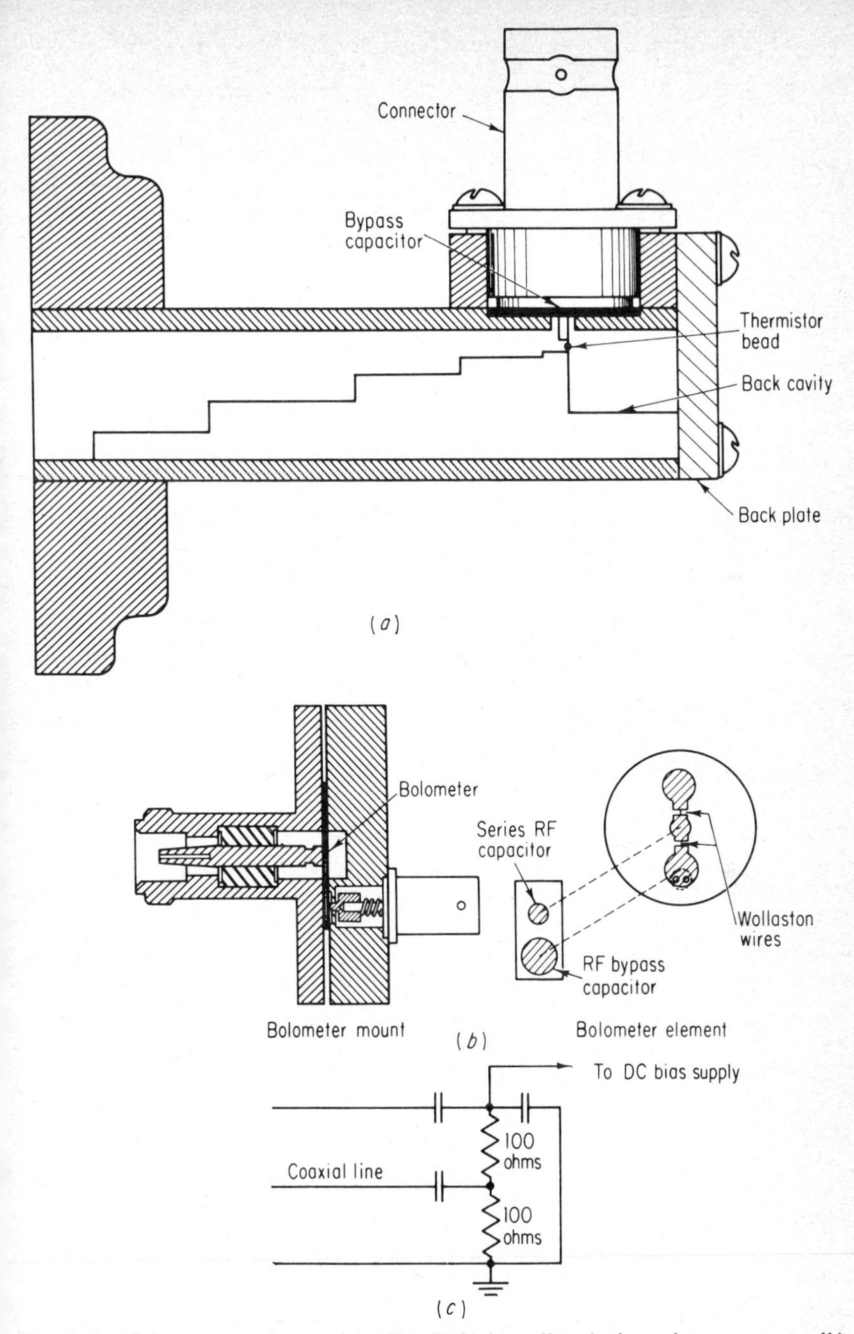

Fig. 8·16 Bolometer mounts. (*a*) The PRD broadband thermistor mount. (*b*) Essential details of the PRD Type 627-A coaxial bolometer mount and bolometer element. (c) The equivalent d-c circuit of the Type 627-A bolometer mount. The d-c bolometer resistance is 200 ohms, and its r-f resistance is 50 ohms. (*PRD Electronics.*)

thermistor by the ridged waveguide in the form of stepped transformers. The impedance can also be decreased to the appropriate value by use of a tapered ridged waveguide. A coaxial barretter mount is shown in Fig. 8·16*b* along with the equivalent circuit. It consists of the Wollaston wires and bypass capacitors mounted on a thin mica disk. The tunable detector mount in Fig. 8·17 accommodates the Sperry 821 or Narda 810-B barretter and crystals such as the type 1N23.

Fig. 8·17 Tunable detector mount. (*Hewlett-Packard Company.*)

In addition to evaluating the error due to loss of power by reflection in thermistor and barretter mounts, it is also necessary to determine to what extent the available power is absorbed by the detecting element and what fraction of the available power is dissipated by stray losses in the mount. This is a measure of the *efficiency* of the mount and is determined by comparing the power values obtained when a known amount of d-c power and a known amount of r-f power are alternately applied to the mount. An accurately calibrated d-c bridge is used to measure the d-c power entering the detecting element. The efficiency would be 100 per cent if all the power entering the mount were used in heating the bolometric element. The efficiency is usually given as

$$\eta = \frac{\text{power absorbed by the bolometer}}{\text{power delivered to the bolometer mount}}$$

The efficiency of commercial bolometer mounts may range from 95 to 98 per cent.

8·12 Short-slot hybrid[1]

The short-slot hybrid junction is a four-port device in which the power input at one port divides equally between two other terminals. If these two

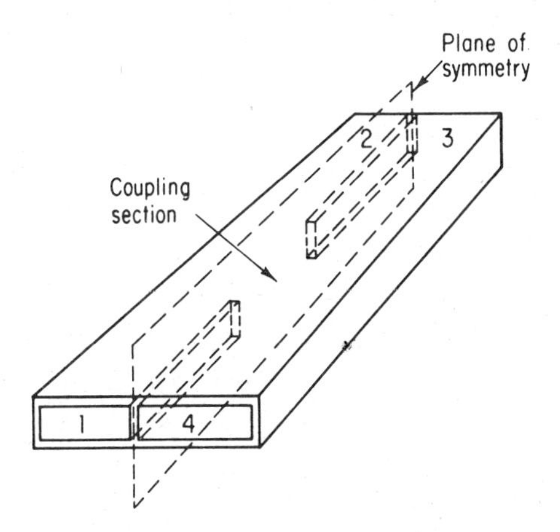

Fig. 8·18 The short-slot hybrid.

output terminals are perfectly matched, the energy reaching the fourth terminal (port) will be zero, and the structure becomes an ideal hybrid junction.

A schematic diagram of the hybrid is shown in Fig. 8·18 in which a plane of symmetry is indicated for the full length of the device. Two waveguide sections are placed side by side with a portion of the common wall removed to permit coupling between the two sections. Under suitable conditions, the power entering port 1 will have divided equally toward ports 2 and 3 by the time the energy reaches the end of the coupling section. The output voltages at port 3 lead the voltage at port 2 by 90°. The output voltage at port 3 leads by 45° what it would be if there were no slot in the waveguide. The short-slot hybrid has many applications in specialized waveguide circuits such as power splitters, phase splitters, phase shifters, couplers, balanced mixers, antenna feeds, bridge circuits, and duplexers.

8·13 Solid-state microwave switches

There are many applications which require rapid switching or modulation of microwave power. The solid-state switching diodes[2] can be used to obtain switching functions at nanosecond speeds. The switches are compact and lightweight and require only milliwatts of driving power. The control of microwave power is obtained by variation of a d-c bias which changes the impedance of the crystal diode. The present low-power applications (less than 1.5 watts) include ON-OFF switching of microwave power, amplitude modulation, electronic controlled attenuators, pulse shaping, and clipping. With improved application techniques and power-handling capabilities, the diode can be employed in phased antenna array systems, transmit-receive devices, etc.

The equivalent circuit of the crystal diode switch is shown in Fig. 8·19. The switching action is achieved by alternately applying forward and reverse bias to the diode. The explanation of switching action is given for the resonant frequency operation with the diode designed so that $X_L = X_C = X_{C_s}$.

Forward Bias. Consider the crystal diode connected across the waveguide. When the diode is forward biased, the nonlinear barrier resistance approaches a very low value and shunts the barrier capacitance. The inductive reactance of the lead is much greater than the spreading resistance. At the operating frequency, $X_L = X_{C_s}$ and the diode appears as a parallel resonant circuit. The crystal diode is in the ON condition, and nearly all the incident power is delivered to the load.

Reverse Bias. When reverse bias is applied to the diode, $X_{C_s} \gg r$ and the nonlinear barrier resistance is much greater than the barrier capacitance ($R \gg X_C$). The barrier capacitance shunts the large resistance and resonates with the lead inductance. The diode impedance is the small spreading resistance r shunted across the waveguide. Most of the incident power is reflected, and nearly all the power that is not reflected is absorbed in r. Very little power reaches the load, and the switch is in the OFF or isolation condition.

In the reverse bias condition OFF, the diode is represented by the spreading resistance r. In the forward bias condition ON, the diode is represented by $r(1 + Q^2)$. $Q = X_L/r$.

The shunt-connected diode is the most convenient waveguide configuration. The approximate insertion loss and isolation for the shunt-connected diode are

$$\text{Insertion loss} = 20 \log \left(1 + \frac{R_L}{2r(1 + Q^2)}\right)$$

$$\text{Isolation} = 20 \log \left(1 + \frac{R_L}{2r}\right)$$

The series-connected diode is the most convenient coaxial configuration. The insertion loss and isolation for the series-connected diode are

$$\text{Insertion loss} = 20 \log \left(1 + \frac{r}{2R_L}\right)$$

$$\text{Isolation} = 20 \log \left(1 + \frac{r(1 + Q^2)}{2R_L}\right)$$

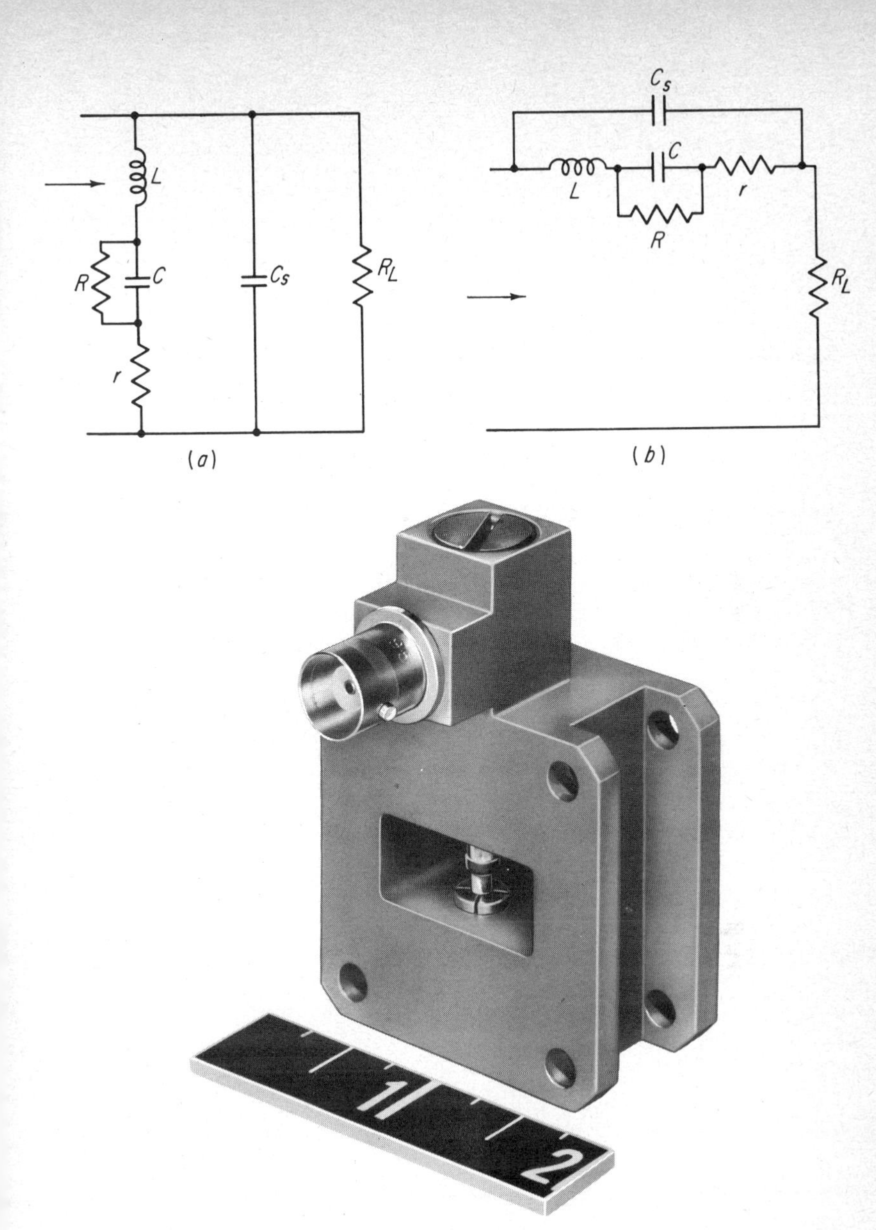

L = inductance of the diode lead (whisker)
r = spreading resistance (fixed series resistance of the diode)
R = diode junction or barrier resistance (nonlinear resistance of the point contact which varies with amplitude and polarity of the bias voltage)
C = diode junction or barrier capacitance (a function of the applied bias voltage)
C_s = shunt capacity of the diode package

Fig. 8·19 Resonant switch operation. (*a*) Shunt switch. (*b*) Series-connected switch. (*c*) Diode switch. (*Hughes Aircraft Company, Solid-state Products, Aerospace Group.*)

8·14 Waveguide joints

A useful form of coupling called a choke joint is shown in schematic form in Fig. 8·20*c*. Choke joints enable two line sections to be joined electrically despite lack of good mechanical contact between sections. The L-shaped channel or slot is a half-wave transmission line. The minimum impedance at

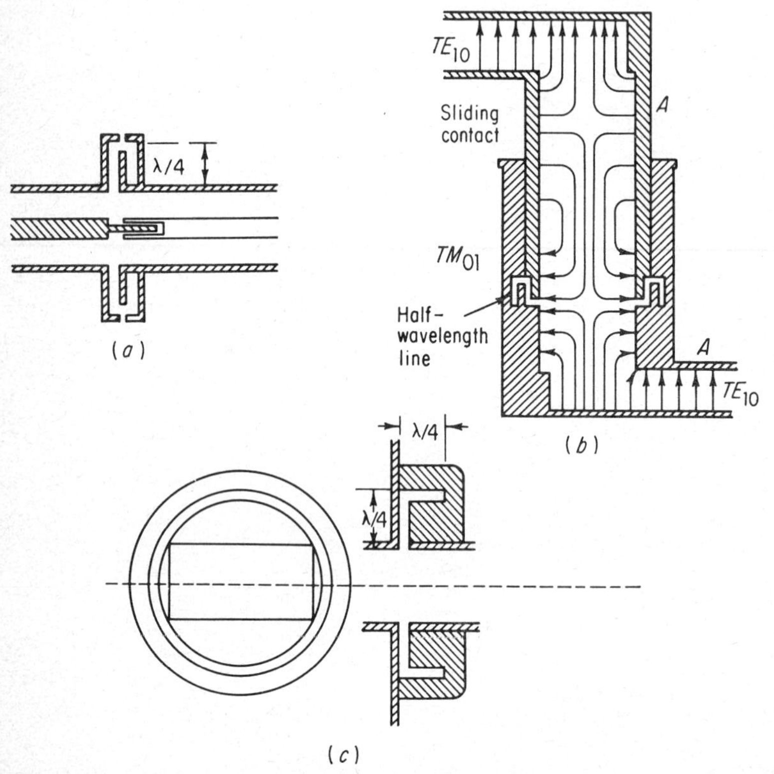

Fig. 8·20 Waveguide joints. (*a*) Coaxial rotating choke joint. (*b*) Waveguide rotating joint. (*c*) Waveguide choke joint.

the shorted end of the slot is transformed to the junction at the waveguide, and the junction behaves as though the adjacent walls were continuous.

A coaxial-line rotating choke joint is shown in Fig. 8·20*a*. Waveguide rotating joints find frequent applications in waveguide systems. A waveguide rotary joint is shown in Fig. 8·20*b*. The rotation is accomplished by using the radial symmetrical TM_{01} mode. The two sections of rectangular waveguide are joined through a section of circular waveguide operating in the TM_{01} mode. This mode is excited directly from the rectangular waveguide mode. Matching partitions, such as inductive windows, are placed in the rectangular waveguide sections to reduce reflections at the junctions of the two guides. Resonant rings may be placed in the circular guide to prevent the formation of the rectangular mode in the circular waveguide.

8·15 Waveguide slotted section

The waveguide slotted section is shown in Fig. 8·21*a*. The slotted section is an accurately machined section of waveguide in which a small longitudinal slot is cut. The section is mounted on a probe-carriage assembly. Since the waveguide sections can be interchanged, measurements can be performed over a number of waveguide bands by interchanging the waveguide slotted sections.

The probe shown in Fig. 8·21*b* is mounted on the carriage and samples the variations of the electric field throughout the length of probe travel along the slot. A crystal detector or barretter is mounted in the probe housing. The detector output is available at the BNC connector and is connected to a standing-wave amplifier. The probe is a thin silver wire which is inserted into the slot by the screw-on knob at the top of the unit. The probe is connected to the inner conductor of the two concentric coaxial lines. Each line is terminated by a variable short circuit. The inner conductor tuner is precisely adjusted by rotation of the inner knob shown in the open slot. The second tuner is a coarse adjustment controlled by the external cylindrical nut. In this way, the standing wave in the probe assembly is varied to obtain maximum output from the detector.

8·16 Antennas

The phenomenon known as *radiation* is the property of a transmission line structure by which wave power detaches itself and continues into space upon reaching the end of the line. An *antenna* or *radiator* is the coupling structure between the guided wave and the free-space wave or vice versa. It is used for either radiating electromagnetic energy into space or collecting electromagnetic energy from space.

Two types of sources are to be considered in a discussion of radiation patterns. The ideal point source radiates equal power in all directions. It is referred to as an *isotropic radiator* or as a *spherical radiator*. The hypothetical isotropic radiator is convenient in theory but is not a physically realizable source. The performance of the ideal source can be calculated and used as a base upon which the performance of a real antenna can be calculated. The power per unit area as a function of the distance from the point source is

$$P_r = \frac{P_t}{4\pi d^2}$$

where P_r is the radial component of average Poynting vector in watts per square meter, P_t is the total radiated power in watts, and d is the distance from the source in meters.

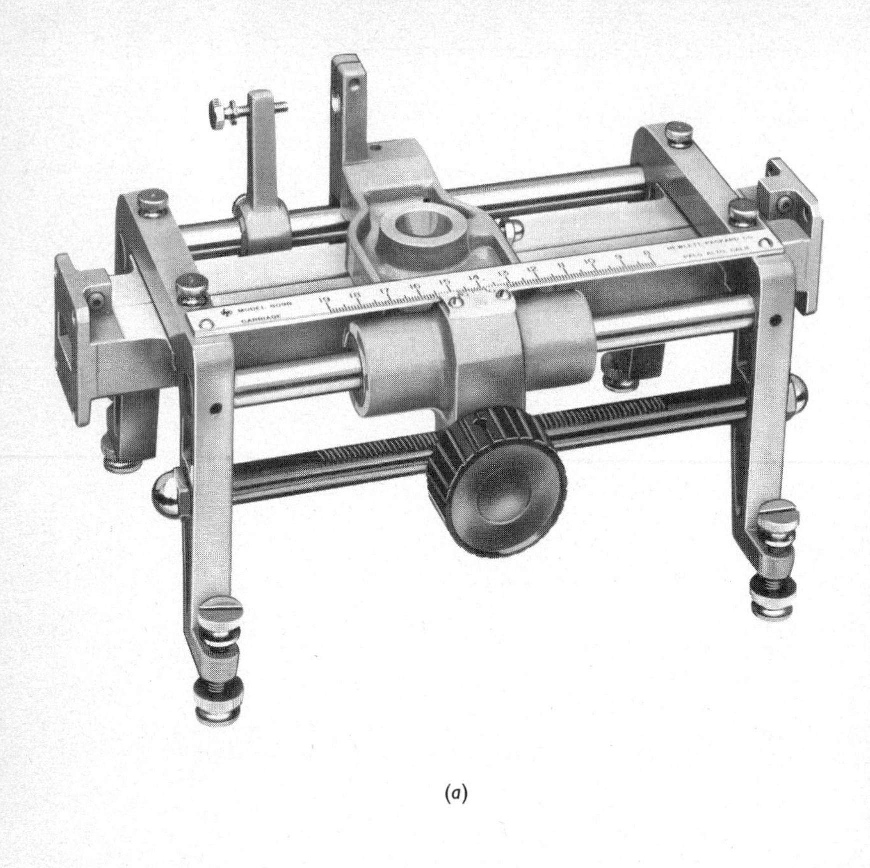

(a)

(b)

Fig. 8·21 **(*a*) Probe carriage and waveguide slotted section. (*Hewlett-Packard Company.*) (*b*) Adjustable broadband probe. (*Sperry Microwave Electronics Company.*)**

The *elementary source*, sometimes referred to as an *oscillating doublet*, may be regarded as an infinitely short linear current element. Ideally, it consists of two closely spaced charges of opposite sign, both oscillating in the same phase. The elementary source differs from the ideal source in that it has an axis. This source can be approximated but cannot be completely realized in practice. The radiation pattern of the elementary source is of considerable interest because it is the pattern produced by a short dipole coincident with

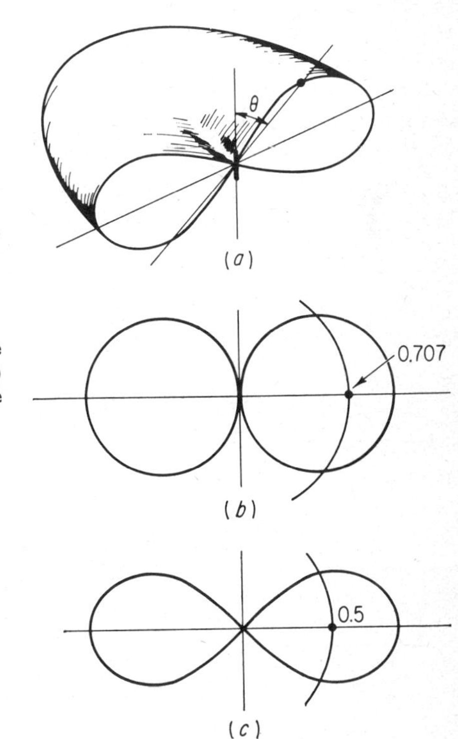

Fig. 8·22 (*a*) Vertical plot of the elementary source pattern. (*b*) Relative signal level. (*c*) Relative power level.

the polar axis. The source radiates but in directions perpendicular to the axis, and there is no radiation along the axis.

There are two kinds of field patterns associated with radiation from the elementary source, the field-intensity patterns and the power-density patterns. The *field-intensity pattern*, sometimes called the *relative field pattern* or *relative signal level*, is a plot of the electric field intensity at a fixed distance from the source as a function of the angle θ formed by the axis and the line joining the source to the point of measurement, as indicated in Fig. 8·22*a*. The relative signal level is proportional to $(\sin \theta)/d$, where d is the distance from the source to the point of measurement. A polar plot of the relative signal level pattern is shown in Fig. 8·22*b*.

The *power-density pattern*, called the *relative power pattern* or *relative power level*, is a plot of the power density at a fixed distance from the source as a function of the angle θ measured from the axis to the line joining the source to the point of measurement. The relative power level is proportional to $(\sin^2 \theta)/d^2$. A plot of relative power level is shown in Fig. 8·22*c*. The power level drops to one-half of its maximum value in directions that are 45° from the direction of maximum power level. The beam width, measured between the two half-power directions, is therefore 90° in a plane containing the axis. The signal level in the half-power directions is 0.707 of its maximum since 0.707 is the square root of 0.5.

The behavior of real antennas can be obtained by considering the line source formed by a large number of ideal isotropic sources which are not directional. The pattern of any antenna can be regarded as having been produced by an array of point sources. A radiating system which consists of a number of point sources arranged in a line is called a *line array* or *linear array*. The simplest linear array consists of two isotropic point sources which are excited in phase. The radiation pattern formed by two isotropic point sources of equal amplitude and the same phase is shown in Fig. 8·23*a*. The vector addition of the fields indicates that the field pattern is a maximum at P_1 and P_2, which are equidistant from the point sources. The path lengths from the sources are different at other points on the pattern such as P_3. The fields from the sources therefore differ in phase, thus producing the field pattern as indicated. This pattern is revolved around the y axis to form the space pattern (doughnut-shaped). The relative field pattern for two isotropic point sources of the same amplitude but opposite phase and spaced one-half wavelength apart is indicated in Fig. 8·23*b*. The low-level beams shown in *c*, *d*, *e*, and *f* are called *side-lobes*. The number and positions of these lobes are determined by the spacing and phase of the sources.

The field patterns in Fig. 8·23*a* and *c* are of the *broadside* type since the maximum radiation is perpendicular to the line joining the sources. The *end-fire* types of array are indicated in *b*, *d*, *e*, and *f*, in which maximum radiation is in the same direction as the line joining the sources. The directive end-fire arrays are illustrated in *e* and *f*.

The Half-wave Antenna. A short linear conductor is often called a short dipole. It is always of finite length even though it may be very short. Current oscillating in the dipole generates electromagnetic waves which travel out into free space at the velocity of light. The dipole has properties that are characteristic of a transmission line, a resonant circuit, and an antenna. The dipole exhibits characteristics of a resonator since energy concentrations oscillate from entirely electric energy to entirely magnetic energy and back twice per cycle.

A common form of the fundamental half-wave radiating element is a conductor of essentially uniform diameter, one-half wavelength long,

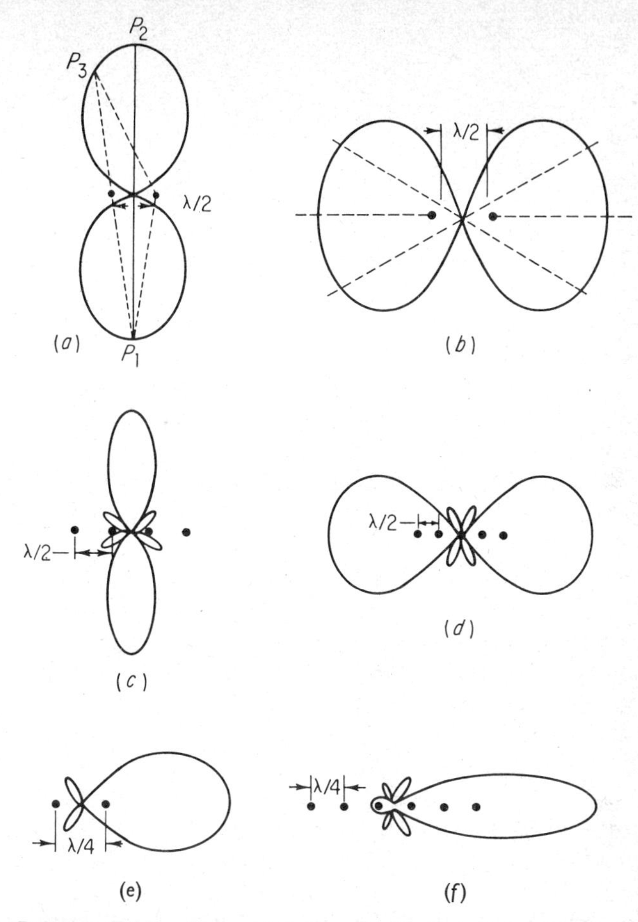

Fig. 8·23 Relative field patterns. (*a*) Two isotropic point sources of the same amplitude and phase spaced one-half wavelength apart. (*b*) Two isotropic point sources of the same amplitude but opposite phase. (*c*) Array of four isotropic point sources of equal amplitude and the same phase spaced one-half wavelength apart. (*d*) Array of four point sources of equal amplitude and opposite phase spaced one-half wavelength apart. (e) Sources equal in amplitude, 90° out of phase. (*f*) Directive end-fire type array.

connected at its midpoint to a two-wire transmission line. This type of connection presents a low impedance to the transmission line and is referred to as *center-fed* or *current-fed*. The *voltage-fed* or *end-fed* method of exciting half-wave antennas presents a high impedance. Both types are illustrated in Fig. 8·24.

The fields which are associated with the stored energy (periodic buildup and collapse of the electric and magnetic fields around the antenna) are called the *induction fields*. The induction fields are principally responsible for

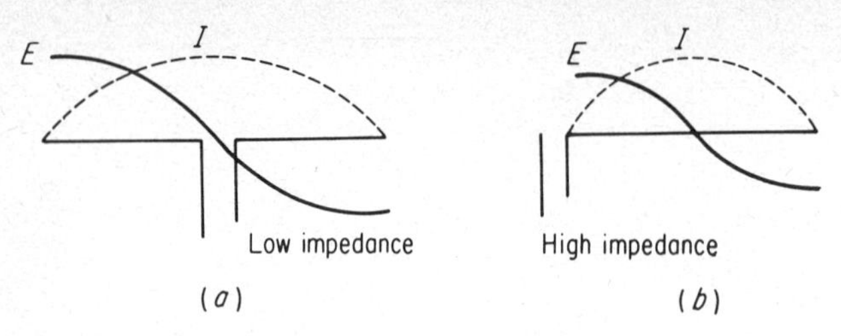

Fig. 8·24 Half-wave antenna feeds.

the behavior of the antenna as a resonant circuit element. The induction field components are large at the antenna and are not detectable beyond a distance of about two wavelengths from the antenna. The resonant behavior of the antenna also includes the movement of charge up and down the length of the antenna at the frequency of the applied wave. At a particular instant of time, maximum negative charge exists on one end of the antenna and maximum positive charge exists on the opposite end. Maximum intensity of electric flux lines exists at this instant (zero current). The flux lines associated with the charges collapse, and the unlike charges come together at another instant of the cycle (maximum current flow). The original lines of force try to return rapidly to the antenna. However, all of the lines of force in the fields cannot return before the new induction field, of opposite polarity, starts to move outward at the beginning of the next half-cycle. The new induction field components encounter the returning fields and force them back away from the antenna. This periodic action around the antenna produces a steady flow of energy into space, as indicated for a particular instant of time in Fig. 8·25.

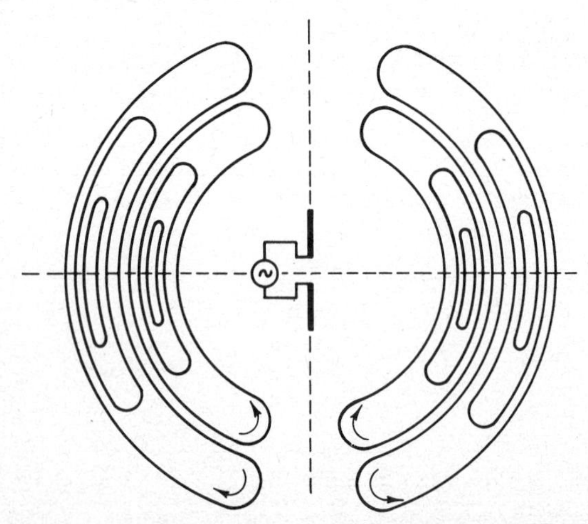

Fig. 8·25 Idealized representation of radiation (electric field) from a dipole.

The portion of the radiated energy which does not return to the antenna is called the *radiation field*. The flux lines expand radially with the velocity of light, and new flux lines are created at the antenna to replace the radiated lines. The frequency of the sinusoidal oscillating fields is the same as the frequency of the current in the antenna. The radiation fields are smaller than the induction fields near the antenna but are much larger in regions removed from the antenna.

The field patterns of the dipole are the same as those shown in Fig. 8·22 for the elementary source. The intensity of the radiated wave depends primarily upon the radiation resistance and the current flow in the antenna. In turn, the radiation resistance depends upon the antenna shape, size, length, height above ground, and the operating frequency.

Antenna Arrays. The limited directive property of the half-wave antenna makes it necessary to use other types of antennas in order to produce a concentration of radiated energy in a specific direction. Directional arrays can be constructed with the aid of elements in which currents are induced by the fields of the driven element. These elements have no transmission line connection to the transmitter or receiver and are usually referred to as *parasitic elements.* A parasitic element can act as a *reflector* or *director.* A parasitic array is perhaps the simplest of all directional arrays. A parasitic element longer than the driven element is called a *reflector* and reinforces radiation in the direction of a line pointing away from itself toward the driven element. This element is usually spaced about 0.15 wavelength from the driven element, as indicated in Fig. 8·26. If the parasitic element is shorter

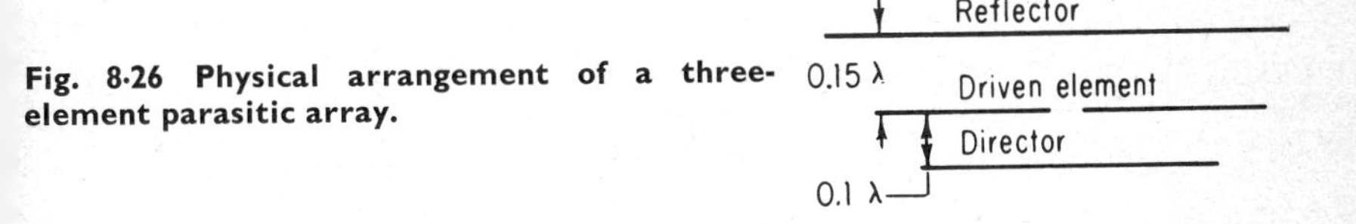

Fig. 8·26 Physical arrangement of a three-element parasitic array.

than the driven element, it is called a *director*. The sharpness of resonance of the multielement array is determined by the lengths of the parasitic elements, and the gain or directional property is chiefly determined by the element spacing. The input resistance of the driven element is usually lowered by the parasitic element since current in the parasitic element causes a voltage to be induced in the driven element. A wide frequency range is obtained by adjusting the director length to resonate at the highest frequency and by adjusting the reflector to resonate at the lowest frequency.

Collinear elements lie in the same plane or axis and are excited in phase. If collinear elements are stacked above and below another set of similar elements, the result is a *broadside array*. The connections of a broadside array consisting of a number of one-half wave antennas are shown in Fig. 8·27*a*. The radiation pattern minus the side lobes is shown in Fig. 8·27*b*. One of the two main lobes can be eliminated by placing a metal surface or metal screen approximately one-quarter wavelength behind the array. One lobe can also be eliminated using a second array of antennas placed behind

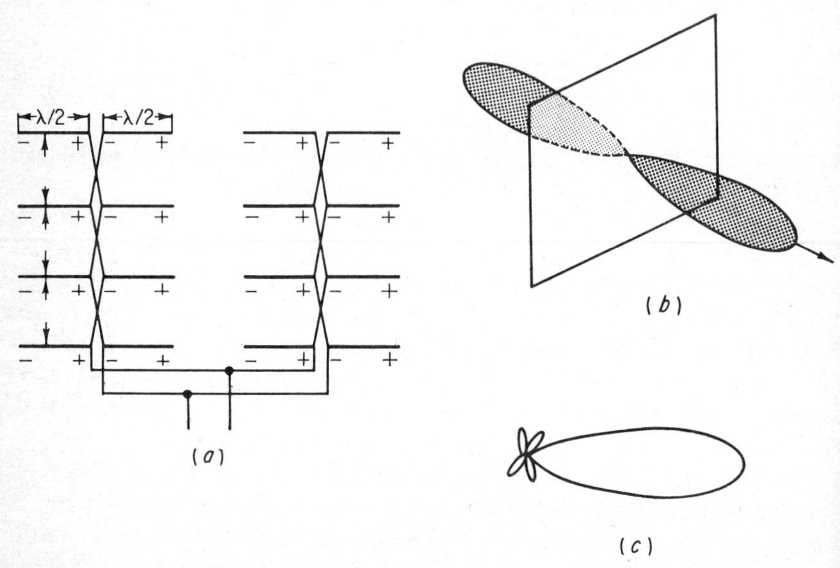

Fig. 8·27 (*a*) Broadside array. (*b*) Radiation pattern of the broadside array. (*c*) Field pattern for broadside array with reflector.

the first array. The resulting pattern is shown at *c*. Additional rows of dipoles in the horizontal and vertical directions provide more narrow beams.

Parabolic Reflector. The parabola is a curve that passes through all points which are equidistant from a point called the focus and a fixed line called the *directrix*. The distance from any point must be measured perpendicular to the directrix, as indicated in Fig. 8·28*a*. Constant-phase surfaces determine the directional characteristics of antennas. The field reflected from the parabola has a single time phase in a plane across the mouth of the parabola as illustrated. Energy from the focal point *P* strikes the parabola surface at points such as *A*, *B*, and *C* and is reflected in a direction parallel to the axis (the angle of incidence is equal to the angle of reflection as shown). The sum of the distance *PA* and *AE* equals the sum of the distances *PB* and *BD*.

Sources of illumination for parabolic antennas are classified as *front feed* and *rear feed*. The feed system may be coaxial or waveguide, as illustrated in Fig. 8·28. The antenna at *b* is a *paraboloid of revolution*, which is sometimes referred to as a *dish*. The antenna at *c* is a *parabolic cylinder*. A schematic diagram of a rear-fed dish antenna is shown at *d*. The parabola-of-revolution antenna is a high-gain antenna used in tracking radar systems where sharp, well-defined beams are required. The parabolic cylinder antenna is used in high-power, search-radar systems applications. High directive

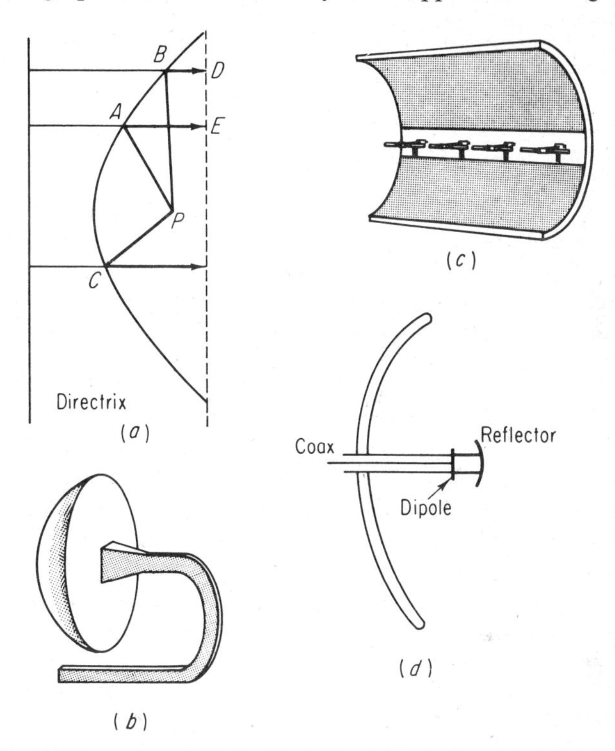

Fig. 8·28 Parabolic antennas. (*b*) Paraboloid of revolution. (*c*) Parabolic cylinder. (*d*) Rear-fed dish.

properties are obtained because of the large dimensions in wavelengths of the plane constant-phase surface.

Electromagnetic Horn. A rectangular waveguide may be flared in either or both of its two dimensions to form a rectangular horn. The radiation pattern is highly directive. The beam is sharp in the plane of the longer sides and comparatively broad in the plane of the shorter sides. The actual beam directivity is determined by horn length and flare angle and is generally increased by an increased horn length. The length and flare angle provide a gradual decrease in waveguide wavelength. As the waveguide wavelength approaches the free-space wavelength, the wave impedance approaches the free-space impedance, and a near perfect impedance match is obtained. If the guide is flared symmetrically in the *a* dimension only, the structure is known as an H-plane *sectoral horn*. The E-plane sectoral horn is flared symmetrically in the b dimension. The *pyramidal horn* is formed by flaring the waveguide in both planes, as shown in Fig. 8·29. Power gain is a function of the flare in both planes. As shown, the constant-phase wavefront is not a plane wave.

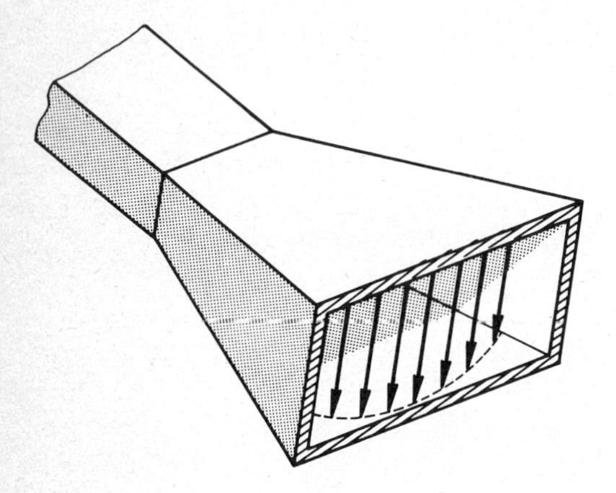

Fig. 8·29 The pyramidal horn.

Metal-plate Lens Antennas. A waveguide lens makes use of the optical properties of microwaves. A constant-phase wavefront can be obtained from a spherical wavefront, as shown in Fig. 8·30. Waves which pass through the thick center of the plane-convex dielectric lens have a lower phase velocity and therefore suffer a decrease in phase shift. Waves which pass through the

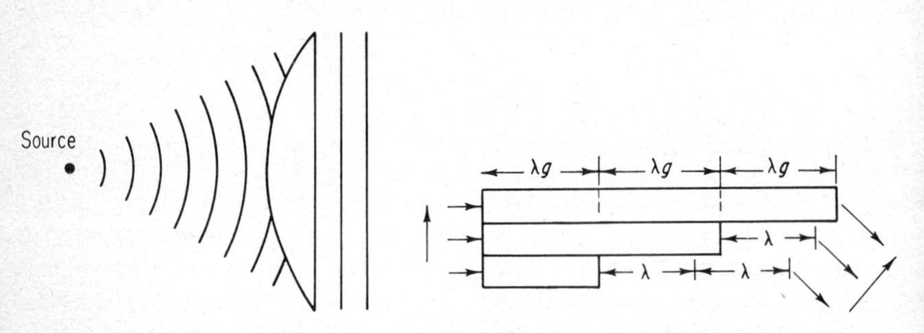

Fig. 8·30 Dielectric lens. **Fig. 8·31 Illustration of phase focusing.**

thin portions of dielectric material have less phase shift and therefore emerge with a relative leading phase. With proper shaping, the lens produces a plane wave at the output side.

The parallel plates of a waveguide lens increase the phase velocity of the electromagnetic wave. Therefore, this type of lens must be concave so that energy at the edges of the lens experiences an increased phase velocity in passing through the longer waveguide section. As an example of focusing, Fig. 8·31 shows a plane-wave incident upon three sections of waveguide. The waveguide wavelength is greater than the free-space wavelength, and the corresponding phase velocity is greater than the free-space velocity. Because of the different velocities, the wavefront is staggered as shown. The waves can be made convergent by proper lengths of waveguides. Alternately, waves originating from a focal point will emerge as a beam of parallel waves, as shown in Fig. 8·32*a*. In the example, it was assumed that waveguide sections

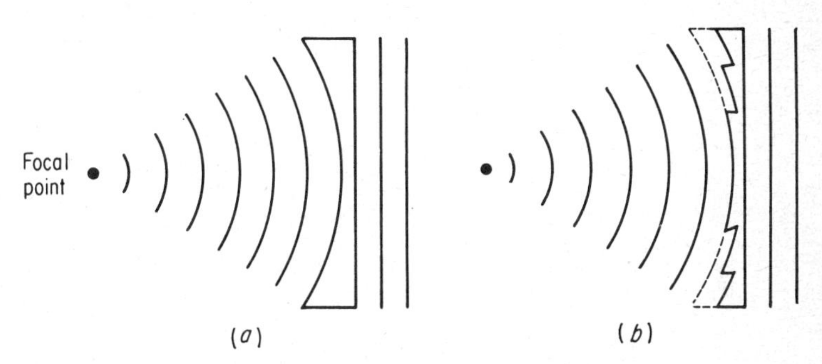

Fig. 8·32 (*a*) E-plane metal-plate lens. (*b*) Stepped or zoned lens.

were placed one on top of the other. However, the phase velocity in a waveguide depends only on the width or *a* dimension of a waveguide propagating the TE_{10} mode. The tops and bottoms can be removed, and the lens is therefore constructed with thin sheets of metal. One form of lens antenna is the stepped waveguide lens shown in Fig. 8·32*b*. It is stepped so that a full-wave path length difference is removed whenever the thickness is increased by a full wave in going from the center of the lens outward.

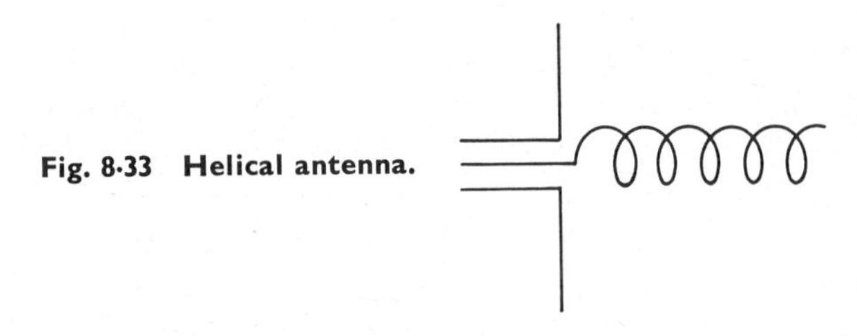

Fig. 8·33 Helical antenna.

The *helical antenna* is shown in Fig. 8·33. This type antenna is useful where a circularly polarized radiation field is desired. Radiation from helices with circumferences of the order of one wavelength and a number of turns is usually a well-defined beam with a maximum radiation in the direction of the helix axis. A wide variety of nonuniform or tapered helices is also possible. The terminal impedance of a helical antenna operating in the axial mode is a pure resistance in the range of 100 to 200 ohms.

Slot Antennas. A linear array antenna which has excellent electrical characteristics and which has mechanical advantages over other types of antennas can be formed by positioning a series of slots along a length of waveguide. The slots radiate electromagnetic energy, and the shape of the radiation pattern depends upon the orientation of the slot with respect to the edges and faces of the guide. The slots may be classified as *resonant*, or *near resonant*, and *nonresonant*. The majority of practical arrays consists of arrays of slots cut in the same face of the waveguide structure. If the field pattern, at a distance from the radiator, is similar for each element or slot, the array is called a *parallel array*.

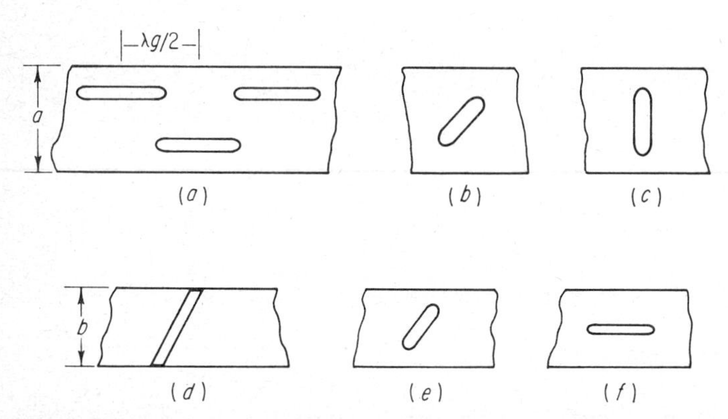

Fig. 8·34 Resonant slots. (*a*) Longitudinal shunt slot array. (*b*) Inclined series slot. (c) Series slot. (*d*) Edge slot (shunt). (e) Shunt inclined slot. (*f*) Longitudinal shunt slot in *b* dimension.

Radiating slots can be located at a number of different positions along the rectangular waveguide, as indicated in Fig. 8·34. The *shunt* type looks like a series resonant circuit in parallel with the line. When this type of slot is arranged in an array, a short circuit is placed an odd multiple of one-quarter waveguide wavelengths from the last slot. The *series* type slot looks like a parallel resonant circuit in series with the line. In a series slot array, a short circuit is placed a multiple of one-half wavelengths from the last slot.

The longitudinal shunt slot and the edge or shunt-inclined slots are most widely used. The longitudinal slot produces polarization transverse to the axis of the guide. The edge slot gives longitudinal polarization, which is the normal requirement. The edge slot can be of the "dumbbell" type; however, the simple edge cut is easier to fabricate.

Nonresonant or traveling-wave arrays have slot spacings other than one-half or one wavelength. The beam emerges at an angle to the array, and the angle varies with frequency. The nonresonant array is matched along its whole length because of the random phase additions of the reflections from the slots.

Antenna Properties. The radiating antenna acts like a network containing resistance and reactance. Some of the resistance loss is due to the heat loss in the antenna, and some is associated with the loss by radiation. The loss due to radiation is referred to as *radiation resistance*. Radiation resistance is defined as the ratio of total power radiated to the square of the effective value of the maximum current in the radiating system. The radiation resistance of a linear center-fed one-half wavelength antenna with sinusoidal current distribution is 73 ohms.

The *directive gain* of an antenna is a measure of its ability to transmit a signal to a distant point. A measure of the ability of an antenna to collect power is called the *effective area*. The *effective area* is the ratio of the power available at the antenna terminals (terminating impedance) to the power per unit area (power density) of the polarized incident wave. That is, the received power is equal to the power flow through an area that is equal to the effective area of the antenna.

From the law of reciprocity, a good transmitting antenna is also a good receiving antenna.

The *gain* of an antenna is the ratio of the maximum radiation intensity to the maximum radiation intensity from a reference antenna with the same power. If the reference antenna is an isotropic point source, then the gain is the *absolute gain*. A simple expression for antenna gain is

$$G = \frac{P}{P_a}$$

where P is the power flow per unit area in the plane, linearly polarized wave which the antenna causes in a distant region and P_a is the power flow per unit area which would have been produced by a reference antenna transmitting the same amount of power. Antenna gain may also be expressed by

$$G = \frac{4\pi A}{\lambda^2}$$

where A is the effective area. From the above formula,

$$A = \frac{\lambda^2 G}{4\pi}$$

Since the gain of the isotropic source is *one*, the effective area is $\lambda^2/4\pi = 0.079\lambda^2$. The simple oscillating doublet has an effective area of $3\lambda^2/8\pi = 0.119\lambda^2$ and a corresponding gain of 1.5. The half-wave antenna has an effective area of $1.31\lambda^2$ and a gain of 1.64. 16.4

Other terms which are commonly used to describe the properties of antennas are the effectiveness, radiation efficiency, radiation intensity, and aperture. The *effectiveness* of an antenna is the ratio of the effective area to the actual area. The *radiation efficiency* is defined as the ratio of the power radiated to the total power supplied to the antenna at a given frequency.[3] *Radiation intensity* is defined as the power radiated from an antenna per unit solid angle in that direction.[3] The *aperture* of an array is defined as the portion of a plane surface near the antenna, perpendicular to the direction of maximum radiation, through which a major portion of the radiation passes.[3]

PROBLEMS

8·1 The magic tee is used in a microwave bridge circuit. Two signals of equal amplitude are applied to arms 1 and 2.

a. Illustrate the electric field orientations in arms 1 and 2 and the **H** arm for a zero output (null) at the output of the **H** arm.

b. Repeat *a* for a null in the **E** arm.

c. Illustrate the electric field orientation in arms 1 and 2 and the **E** arm for a zero output (null) at the output of the **E** arm.

8·2 Plot the attenuation characteristic of the type 382A attenuator in 10° increments from 0 to 90°.

8·3 What is the directivity of the coupler in Fig. 8·9*c* if P_{0i} is -10 dbm and P_{0r} is -42 dbm? If P_i is 0 dbm, what is the coupling value?

8·4 The directivity of the coupler in Fig. 8·9*c* is 45 db. With a matched load connected to the main line input, the power output P_{0r} is 63 db below the input power P_{ir}. What is the coupling value of the coupler? If P_{ir} is 1 watt, what is P_{0r}?

8·5 If a short circuit is placed on the normal input line of the coupler in Prob. 8·4, what is the output power P_{0i}, assuming no loss in the main line?

8·6 Illustrate the expected detector output waveforms if a 1,000-cps square-wave signal is alternately applied to a crystal, barretter, and thermistor.

8·7 Make a comparison summary between crystal and barretter characteristics and applications.

8·8 Make a comparison summary between barretter and thermistor characteristics and applications.

8·9 Typical values for the Philco 1N3482 diode at 9 Gc are $L = 3$ nanohenrys, $C = 0.1$ picofarads, $r = 10$ ohms, and $R_L = 400$ ohms. Calculate the insertion loss and isolation for shunt switch operation.

8·10 What are the insertion loss and isolation values of the same diode used in a series 50-ohm coaxial line application?

8·11 The power input to the normal output terminal of a 20-db directional coupler is 300 mw. The directivity is greater than 60 db and can be ignored in the calculations.

a. How much power appears at the auxiliary arm output if the main line input is terminated with a short circuit?

b. How much power arrives at the auxiliary arm when the main line input is terminated with loads having the following VSWR values: 1.5, 2.0, 3.6, and 6.0?

REFERENCES

1. Henry J. Riblett, The Short-slot Hybrid Junction, *Proc. IRE*, vol. 40, no. 2, February, 1952.
2. M. R. Millet, Microwave Switching by Crystal Diodes, *IRE MTT*-6, no. 3, pp. 284–290, July, 1958.

   R. V. Garver, E. G. Spencer, and M. A. Harper, Microwave Semiconductor Switching Techniques, *IRE MTT*-6, no. 4, pp. 378–383, October, 1958.

   R. V. Garver, Theory of TEM Diode Switching, *IRE MTT*-9, no. 3, pp. 224–238, May, 1961.
3. "Standards on Antennas," Definitions of Terms, The Institute of Radio Engineers, New York, 1948.

CHAPTER

9

FREQUENCY MEASUREMENT DEVICES

Introduction. The measurement of frequency or wavelength is one of the primary requirements in most microwave problems. The operating principles of frequency measuring devices are explained in this chapter.

Frequency is the more fundamental quantity because the frequency of oscillation is the same at all parts of the system under steady-state conditions. Wavelength, on the other hand, depends upon the configuration of the electric and magnetic fields as determined by the geometry of the measuring device. Even though the resonant frequency and wavelength can be computed from the dimensions of the frequency-measuring structure, the resonant system is usually calibrated against a frequency standard of some type.

9·1 Transmission line resonators

The resonance properties of short-circuited and open-circuited transmission lines were discussed in Chap. 2. The relations between line length and frequency or wavelength for transmission lines terminated in a short circuit will be reviewed briefly by referring to Fig. 9·1. An input voltage wave to the quarter-wave line at *a* will have a phase change of 90° in traveling from the input (open) to the shorted end, 180° phase change at the short, and another 90° in returning to the open end. The total phase change is 360°, and this reflected wave is in phase with the applied wave. By referring to the illustrations, it can be seen that resonance occurs when the input terminals are shunted with a reactance that is equal in magnitude and opposite in sign to the input impedance. In the case of the open input line, the shunting reactance is infinite and resonance occurs at frequencies that make the input impedance infinite. These frequencies are called the *natural* frequencies of oscillation. The lowest natural frequency f_1 corresponds to the quarter wavelength voltage distribution, and other resonances occur at $3f_1$, $5f_1$, etc. The natural frequencies of the line shorted at both ends are observed to be $2f_1$, $4f_1$, $6f_1$, etc.

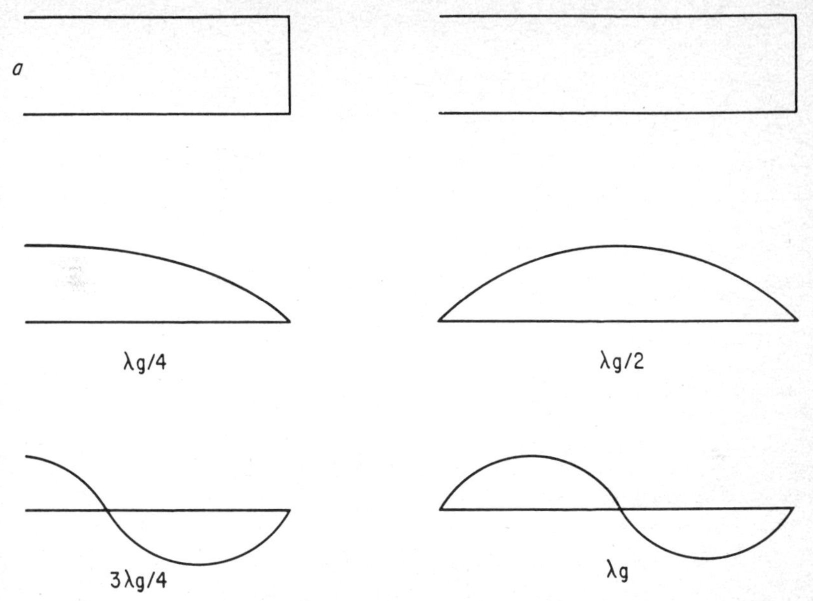

Fig. 9·1 Voltage distributions on sections of resonant transmission lines.

The different voltage and current patterns are referred to as *modes*. The $\lambda_g/4$, $3\lambda_g/4$, $\lambda_g/2$, and λ_g modes of resonance are shown in the illustration. The resonant mode excited by the lowest possible frequency is called the *fundamental* mode or the *dominant* mode.

9·2 Cavity resonators

A cavity resonator, sometimes called a *resonant cavity* or *wavemeter*, is a dielectric region completely surrounded by conducting walls. It is capable of storing energy and is analogous to the low-frequency resonant circuit.

The cavity resonator is an essential part of most microwave circuits and systems. Cavities can be listed by three general types according to physical construction.

1. Sections of uniform transmission line (such as coaxial transmission line)
2. Irregular shapes such as reentrant cavities
3. Geometrical shapes such as right circular cylinder, rectangular boxes, spheres, etc.

Some of the uses of cavities are listed below.

1. Measurement of frequency (a primary requirement in most microwave problems)
2. Selective filters
   a. Bandpass
   b. Band rejection

3. Applications in accelerators, oscillators, klystrons, ammonia masers, and parametric amplifiers
4. Concentration of electric fields for power breakdown
5. Applications in measurement of properties of materials

An essential property of the above structures is that every enclosed cavity, with a highly conducting boundary, can be excited in an infinite sequence of resonant modes. Each mode is characterized by a particular standing-wave distribution of surface current. The frequencies at which resonance occurs depend upon the shape and size of the enclosed cavity.

9·3 The coaxial cavity

A coaxial cavity resonator can be formed by closing the ends of a section of coaxial transmission line or by closing one end and leaving one end open in the same manner as shown for the parallel two-wire transmission line.

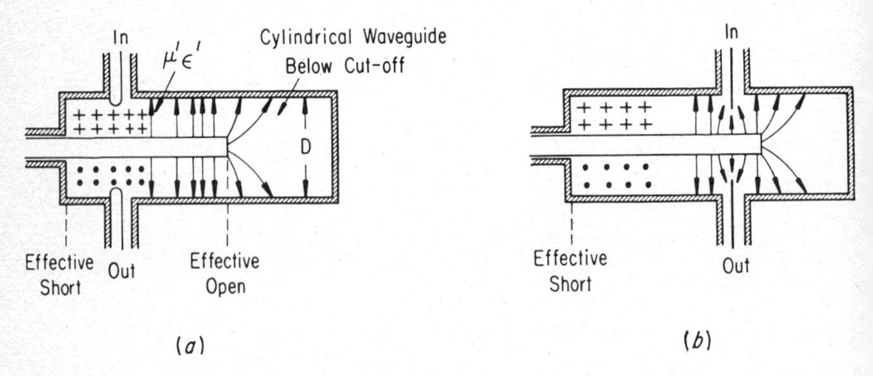

Fig. 9·2 Sectional views of coaxial resonators operated in odd multiples of a quarter wavelength. (*a*) Magnetic (loop) coupling. (*b*) Electric (probe) coupling.

The spectrum of the resonant modes in the coaxial cavity is the same as that described for the parallel-wire lines and comprises natural frequencies at integral multiples of the lowest natural frequency. Although the characteristics of the coaxial line resonator are in many respects similar to those of the parallel-wire resonators, coaxial resonators have the advantage that the electromagnetic field may be entirely confined within the outer conductor, thereby eliminating the undesirable radiation loss.

Coaxial cavities have been constructed using the TEM mode so that $\lambda_g = \lambda$, the free-space wavelength. This is strictly true only for a lossless cavity. For a practical cavity the velocity of propagation is slightly slower in the coaxial line than it would be in free space.

The quarter-wave type of coaxial cavity is shown in Fig. 9·2. The coupling devices for magnetic coupling are most effective near the short-circuited or high current end. The electric coupling devices are located an odd number of quarter wavelengths from the short-circuited end in the region of intense electric field for this type of cavity. The coupling methods are similar for a line any number of quarter wavelengths long. The effective open is obtained with a circular waveguide-below-cutoff.

The $\lambda/4$ cavity of Fig. 9·2 resonates in the TEM mode when the effective length is an odd multiple of a quarter wavelength as given by

$$L = \lambda \frac{2K - 1}{4} \qquad \text{where} \qquad \lambda = \frac{2.998 \times 10^{10}}{f\sqrt{\mu'\epsilon'}} \quad \text{cm}$$

K is an integer and applies so long as the cylindrical waveguide section is below cutoff; $K = 1$ for $\lambda/4$, and $K = 2$ for $3\lambda/4$, etc.

The half-wave coaxial cavity (shorted at both ends) resonates when the following relation is satisfied:

$$L = K\frac{\lambda}{2}$$

where $K = 1$ for $\lambda/2$, and $K = 2$ for λ, etc.
Sectional views of this type of cavity are shown in Fig. 9·3.

9·4 Types of wavemeters

Wavemeters are classified as *transmission*, *reaction*, or *absorption* types. The two-port cavity in Fig. 9·3*a* is of the transmission type. Maximum energy passes through the cavity only when the frequency of the input signal is equal to the resonant frequency setting of the cavity. The cavity is effectively a short circuit except at frequencies near resonance; these bandpass characteristics are shown in Fig. 9·3, where it is noted that maximum signal through the cavity is obtained at the resonant frequency f_0.

The *reaction* type frequency meter is illustrated in Fig. 9·3*b*. The cavity absorbs energy from the transmission line at the resonant frequency, as shown by the frequency meter reaction as it would appear on a swept-frequency klystron mode displayed on an oscilloscope.

The *absorption* type wavemeter or frequency meter is represented by the one-entry coaxial wavemeter shown in Fig. 9·3*c*. Viewed from the input, this cavity exhibits the same characteristics as the transmission type.

The transmission and absorption type wavemeters should be used in a matched system and may require isolation when the operation of the particular system is affected by the variations of cavity impedance when the wavemeter is tuned.

9·5 Cavity Q

The power dissipation in devices which store energy is usually described by the quality factor Q. This useful figure of merit is the ratio of energy stored in a device to the energy dissipated in a specified time interval. For

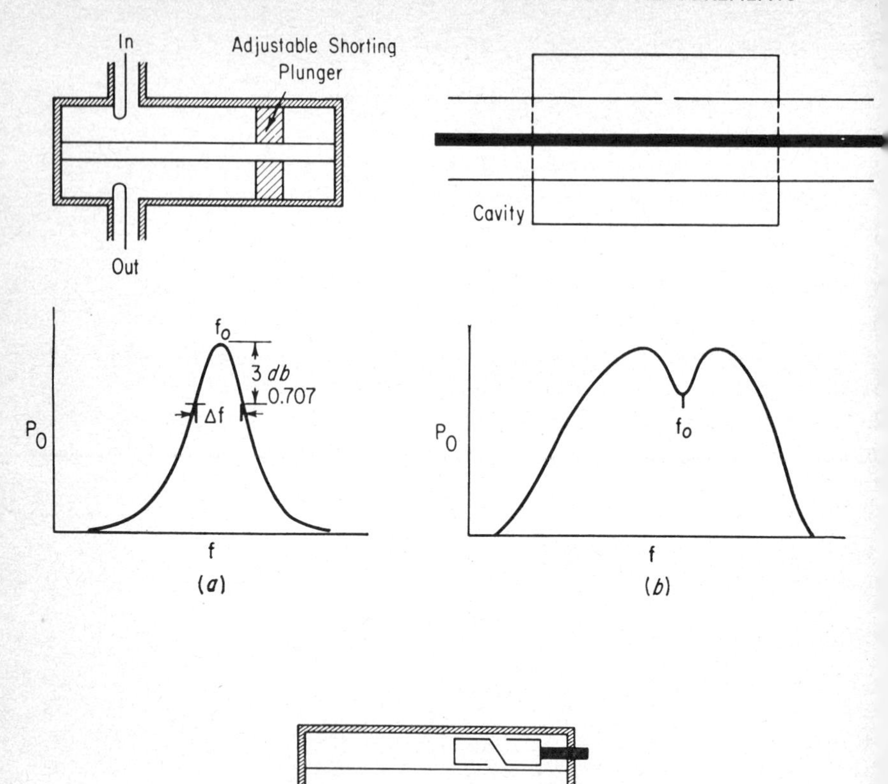

Fig. 9·3 The half-wave coaxial cavity. The power output versus frequency of the reaction type is shown as it would appear on a swept-frequency klystron mode. (a) Transmission type. (b) Reaction type. (c) Absorption type.

the time interval $1/2\pi$ times the period of oscillation

$$Q = 2\pi \frac{\text{energy stored}}{\text{energy dissipated per cycle of oscillation}}$$

An equivalent relation is

$$Q = 2\pi f \frac{\text{energy stored}}{\text{average dissipated power}}$$

The Q is also defined as the ratio of the mean passband frequency to the 3-db bandwidth $f_0/\Delta f$, that is, the bandwidth between the frequencies for which the amplitude of the surface current, the surface charge, the electric field, and the magnetic field are reduced to 0.707 of their value at the resonant frequency f_0, as illustrated in Fig. 9·3*a*.

In practical resonators the amount of energy dissipated per cycle is small compared to the energy stored in the system. The dissipated power is lost internally in the resonant circuit itself and also in the equivalent resistances coupled into the circuit from the external input and output circuits. In general, Q becomes greater as the conductivity of the metal is increased, or the ratio of volume to enclosed metal surface is increased, and as the wavelength is increased. The Q of coaxial cavities is limited at the high frequencies because of the necessity of restricting the radii of the conductor in order to prevent the excitation of the next higher mode which would cause a spurious response. The first higher mode is the TE mode which occurs when the average circumference is approximately equal to the wavelength. The TE mode can propagate in any coaxial line when the frequency in megacycles is approximately equal to

$$f = \frac{3 \times 10^4}{\pi(a + b)}$$

where a and b are the conductor radii.

In general, the cutoff wavelength for the TE mode is approximately given by

$$\lambda_c = \frac{2\pi}{n} \frac{a + b}{2}$$

where $n = 1, 2, 3, \ldots$, and the inner and outer radii are a and b, respectively.

The actual measured Q of a resonant cavity includes all losses associated with the cavity and is called the loaded Q (designated Q_L). The Q of the cavity without any coupling devices is the theoretical or unloaded Q (designated Q_0). Commercial cavities in the 10- and 3-cm range have loaded Q values in the range of 4,000 to 12,000.

9·6 Tuning plungers

In the previous discussion of resonant cavities, a good short circuit between the inner and outer conductor was assumed at one or both ends of the cavity. It is difficult to obtain an acceptable short circuit since the short circuit must be provided by an adjustable conducting plunger.

The required low impedance is provided by contacting or noncontacting types of plungers. Finger contacts are usually provided in the metal-to-metal contacting types. Insulating films on the finger contacts lower the Q of this type of cavity, and small metal particles along with poor finger contact cause noise and erratic tuning. The noncontacting plungers can be constructed to closely simulate a short circuit over a frequency range of 5 to 1 or greater. Two low impedance joints are shown in Fig. 9·4. The double choke joint (Z plunger) is most satisfactory. The low impedance coaxial lines

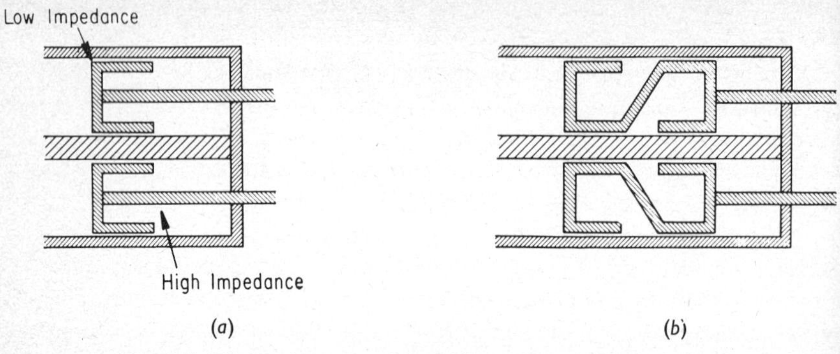

Fig. 9·4 Tuning plungers. (a) Choke plunger. (b) Z plunger.

formed by the gaps between the plunger and the line conductors are terminated in a much higher impedance of the section of the resonator beyond the plunger. The high impedance is transformed to a low impedance since the plunger is a quarter wavelength long.

9·7 Reentrant cavities

Small cavities as illustrated in Fig. 9·5 are called reentrant cavities. The cavity is generated by rotating each of the figures in the illustration about the axis shown.

This type of cavity is widely used as the resonant element of klystron oscillators where it is used in the process of velocity modulation of an

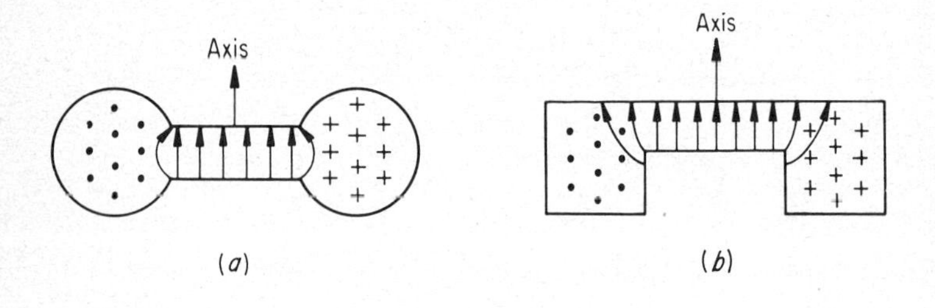

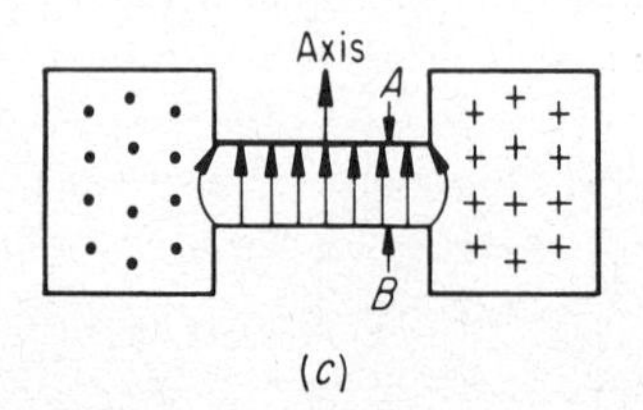

Fig. 9·5 Illustration of reentrant cavities. (a) Toroidal resonator. (b) End gap (c) Center gap.

electron stream. In this application the walls of the cavity at A and B (the gap region) are actually grids through which the electron stream passes. This active region must be as small as possible since the transit time across the gap must be small for efficient energy transfer.

Methods of tuning the reentrant cavity can be understood more easily if the resonator is thought of in terms of lumped elements. This is possible because the electric and magnetic fields for the most part occupy different regions of the cavity. The gap region occupied by the electric field may be considered as the capacitive element, and the annular magnetic field may be considered as the inductive element of the lumped constant circuit. From this equivalence it can be seen that the frequency can be increased by increasing the separation between the grids of the gap (thus decreasing the capacitance) or by reducing the inductance by decreasing the cross-sectional area of the cavity. A tuning screw can be introduced into the cavity to decrease the inductance, thereby increasing the resonant frequency.

9·8 Rectangular waveguide resonators

The firmest restriction on the shape of an electromagnetic field is given by a closed conducting cavity. The hollow rectangular cavity shown in Fig. 9·6 may be analyzed as a length of rectangular waveguide short-circuited at

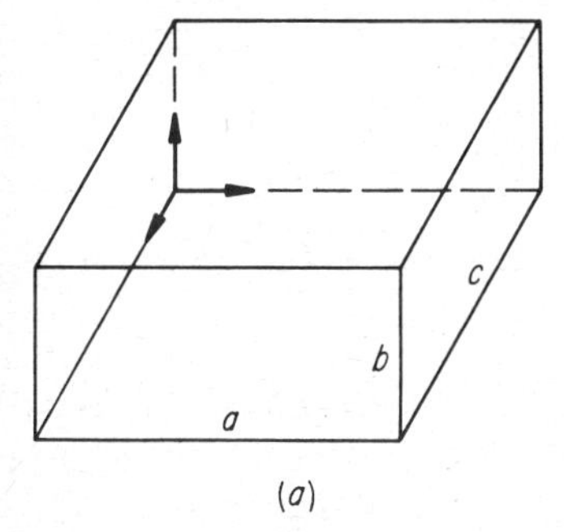

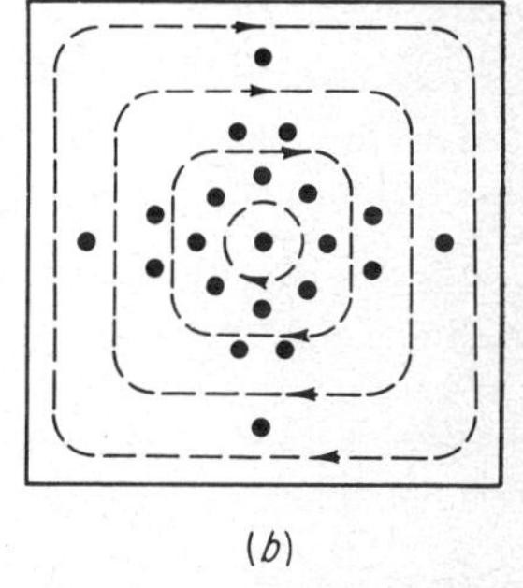

Fig. 9·6 Rectangular waveguide resonator. (*a*) Rectangular cavity. (*b*) Top view of the electric and magnetic fields of the TE_{101} mode in the rectangular waveguide cavity.

both ends. Such a cavity has an infinite number of modes of propagation, characterized by the configuration of the electric and magnetic fields, and the only frequency requirement set up is that of a minimum frequency.

A stable field pattern is obtained by the reflections from all surfaces of the structure. The electric field has the same phase throughout the cavity, and the magnetic field increases and decreases, as a unit, 90° out of phase with the variations of the electric field. The electric field is zero at the end and sides of the rectangular structure, while the magnetic field is zero at the center and maximum at the sides and ends of the structure.

The free-space wavelength at which the cavity is resonant is given by

$$\lambda = \frac{2}{\sqrt{(m/a)^2 + (n/b)^2 + (p/c)^2}}$$

where m = number of half-wave variations of the electric and magnetic fields along the a dimension

n = number of half-wave variations of the electric and magnetic fields along the b dimension

p = number of half-wave variations of the electric and magnetic fields along the c dimension

m, n, and p are positive integers, one of which may be zero; a, b, and c are measured in centimeters

9·9 Cylindrical cavity wavemeters

A circular cylindrical cavity is the most suitable shape because it lends itself readily to accurate machining and mechanical construction of moving parts. As in the case of the rectangular waveguide, a large number of resonances are possible because each TE and TM wave of the circular waveguide gives rise to a set of modes of oscillation. Actually, there is a triple infinity of these modes as indicated by the three subscripts of the mode designations. For a specific application, the cavity is usually designed and excited in such a manner that only one mode of resonance is obtained over a limited frequency range.

The subscripts which describe the particular mode (TE_{mnp} or TM_{mnp}) are defined by

m = number of full-period variations of the radial electric field with respect to θ (full-period variations around the circumference of the cavity)

n = number of half-period variations of **E** with respect to the radius. Except for the TE_{0np} cavity modes, n can be defined as the number of planes that can be passed through the axis perpendicular to all electric field lines

p = number of half-period variations of the electric field along the z axis.

The configurations of the electric and magnetic fields associated with particular TE and TM modes of propagation in cylindrical waveguides are shown in Fig. 7·10 in Chap. 7. The cross-sectional details of the most often used cylindrical cavity resonators are shown in Fig. 9·7. A simple noncontacting end plate is used in a TE_{0np} cavity because this mode has no conduction currents across the junction between the end plates and the cavity walls. This type of plunger end plate is not a perfect short for modes which require current flow between the plunger and the end plate. Fields of these modes couple through the gap into the section behind the plunger. A lossy material, such as polyiron, is placed in the region as shown in a in order to absorb the power from these unwanted modes. The highest Q for practical cavity size can be obtained using the TE_{0np} modes.

The TE_{111} mode cavity uses a nonconducting choke to provide the required short circuit between the plunger and the side wall of the cavity. This mode is selected because it is the dominant mode of the circular waveguide and permits the largest tuning range without interfering modes. This can be verified by examination of the mode chart of Fig. 9·9.

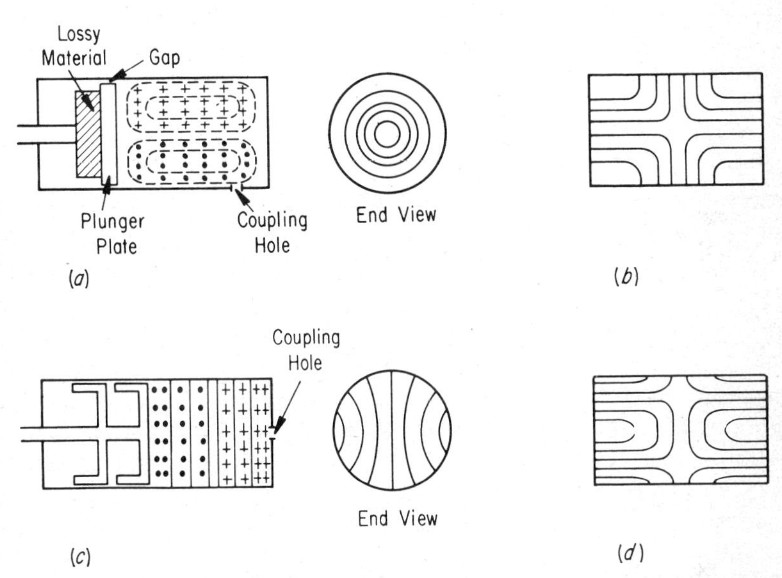

Fig. 9·7 Cross-sectional views of (a) TE_{011} mode and (c) TE_{111} mode cavity wavemeters. The electric field configurations in a cylindrical cavity are shown at (b) for the TM_{011} mode and at (d) for the TM_{111} mode.

The magnitude of coupling to the cavity is determined by the diameter and thickness of the coupling iris which is positioned in the waveguide to couple to the field component of the desired mode. The coupling is accomplished by cutting holes in the common wall between the cavity and the section of waveguide on which the cavity is mounted. Since the electric field is zero everywhere at the boundary surface of the TE_{0np} cavity, the coupling to the mode is magnetic. Maximum coupling is at the side of the cavity an odd quarter λ_g from the end or at the end about 48 per cent out from the center to the edge of the cavity.

In order to prevent excessive changes in frequency indication with changes in temperature and environmental conditions, invar is frequently used in the construction of the cavity and the cavity is filled with nitrogen and hermetically sealed. The invar cavity is carefully finished, copper plated, then silver plated and polished to provide a mirror-like finish. The Q of such cavities may range from a low Q of 1,000 to a high Q greater than 50,000. As an example, commercial wavemeters in the range of 3.95 to 10 Gc have loaded Q values in the range of 4,000 to 8,000; from 8.2 to 12.4 Gc the loaded Q ranges between 2,000 and 12,000. The accuracy obtainable using these wavemeters varies between 0.005 and 1 per cent.

9·10 Wavemeter circuits

The equivalent circuit of the transmission type wavemeter near resonance and at particular reference planes is illustrated in Fig. 9·8*a*. When the cavity is tuned to the frequency of the incident waves, power is transmitted through the cavity. At frequencies far from resonance the wavemeter acts as a short circuit, and essentially no power is transmitted through the cavity.

By proper selection of the reference plane, the equivalent circuit can be reduced to a simple series or parallel *RLC* circuit. For each case the input impedance of the cavity at resonance is a *pure* resistance. The maximum and minimum of the VSWR at resonance fall on the pure resistance axis when drawn on a Smith chart. Also, the maximum or minimum of the standing-wave pattern is at the reference terminal, and there is no frequency pulling of the cavity. As an example, the transmission type wavemeter appears as a short circuit when tuned off resonance, and the detuned short position (series *RLC*) can be obtained with the slotted section probe. If the wavemeter is now tuned to resonance, the minimum or maximum of the standing-wave pattern should be located at the detuned short position which had been located. If the maximum or minimum does not occur at the located reference plane, then there is a source or load mismatch, and the resonant frequency of the cavity has been changed (pulled). This indicates that a wavemeter frequency indication can be in error when connected to a reactive load. Therefore it is frequently necessary to place attenuation between a wavemeter and any associated component which has a high VSWR.

The reaction type wavemeter and its equivalent circuit are shown in Fig. 9·8*b*. The power absorbed by a load on this type meter is decreased when the wavemeter is tuned through resonance since the cavity absorbs a part of the power from the main line at resonance. At frequencies far from resonance the cavity presents an effective short circuit to the waveguide wall. A reaction type display is illustrated in Figs. 9·3 and 12·4 where a reaction wavemeter absorbs power from a swept frequency signal.

9·11 Coupling parameter and frequency pulling

The coupling parameter is defined as the normalized admittance *at resonance*.

$$\beta = \frac{G_0}{G_c} \tag{9·1}$$

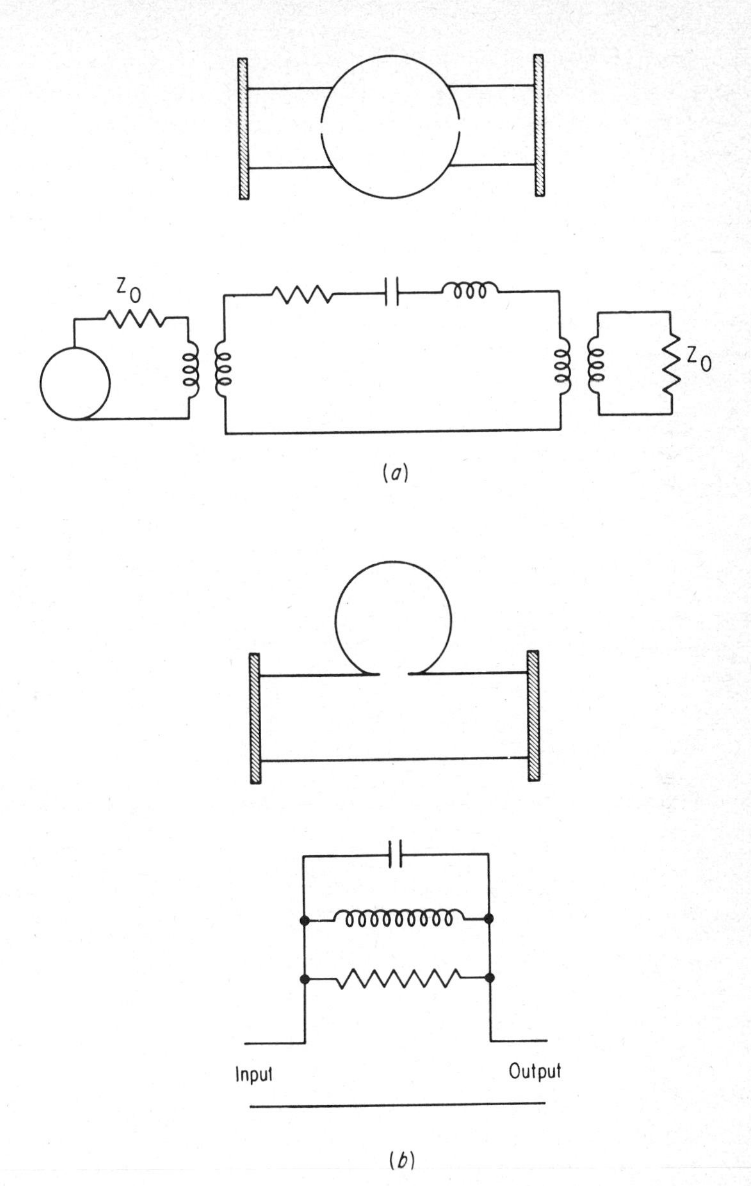

Fig. 9·8 Transmission and reaction type wavemeter circuits. (*a*) Transmission type wavemeter and its equivalent circuit. (*b*) Reaction type wavemeter and its equivalent circuit.

where G_0 is the characteristic admittance of the transmission line and G_c is the cavity admittance. Using the previous example of the detuned short reference plane, if the standing-wave pattern is located at a minimum at the detuned short position when the cavity is tuned to resonance, $G_0/G_c = 1/\text{VSWR}$ and the cavity is said to be *undercoupled*. If the standing-wave pattern is at a maximum at the reference plane, $G_0/G_c = \text{VSWR}$ and the cavity is *over-coupled*. Phase reflections from an undercoupled resonator remain the same at resonance as at frequencies far above or below resonance. An overcoupled system exhibits a continuous phase shift through π rad as the frequency changes from a frequency below resonance to a frequency above resonance.

The unloaded Q of the wavemeter can be computed after determination of the loaded Q, coupling parameter, and minimum or maximum of the standing-wave pattern.

$$\begin{aligned} Q_0 &= (1 + \beta)Q_L \qquad \text{single-port cavity} \\ Q_0 &= (1 + 2\beta)Q_L \qquad \text{two-port equal-coupled cavity} \end{aligned} \tag{9·2}$$

An approximate expression[1] for the maximum possible frequency shift for each of two equal coupling holes of a transmission type cavity is

$$\frac{\Delta f}{f} = \frac{Q_0 - Q_L}{8Q_0Q_L}\frac{\rho^2 - 1}{\rho} \tag{9·3}$$

A similar expression for the reaction type wavemeter matched in one direction is

$$\frac{\Delta f}{f} = \frac{Q_0 - Q_L}{8Q_0Q_L}\frac{\rho - 1}{\rho + 1} \tag{9·4}$$

9·12 The mode chart

The *mode chart* is a plot of the formula relating resonant frequency to the mode, shape, and dimensions of the cylindrical cavity. The equation is written as

$$(fD)^2 = \left(\frac{cr}{\pi}\right)^2 + \frac{1}{\mu'\epsilon'}\left(\frac{c}{2}\right)^2\left(\frac{D}{L}\right)^2 p^2 \tag{9·5}$$

$$(fD)^2 = A + Bp^2\left(\frac{D}{L}\right)^2 \tag{9·6}$$

where A = a constant depending upon the mode; this is the Y intercept on the mode chart

B = a constant depending upon the velocity of electromagnetic waves in the dielectric ($B = 0.34799 \times 10^8$ for air at 25°C and 60 per cent relative humidity)

p = number of half wavelengths along the cavity axis

f = frequency, Mc per sec

L = length of the cavity, in.

D = diameter of the cavity, in.

$c = 1.17981 \times 10^{10}$ in. per sec (velocity of light)

r = root of the Bessel function (p_{mn})

Bp^2 is the slope of the straight line representing the mode. The mode chart for the right circular cylinder resonant cavity is shown in Fig. 9·9, and a table of the constant A for several modes is given in Table 9·1.

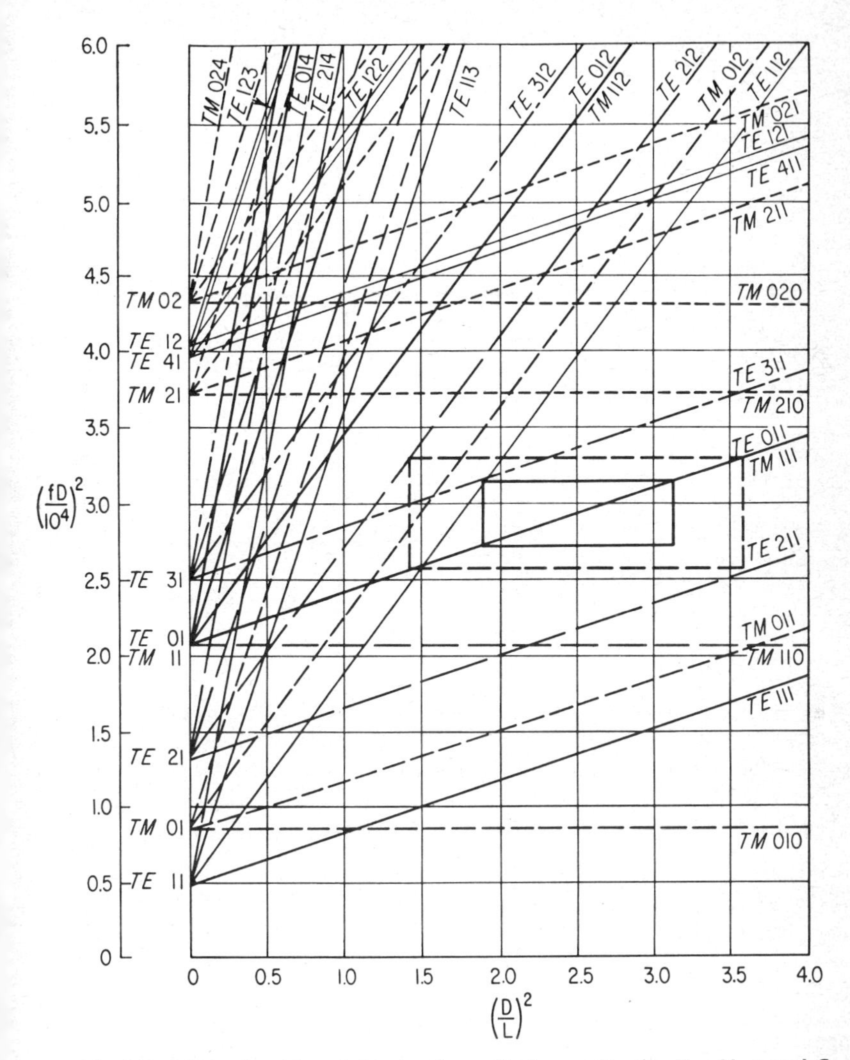

Fig. 9·9 Mode chart for right circular cylinder cavity. (*D. Van Nostrand Company.*)

The diagonal of the rectangle representing the operating area of the cavity is the locus of the mode that the cavity was designed for. This rectangle must be clear of loci of other modes if it is to be free of spurious responses. The solid rectangle represents a design free of spurious responses.

A *crossing mode* is one which crosses the main mode within the operating rectangle. The crossing mode causes least trouble when it does not couple electromagnetically to the main mode. If two modes couple to each other, they do not actually intersect; on a swept-frequency-reaction type display, two equal reactions appear in the area of resonance. This coupling problem is usually found in the area of main mode crossing with TM_{mn0} modes which are indicated on the mode chart where it is noted that they are independent of the cavity length. These modes are a function of diameter only. If a wavemeter is tuned to the area of TM_{mn0} mode interference, two equal reactions

Table 9·1 Constants for use in computing resonant frequencies of circular cylinders

Mode	p_{mn}	A
TE_{01}	3.83171	2.0707×10^8
11	1.84118	0.47810×10^8
12	5.33144	4.0088×10^8
21	3.05424	1.3156×10^8
31	4.20119	2.4893×10^8
41	5.31755	3.9879×10^8
TM_{01}	2.40483	0.81563×10^8
02	5.52008	4.2975×10^8
11	3.83171	2.0707×10^8
21	5.13562	3.7197×10^8

would be shown on a swept-frequency display. If the source and wavemeter are tracked upward in frequency, one reaction disappears while the other is tracked as the main mode. If the source is now tuned downward from the interference area, the same effect is observed with the opposite reaction becoming the main mode reaction.

When the crossing mode couples to the main mode, the Q of the main operating mode is lowered. This is particularly true if the Q of the crossing mode is lower than the Q of the main mode. The Q of the crossing mode is sometimes lowered by placing lossy material in the cavity, and this can also lower the Q of the main operating mode. If the dotted rectangle is considered as the design rectangle for a particular cavity, then the TE_{112} mode is a crossing mode.

Interfering modes may be defined as modes which cause undesired frequency indications within the operating area of a frequency meter. A search for interfering modes could commence from the high frequency end of the cavity which is the top right-hand corner of the design rectangle. The frequency meter reaction is presented on a swept-frequency display when the source frequency is set to the same frequency. If the source frequency is now kept constant and the frequency meter is tuned downward in frequency, the search is being made along the top horizontal line of the design rectangle. For the case of the dotted rectangle, a reaction would be noted when the length of the cavity reaches $(D/L)^2$ in the range of 2.25, and this reaction is the TE_{311} mode, as indicated in Fig. 9·9. If the frequency meter is tuned downward in frequency, the next mode encountered is the TE_{112} and then the TM_{012} mode. When the top left corner is reached $(D/L)^2 = 1.4$, and the cavity is tuned to its maximum length. The source frequency is now tuned downward without moving the cavity, the left vertical line is searched, and the TE_{311} and TM_{012} modes are observed on the swept-frequency display. In order to check for *crossing modes*, the source frequency and cavity frequency are again tuned to the high frequency end (the top right-hand corner of the design rectangle). If the source and cavity frequencies are tracked downward together in such a way that the reaction is observed continuously, any crossing modes that exist in the cavity can be found. The TE_{112} mode would be observed in the dotted rectangle. The bottom horizontal of the design rectangle is traversed by varying the cavity length over its complete range when the source frequency is kept at the low frequency setting. The remaining side of the rectangle is the right vertical line which is traversed by setting the cavity to the high frequency setting and varying the source from the high to low range limits. If each of the interfering modes is plotted, a mode chart constructed from the known cavity dimensions can be used to identify the interfering modes.

9·13 Filters

Microwave filters are frequency-selective devices. The theory and design of low-pass, high-pass, and bandpass microwave filters of many types have been covered extensively in the literature.[2] A treatment of the subject requires a knowledge of resonant type structures. The transmission type resonant cavity forms the simplest symmetrical unit that embodies the principal properties of a filter. When the transmission type cavity is placed in the line, energy is transmitted through the cavity at resonance and is reflected at other frequencies. A flat passband and steep attenuation characteristics outside the passband are required in order to obtain the most efficient use of frequency space. The single cavity filter does not fulfill this requirement. However, a combination of sections in cascade can be made to provide almost any kind of response that may be desired.

If two identical resonant-waveguide cavities are placed in close proximity, the adjacent coupling irises form a single reflection and can be considered as a single iris. Typical response curves of this *iris-coupled filter* are shown in Fig. 9·10*a* and *b*. In general, as the number of resonant cavities is increased, the passband has more uniform response, and the slope of the attenuation characteristic increases, as illustrated in Fig. 9·10*c*. An increase in the number of resonant cavities also increases the insertion loss in the passband.

When several cavities are cascaded to obtain certain design characteristics, the coupling between cavities can be either direct coupled or quarter-wave coupled. In the *direct-coupled* filter, each resonant-waveguide cavity is coupled to an adjacent cavity by means of an aperture or iris. The *quarter-wave–coupled* filter is formed by separating adjacent cavities by an odd number of quarter-wavelength sections. This distance is usually made three-quarters of a wavelength.

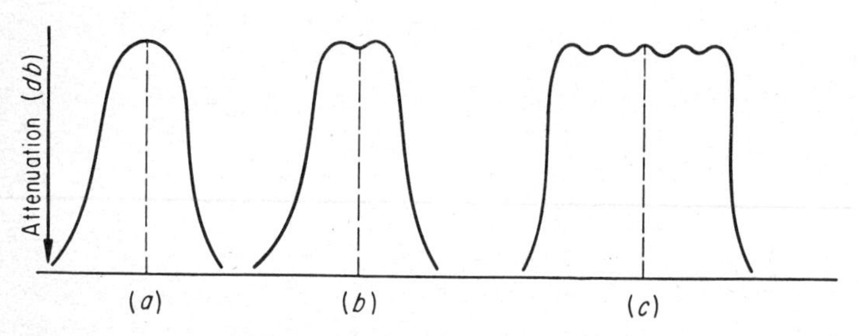

Fig. 9·10 (*a*) Optimum flatness of two-cavity filter. (*b*) Overcoupled two-cavity filter. (*c*) Bandpass response of a multicavity coupled filter.

A typical waveguide type direct-coupled filter is shown in Fig. 9·11*a*. Capacitive screw and lock-nut assemblies are usually provided for tuning the cavities to the proper resonant frequencies. The cavity is made slightly shorter than required for resonance at the design frequency. The resonant frequency of the cavity can be varied above and below the design frequency with the capacitive tuning screw. Insertion of the screw into the cavity effectively lengthens the cavity. The penetration of the screw is kept at a minimum in order to prevent a reduction of cavity Q. The screws are silver-plated to keep the insertion loss at a minimum.

The resonant cavities usually employ inductive irises as coupling elements. Four of the most practical coupling irises are shown in Fig. 9·12. Higher Q is obtained by using the round-hole coupling iris. In practice, the desired frequency response of the filter is obtained by changing the size and positions of the irises.

Circular waveguide filters can be constructed in a similar manner using circular disks spaced along the line.

Short-circuited coaxial stubs, spaced at intervals along the line, can be used to form high-pass and bandpass filter characteristics. A low-pass coaxial filter employing series-capacitance disks is shown in Fig. 9·11*b*.

Tunable multiple-cavity filters are usually designed in the form of right circular cylinders such as the TE_{111} and TE_{011} mode cavities. By referring to the mode chart it is noted that these cavities have a greater frequency range without crossing or other interfering modes.

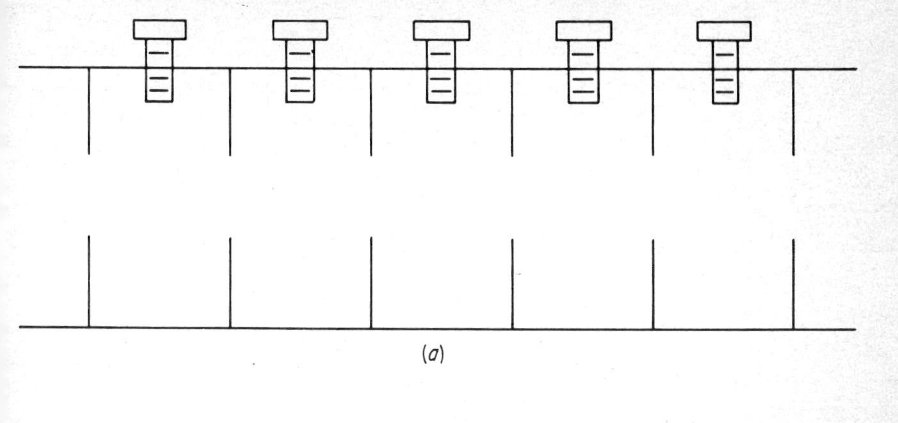

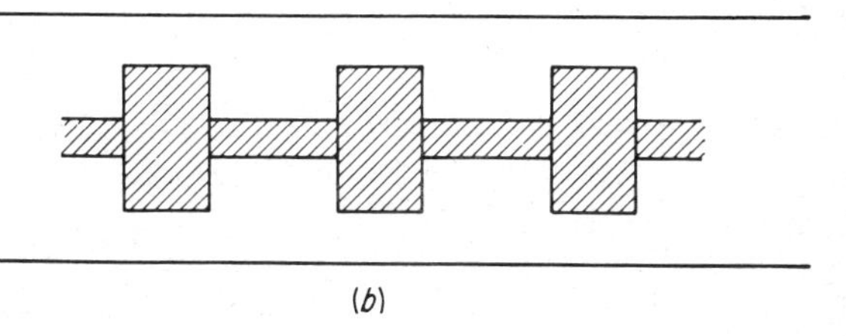

Fig. 9·11 **(*a*) Multicavity coupled filter. (*b*) Low-pass coaxial filter.**

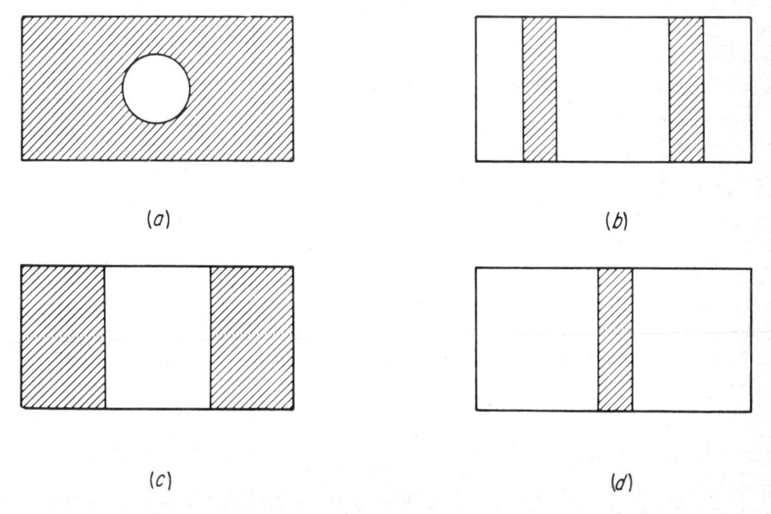

Fig. 9·12 **Inductive irises used as coupling elements.**

PROBLEMS

9·1 What is the effective length of a three-quarter wavelength coaxial cavity operating in the TEM mode at a frequency of 2 Gc?

9·2 What is the effective length if the above cavity is filled with a material which has a dielectric constant of 3.4?

9·3 Calculate the free-space wavelength of a rectangular cavity operating in the TE_{103} mode if the length of the cavity is 2.8 in. and the a and b dimensions are 0.9 and 0.4 in., respectively. What is the operating frequency?

9·4 Express the coupling parameter β in terms of the characteristic impedance of the transmission line and the cavity impedance Z_c at resonance.

9·5 Express undercoupling and overcoupling in terms of the characteristic impedance Z_0 of the transmission line and the cavity impedance Z_c.

9·6 Develop a mode chart for a right circular cylinder cavity operating in the TE_{011} mode. The cavity diameter is 2.0 in.; the length is 1.25 in. at the low frequency end and 0.7 in. at the high frequency end.

a. Plot the following modes on the chart: TE_{111}, TM_{011}, TE_{211}, TE_{311}, TE_{121}, TE_{411}, and TM_{211}. *Note:* Use $(D/L)^2 = 7$ in order to find the point $(fD)^2$ using Eq. (9.6).

b. Plot the following modes on the chart using $(D/L)^2$ equal to 3 in Eq. (9.6) to find the point for $(fD)^2$ on the chart: TE_{212}, TM_{012}, and TE_{112}.

9·7 What are the approximate frequencies at which the TM_{020} and TM_{210} modes cross the main operating TE_{011} mode of the cavity in Prob. 9·6?

REFERENCES

1. *PRD Report*, vol. 2, no. 2, July, 1953.
2. Herbert J. Reich, ed., "Very High Frequency Techniques," vol. II, chaps. 26 and 27, McGraw-Hill Book Company, New York, 1947.

   G. L. Ragan, "Microwave Transmission Circuits," M.I.T. Radiation Laboratory Series, vol. 9, chaps. 9 and 10, McGraw-Hill Book Company, 1948.

CHAPTER

10

FERRITE DEVICES

10·1 Microwave ferrite phenomena

Ferrites are magnetic materials which are formed by replacing some of the iron atoms in magnetite, an iron oxide, by other metallic atoms such as zinc, nickel, cobalt, manganese, and aluminum. The ferrite exhibits good magnetic properties, has high resistivity (specific resistance), and is transparent to electromagnetic waves. An electromagnetic wave propagating through the ferrite encounters strong interaction with the spinning electrons which cause the magnetic properties of the ferrite.

If a material has paired electrons, the electrons rotate in opposite directions and the magnetization effects cancel. However, if unpaired electrons are present, the unpaired electrons have a magnetic moment along the axis of rotation, thus producing the magnetic properties of the material.

The behavior of a ferrite material in a microwave field can be illustrated by a description of the interaction between the electromagnetic properties of the ferrite and the microwave energy.[1]

A graphic picture of the behavior of the ferrite material is obtained by considering the spinning electron as a tiny magnet which is magnetized in the direction shown in Fig. 10·1*a*. Since the electron has mass, rotation causes it to have angular momentum and a magnetic moment along the axis of rotation. The electron acts like a gyroscope, and, instead of aligning itself with the applied magnetic field **H**, it precesses about the axis in a clockwise direction, as shown in Fig. 10·1*b*. The frequency of precession, called the *gyromagnetic resonant frequency*, is a function of the applied magnetic field. If the applied magnetic field is great, the torque on the electron is large and the precession frequency is high.

The ferromagnetic resonance induced by the external d-c magnetic field occurs at

$$f_r = \gamma \mathbf{H}_r$$

where f_r is expressed in megacycles per second, $\mathbf{H}_r$ is the external field in oersteds, and γ is the gyromagnetic ratio. The natural gyromagnetic resonant frequency is 2·8 Mc per oersted of magnetic field.

When a rotating magnetic field (circularly polarized microwave energy) is applied in a plane perpendicular to the aligning d-c field, the magnetic moment vector has an instantaneous component of torque, parallel to the

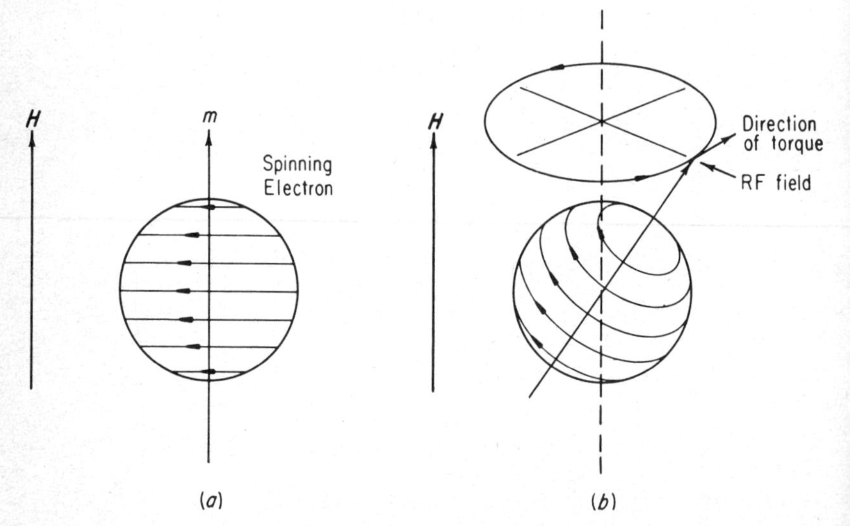

Fig. 10·1 The electron aligns itself with the applied static field at (*a*). The instantaneous component of torque perpendicular to the instantaneous r-f magnetic field and the electron magnetic moment is shown at (*b*).

direction of the magnetic moment, which adds to the energy of the precessional motion. If the rotating magnetic field is in the same direction and at the same frequency as the precession, more and more energy is delivered to the electron, causing it to go into violent oscillation. The energy received from the microwave field is dissipated in the crystalline lattice structure in the form of heat. There is maximum absorption of microwave energy at this frequency.

The second quantity, characteristic of the absorption, is the *ferrimagnetic resonant line width* Δ**H**. The resonant line width represents the field strength over which resonance occurs at a given frequency. Line width is defined as the separation of the two magnetic-field values at which the power that is absorbed by the ferrite material is one-half the maximum absorption value. A broad line width indicates broadband capabilities.

If the magnetic field is rotating in a direction opposite to that of the precession, microwave energy is alternately delivered to and extracted from the electron with little net effect on either the ferrite material or the microwave energy. The difference in the loss due to the two senses of circular polarization is most apparent at the resonant frequency. Less microwave energy is delivered to the electron as the frequency difference between the precessional and rotational frequencies becomes greater. At frequencies far removed from resonance, the major difference is in the velocities of the two circularly polarized waves. The difference in velocity causes different phase-shift characteristics for the two senses of circular polarization. This difference in

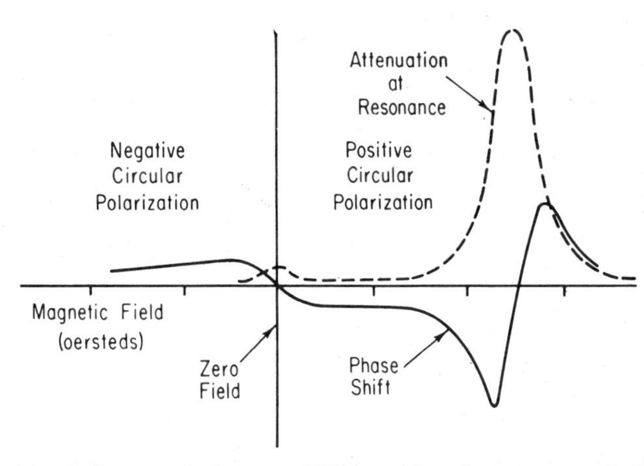

Fig. 10·2 A typical curve of r-f permeability as a function of magnetic field and for the two senses of circular polarization.

the velocity of propagation is due to the difference in permeability encountered by the clockwise- and counterclockwise-rotating, circularly polarized magnetic fields.

Typical phase shift and attenuation characteristics as a function of magnetic field are shown in Fig. 10·2. Rapid phase shift and maximum attenuation occur at the resonant frequency.

Polarization of the magnetic field intensity is the important consideration when considering ferrite phenomena. Circular and linear polarization characteristics determine the operating characteristics of the devices to be considered. A circular-polarized magnetic field intensity exists for almost all modes at some points in the transmission line but never at all points. A vertical plane of circular polarization exists at approximately one-half the distance between the center and side wall of a waveguide propagating the TE_{10} mode. Circular polarization exists along the center axis of a circular waveguide propagating the TE_{11} mode.

10·2 Faraday rotation isolator

The previous presentation makes it possible to explain the Faraday rotation effect. In 1845 Michael Faraday demonstrated that the plane of polarization of a linearly polarized wave is rotated when plane-polarized light is sent through certain materials in a direction parallel to an applied magnetic field. This effect is also evident at microwave frequencies. A linearly polarized wave can be considered to be composed of two circularly polarized waves equal in magnitude but rotating in opposite directions. These two circularly polarized waves can, at all times, be combined to yield the linearly polarized

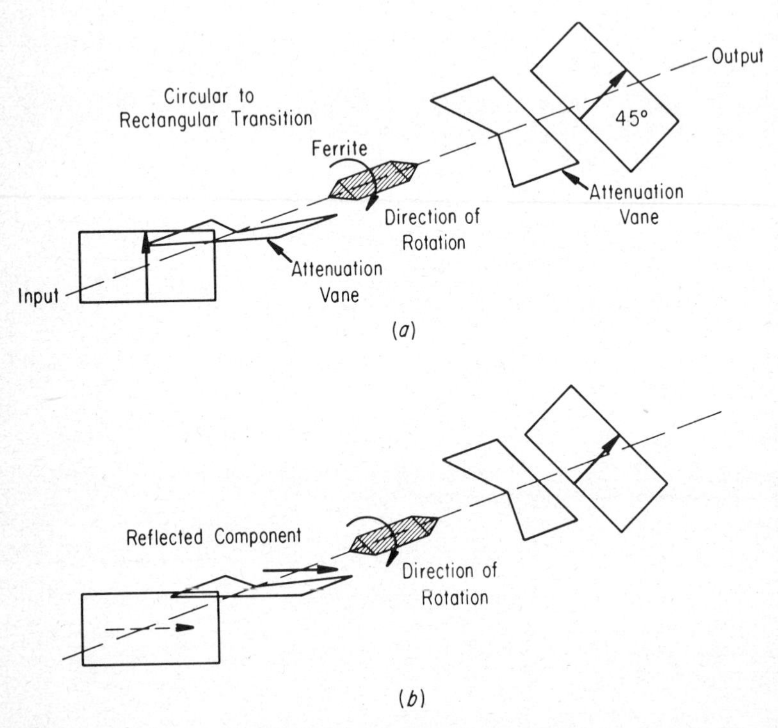

Fig. 10·3 Schematic illustration of the Faraday rotator. The permanent magnet is located in the vicinity of the ferrite. (*a*) Forward direction. (*b*) Reverse direction

wave. If the r-f field applied to the ferrite is well below the resonant magnetic field, the two circularly polarized components of the input linear wave pass through the ferrite with negligible loss, but the two components experience different phase shifts since one wave aids precession of the spinning electron while the other wave opposes the precession. The phase rotation is equal to one-half the difference between the phase shifts of the circularly polarized waves. The direction of the resultant phase rotation is determined by the direction of the magnetic field and not by the direction of propagation of the microwave energy. This phenomenon is applied to produce a non-reciprocal device called an *isolator*.

The schematic representation in Fig. 10·3 is used to explain the basic operation of the isolator. A pencil-shaped ferrite material is mounted in the circular waveguide section. A permanent magnet is placed outside the waveguide to provide the longitudinal magnetic field through the ferrite. The circular waveguide is transformed to rectangular waveguide at both ends. Resistive attenuator vanes are placed parallel to the wide dimension of the waveguide as shown. The input electric vector is perpendicular to the input attenuator vane and is not affected by it. The wave is transformed into a TE_{11} circular waveguide mode, and the polarization of this wave is shifted clockwise 45° in passing through the ferrite. The emerging electric field is also perpendicular to the second resistive vane and passes out of the device through a rectangular waveguide which is oriented to correspond to the emerging polarization, as shown in Fig. 10·3*a*. Figure 10·3*b* illustrates the operation of the device for a signal reflected back through the device. The reflected wave is rotated 45° *in the same direction* in which the forward wave was shifted and is absorbed in the resistive vane since the electric field is now in the plane of the vane. Practical isolators of this type have forward loss up to 1 db and reverse isolation upward of 20 db. The Faraday rotator is not suitable for high-peak-power operation since nonlinear effects occur at relatively low peak powers.

10·3 Resonant absorption isolator

The basic form of this type of isolator is shown in Fig. 10·4. There are many other configurations of such isolators. The ferrite is located approximately one-half the distance from the side to the center of the guide where the plane of the r-f magnetic field is circularly polarized. Resonant absorption takes place as described previously and depends on the strength of the d-c magnetic field, the direction of propagation, and the frequency of the microwave energy.

10·4 Phase changers

The previous discussion of microwave ferrite phenomena indicated that phase shift is the most predominant characteristic at frequencies far removed from magnetic resonance where the ferrite is biased into the low field or high field region. The phase change can be either reciprocal or nonreciprocal in manner. The phase shift is reciprocal when the ferrite is placed to one side of the waveguide and is nonreciprocal when the ferrite is centrally located in the waveguide.

10·5 Circulators

A very useful device which incorporates the nonreciprocal phase shift of ferrites is the circulator. There are many applications of the various types of

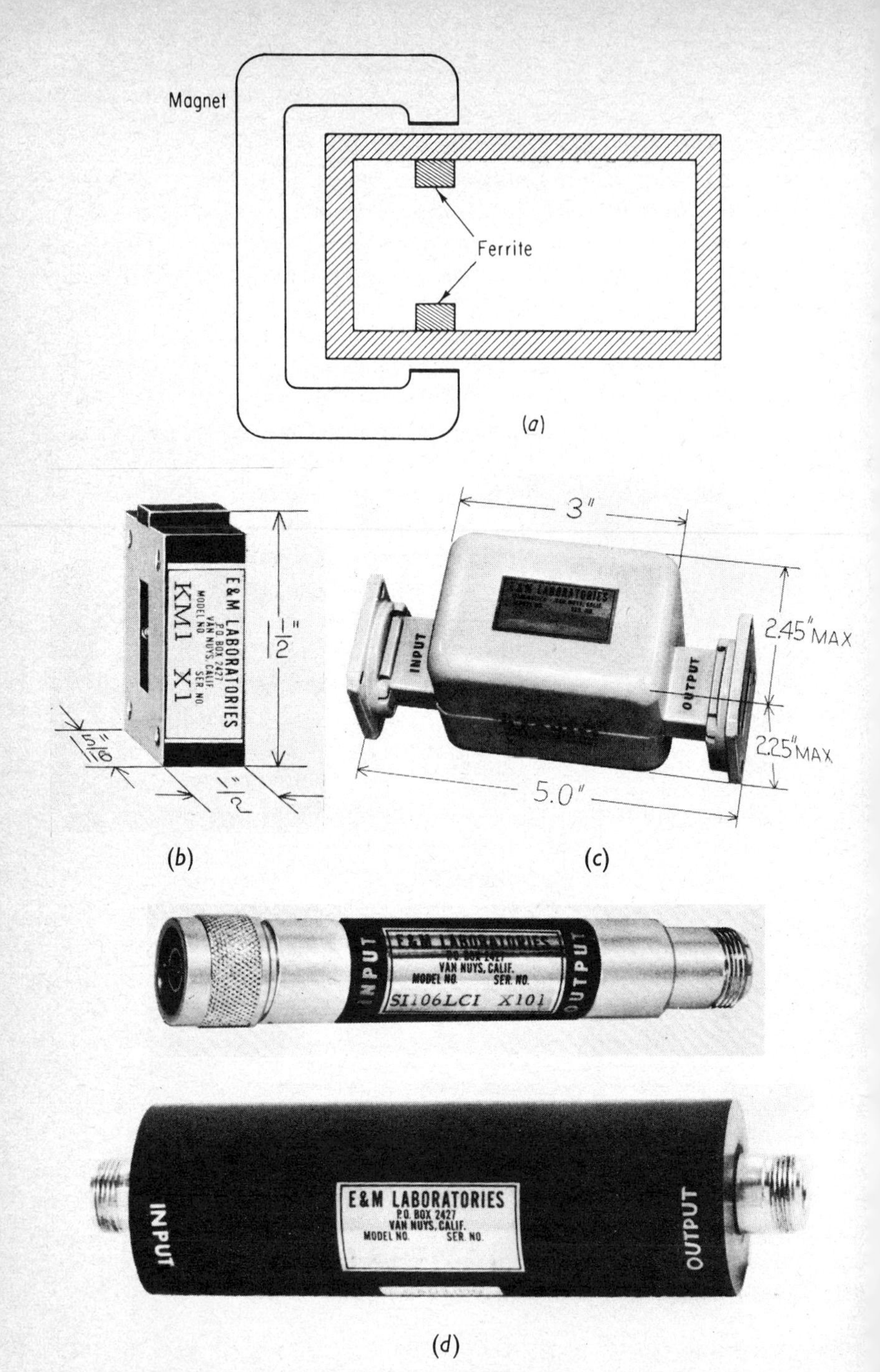

Fig. 10·4 Resonance absorption isolator. (*a*) Schematic representation of a waveguide ferrite isolator. (*b*) Miniature X-band isolator. (*c*) Broadband isolator. (*d*) Broadband coaxial isolators. (*E & M Laboratories.*)

isolators. A typical application of a four-port circulator used as a duplexer in a radar system is shown in Fig. 10·5. Power entering port 1 emerges from port 2 only, power entering port 2 leaves through port 3, power entering port 3 leaves through port 4, and power entering port 4 leaves through port 1. Two of the many types of circulators are shown in Fig. 10·6*a* and *b*.

A 45° Faraday rotator type circulator is shown in Fig. 10·6*a*. The vertical polarized wave enters port 1, passes into the circular section where it is rotated 45° by the ferrite, and leaves through port 2. A wave entering port 2 will be horizontally polarized after passing through the ferrite where it is

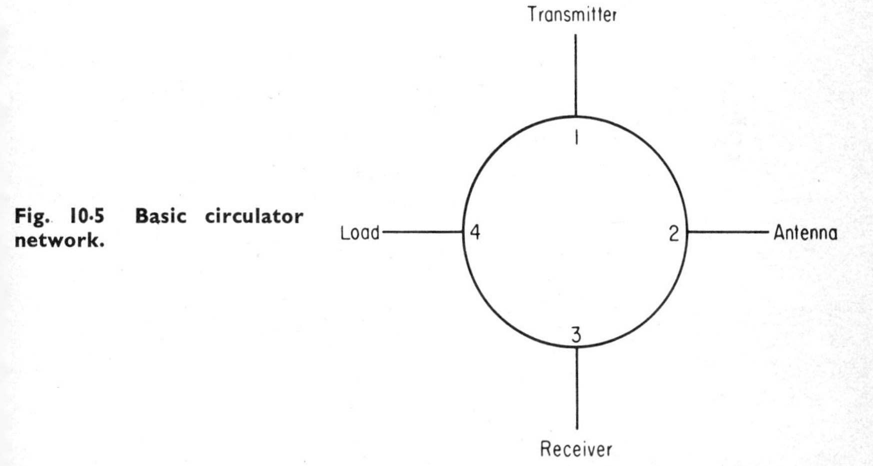

Fig. 10·5 Basic circulator network.

rotated 45°. Consequently, it appears at port 3. Similarly, a wave entering port 3 has its polarization rotated by 45° and leaves only through port 4.

A typical waveguide-circulator network is illustrated in Fig. 10·6*b*. Energy entering arm 1 is divided equally into each waveguide arm by the first 3-db hybrid. The energy traveling down arm 1 undergoes a net differential phase shift of 180° with respect to the energy traveling down the other waveguide and then recombines with the energy of the opposite arm in such a way that all energy leaves through arm 2. Energy entering arm 2 is equally divided by the 3-db hybrid, undergoes zero net differential phase shift since it is traveling in the opposite direction, and is recombined at the second 3-db hybrid, and all energy leaves through arm 3. Likewise, energy entering arm 3 leaves at arm 4, and energy entering arm 4 leaves at arm 1.

10·6 Electronic controlled ferrite devices

If the necessary magnetic field is supplied by current flow through an external magnet coil, the characteristics may be varied according to the

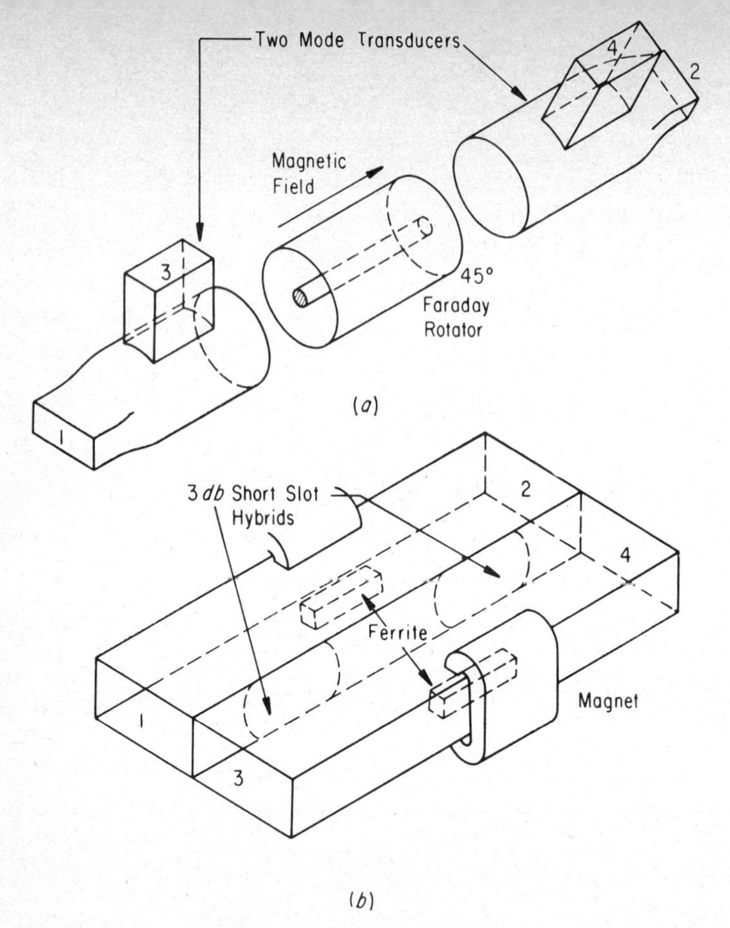

Fig. 10·6 Four-port circulators. (*a*) Four-port Faraday rotator circulator. (*b*) Waveguide circulator network.

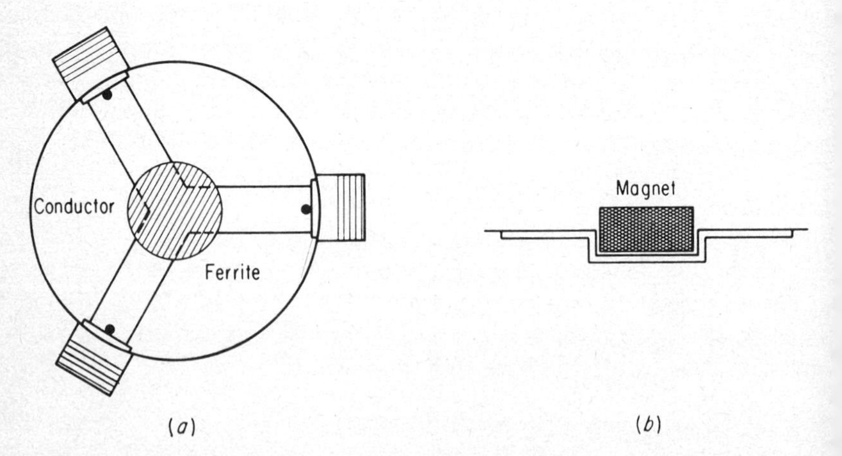

Fig. 10·7 The wye circulator. (*a*) Top view. (*b*) Side view of top plate and magnet location.

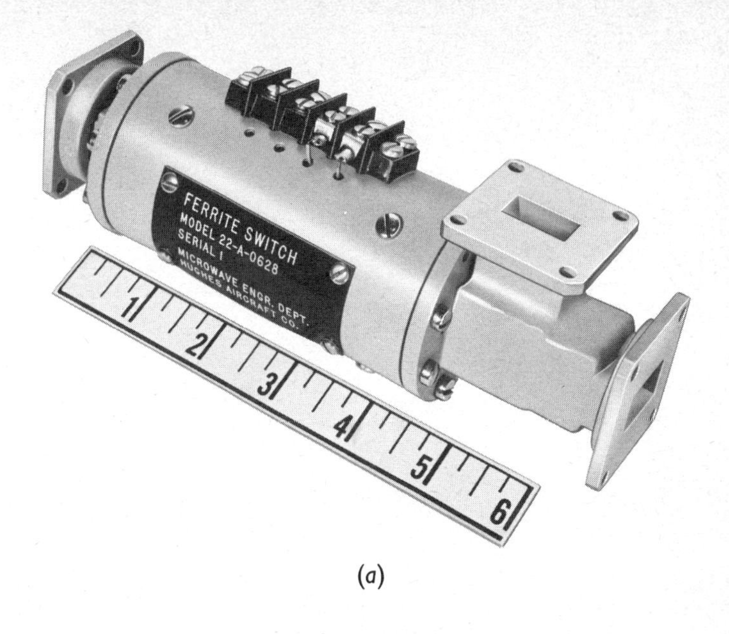

(*a*)

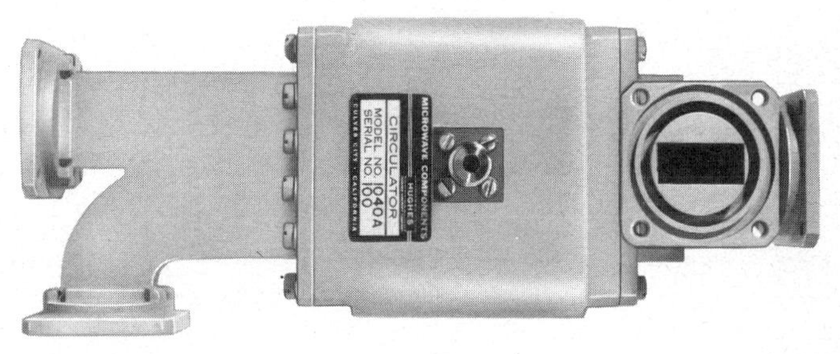

(*b*)

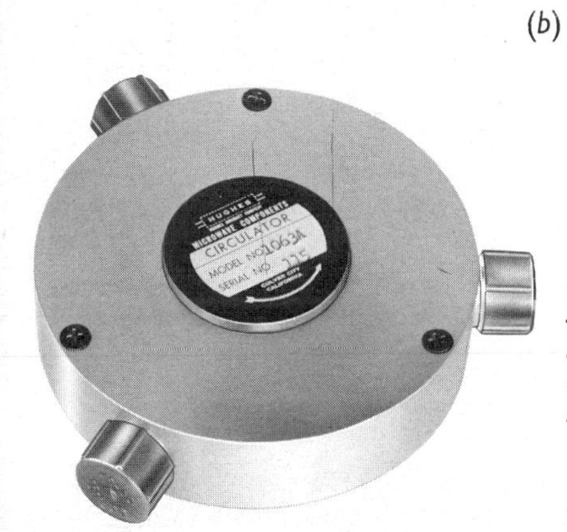

(*c*)

Fig. 10·8 Ferrite devices. (*a*) X-band ferrite switch. (*b*) Four-port waveguide circulator. (*c*) Three-port coaxial circulator. (*Hughes Aircraft Company, Solid-state Products, Aerospace Group*.)

control signal applied to the magnet coil. Variable ferrite attenuators, phase shifters, switches, and modulators find wide application in microwave techniques.

REFERENCE

1. B. Lax and K. J. Button, "Microwave Ferrites and Ferrimagnetics," McGraw-Hill Book Company, New York, 1962.

CHAPTER

11

ACTIVE MICROWAVE DEVICES

Introduction. The active microwave devices discussed in this chapter are separated into two groups. The first group consists of traveling-wave-tube amplifiers, backward-wave oscillators, klystrons, and magnetrons. These amplifiers and oscillators use the properties of free electrons and use d-c energy as a source of power. The basic mechanism of operation is a *velocity modulation* process which produces a density-modulated stream of electrons. The velocity modulation is obtained by acceleration and retardation of electrons by an alternating voltage impressed between electrodes that have relatively short spacing or by means of a moving electric field that has approximately the same velocity as the electron beam.

The second group consists of the parametric amplifier, maser, and laser. These amplifiers and oscillators differ from the first group in that they use the properties of bound electrons and they use r-f energy, instead of d-c energy, as a source of power for amplification or oscillation. The electrons are bound to the atoms of a solid, and the density of potentially usable electrons per cubic centimeter is up to 15 times that of an electron beam.

11·1 The traveling-wave-tube amplifier[1]

Wide band amplification at microwave frequencies is necessary in numerous laboratory applications. Amplification over wide frequency ranges is obtained by means of the traveling-wave-tube principle.

The basic traveling-wave-tube amplifier consists of an electron gun which projects a focused electron beam through a helically wound coil to a collector electrode, as shown in Fig. 11·1. The focused electrons are held in a pinlike beam, through the center of the helix, by a powerful magnetic field around the full length of the tube.

A c-w signal, coupled into the gun end of the helix, travels around the turns of the helix and thus has its lineal velocity reduced by an amount equal to the ratio of the length of wire in the helix to the length of the helix itself. The electron velocity, determined by the potential difference between the cathode and helix, is adjusted so that the electron beam travels a little faster than the c-w signal. The electric field of the c-w signal on the helix interacts with the electric field created by the electron beam and increases the amplitude of the signal on the helix, thus producing the desired amplification.

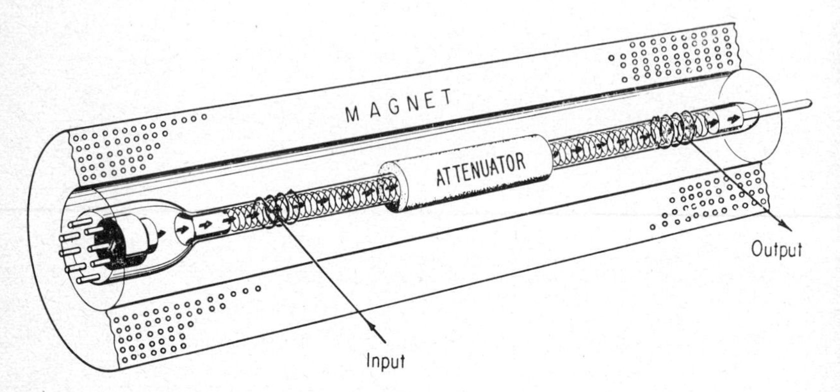

Fig. 11·1 Cutaway view, showing important elements of a traveling-wave-tube amplifier. (*Hewlett-Packard Company.*)

The principal elements of a typical traveling-wave tube and the important steps in the amplification process are shown in Fig. 11·2. The following steps should be followed by referring to the numbered captions on the diagram.

1. The electron beam is directed through the center of the helix.
2. The r-f signal is coupled into the helix. The direction and magnitude of the instantaneous distribution of the axial electric field of the traveling wave at the beginning of the tube is shown in the detail.
3. The electron charge density is modified by the continuous bunching process described in the detail.
4. Amplification of the signal on the helix begins as soon as the electron stream exhibits concentration of the space charge in bunches. The electron bunches interact with the field of the c-w signal and induce r-f current in the helix (propagating circuit), thus adding a small amount of voltage to the c-w signal on the helix. The growing or amplified c-w signal then produces a denser electron bunch which in turn adds a still greater voltage to the c-w signal, and so on.
5. Amplification increases as the greater velocity of the electron beam pulls the electron bunches more nearly in phase with the electric field of the c-w signal. The additive effect of the two fields, exactly in phase, produces greatest resultant amplification.

6. Attenuators placed near the center of the helix reduce all the waves traveling along the helix to nearly zero. This prevents undesired waves, such as waves reflected from mismatched loads, from returning to the tube input and causing oscillation.
7. The electron bunches travel through the attenuator unaffected.

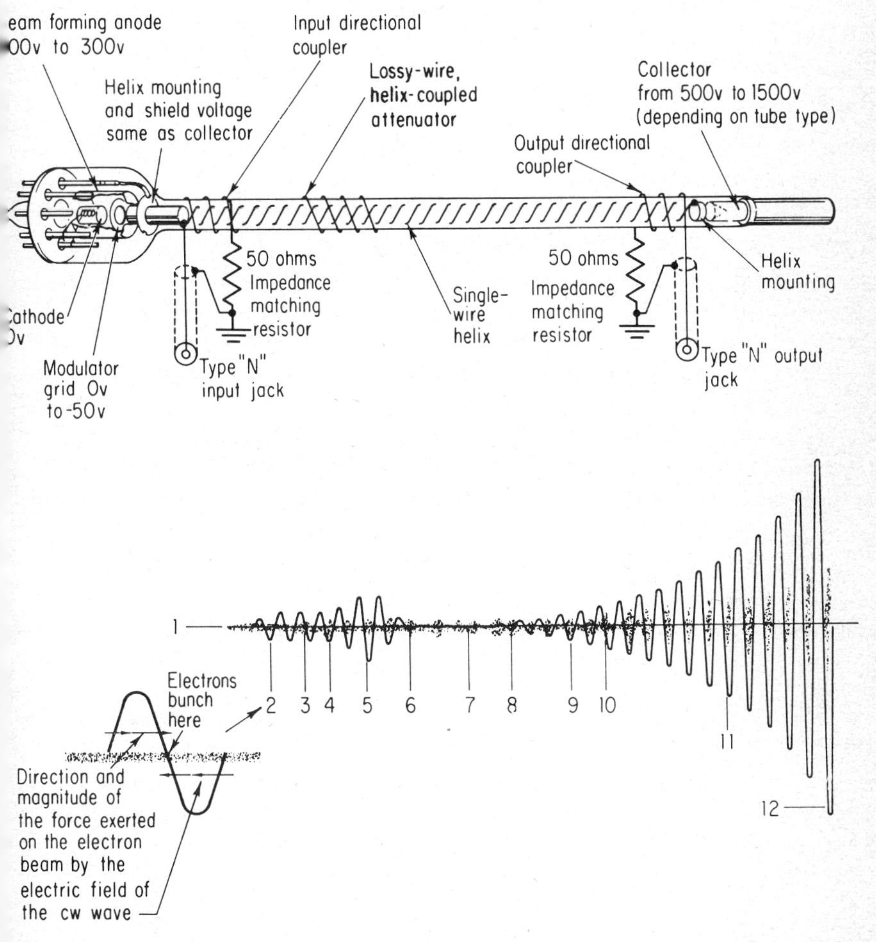

Fig. 11·2 The traveling-wave tube and how it works. (*Hewlett-Packard Company*.)

8. The electron bunches emerging from the attenuator induce a new c-w signal on the helix. This new c-w signal is the same frequency as the original c-w signal.
9. The field of the newly induced c-w signal interacts with bunched electrons to begin the amplification process again.
10. For a short distance the velocity of the electron bunches is reduced

slightly because of the large amount of energy absorbed by the formation of the new c-w signal.

11. Amplification increases as the greater velocity of the electron beam pulls the electron bunches more nearly in phase with the electric field of the c-w signal. The final wave increases exponentially with distance.
12. The amplified c-w signal is coupled out of the helix. *Note that the amplified c-w signal is a new signal the energy of which is wholly supplied by the bunched electron beam.*

11·2 The backward-wave oscillator[2]

The backward-wave oscillator provides a flexible source of microwave energy that can be voltage tuned over bandwidths from 1.5:1 to as high as 5:1. The output frequency of the backward-wave oscillator is determined by a frequency-selective feedback and amplification process rather than by resonant circuits as used in conventional microwave oscillators.

The backward-wave oscillator tube consists of an electron gun, a helix structure, and a collector at the far end of the helix, as shown in Fig. 11·3. Physically, the backward-wave oscillator resembles the traveling-wave amplifier tube, although for comparable frequencies it is larger in diameter and somewhat shorter in length. Another difference, not apparent from a visual inspection of the tube, is that the helical backward-wave oscillator uses a hollow electron beam with a strong concentration of the electrons near the helix. This hollow electron beam is focused along the length of the helix by a strong magnetic field supplied by an axial solenoid surrounding the tube.

The r-f output of the backward-wave oscillator is a result of the interaction between the electron beam and the electric fields accompanying a microwave signal present on the helix. The term *backward-wave oscillator* is quite appropriate for this tube since the r-f energy moves and builds up in a direction opposite to that of the electron beam and is coupled out at the gun end of the tube via the helix terminal.

A cross section of the helix and a portion of the electron beam are shown in Fig. 11·4. The helix structure consists of a cylindrically wound flat-wire tape; the electron beam is hollow and passes very close to the helix turns. The strong axial magnetic field focuses the electrons in the beam and allows movement only in the direction of the axis of the tube. The lines of force of the electric fields associated with an r-f wave traveling along the helix are also shown in Fig. 11·4. Although these fields rotate around the helix at the velocity of light, the axial velocity is equal to the ratio of the turn-to-turn spacing of helix divided by its circumference. The axial electric fields will be strong between helix turns and very weak under the turns since electric fields cannot exist parallel to a conductor. The strong effect of these fields, between helix turns, on the velocity of the electrons in the beam produces an interaction process which may be regarded as equivalent to a feedback amplifier.

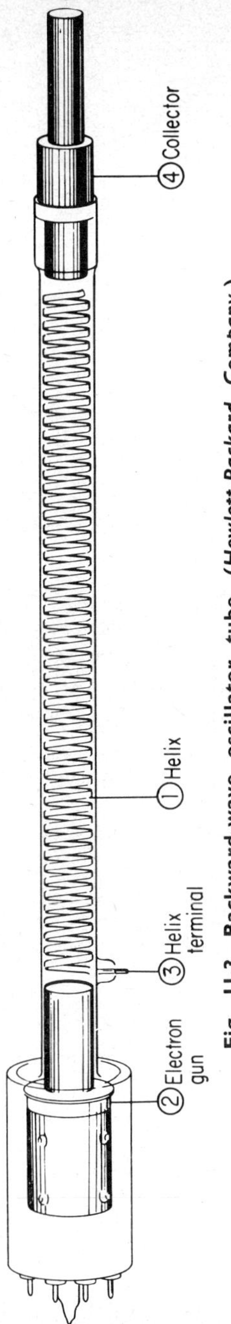

Fig. 11·3 Backward-wave oscillator tube. *(Hewlett-Packard Company.)*

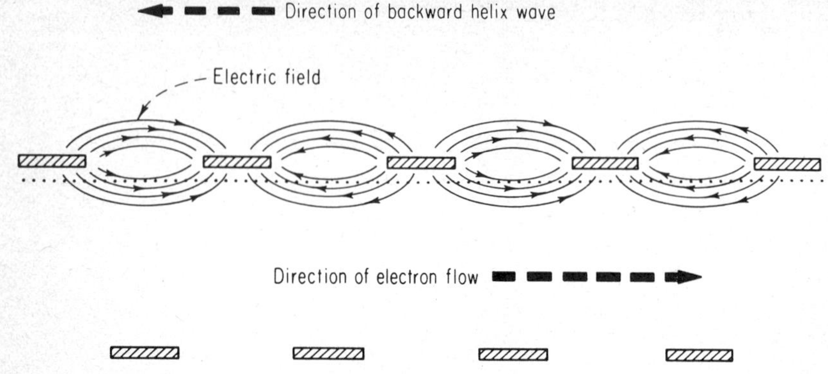

Fig. 11·4 Cross section of helix, electron beam, and helix wave. (*Hewlett-Packard Company.*)

In this way, feedback loops are formed between the midpoints of adjacent helix gaps. If the percentage feedback is sufficiently great and phase relations are properly adjusted, the amplifier will oscillate.

Although the concept of discrete feedback loops is a useful device for explanation, the backward-wave interaction is actually a continuous process. The maximum coupling between the helix wave and the beam will occur midway between gaps and gradually taper off to a minimum directly under the helix turns. One of these regenerative loop chains exists at each angular position around the helix. Each of these regenerative loop chains is independently coupled to the helix transmission line, so the net effect is a continuous

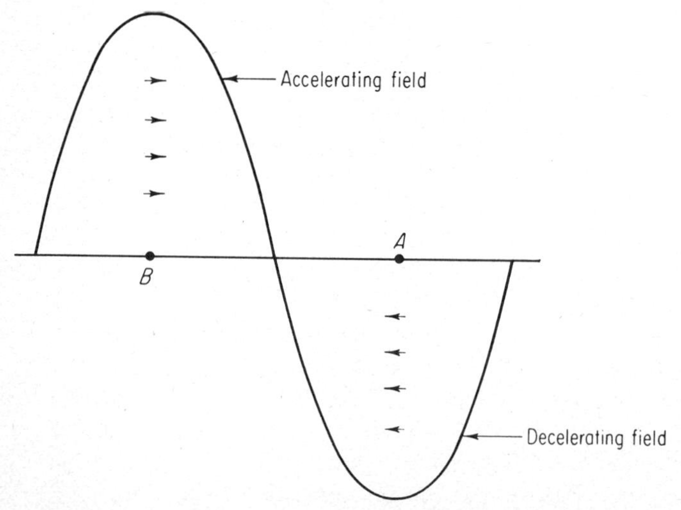

Fig. 11·5 Axial electric field of helix wave which produces velocity modulation and bunching of electron beam.

amplification and feedback process occurring down the entire length of the tube.

The basic mechanism of amplification is a velocity modulation process which causes the electrons to bunch in the beam. Figure 11·5 shows the sinusoidal variations in amplitude of the electric field at the midpoint between helix turns. The phase relationship between the backward wave on the helix and the velocity of the electron beam is such that each specific portion of the electron beam will be affected by an electric field of the same phase as it passes successive gaps down the helix. Referring to Fig. 11·5, an electron at point A experiencing the decelerating effect of the field at the first gap in the helix will experience a continuous decelerating effect, caused by fields of the same phase and direction of force, as it proceeds down the tube. In a like manner, an electron at point B will be continuously accelerated in its journey down the tube. In this way, some parts of the electron beam are slowed down while others are advanced, and the net effect is a bunch formed at the midpoint of Fig. 11·5 between the accelerating and decelerating fields. This situation is shown in Fig. 11·6. The spiral form of the bunched electron beam is due to velocity modulation which occurs at different r-f phases at various angular positions around the spirally wound helix.

At this point it should be mentioned that the average electron velocity of the beam is slightly faster than the effective phase velocity of the amplifier chain. This means that the electron bunches will advance a quarter of a cycle as they approach the collector end of the tube, and thus encounter the full decelerating effects of the electric field and give up a maximum amount of kinetic energy to the wave on the helix.

Figure 11·5 shows that the density of the electron bunches increases according to a sine-wave relationship. Figure 11·7 shows the envelope of the bunching rather than instantaneous amplitudes since many r-f cycles exist along the length of the backward-wave oscillator tube. The wave on the helix moves from right to left toward the gun end of the tube and gains amplitude between each turn according to the degree of electron bunching in the beam. In this way, the envelope of the wave on the helix shown in Fig. 11·7 is the integral of the bunching envelope so the maximum energy transfer from the beam to the wave on the helix occurs at the collector end of the tube.

Now that a correspondence between the chain of lumped regenerative loops and the helix backward-wave oscillator tube has been established, it can be seen that if the velocity of the electron beam is varied, the phase delay around each of the regenerative loops will be changed and, if the electron beam current is high enough, the chain of regenerative loops will oscillate at a frequency where the phase delay of each loop is equal to one cycle.

Oscillations begin in the backward-wave oscillator in much the same manner as they begin in other oscillators. Noise waves are established on

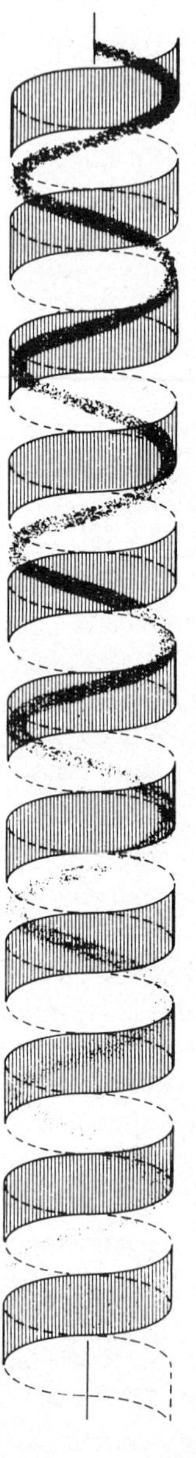

Fig. 11·6 Helix showing bunching of electron beam. *(Hewlett-Packard Company.)*

the helix from the shot noise coupled from the electron beam and from thermal energy developed in the termination at the input end of the tube. The waves traveling backward on the tube (to the left) velocity modulate the beam. Velocity modulation causes the electrons to bunch and in turn reinforce the wave that exists on the helix at the frequency where the single loop delay is equal to one cycle. In this way, oscillations are built up at a single frequency determined solely by the electron beam velocity, which is a function of the cathode-to-helix voltage on the tube.

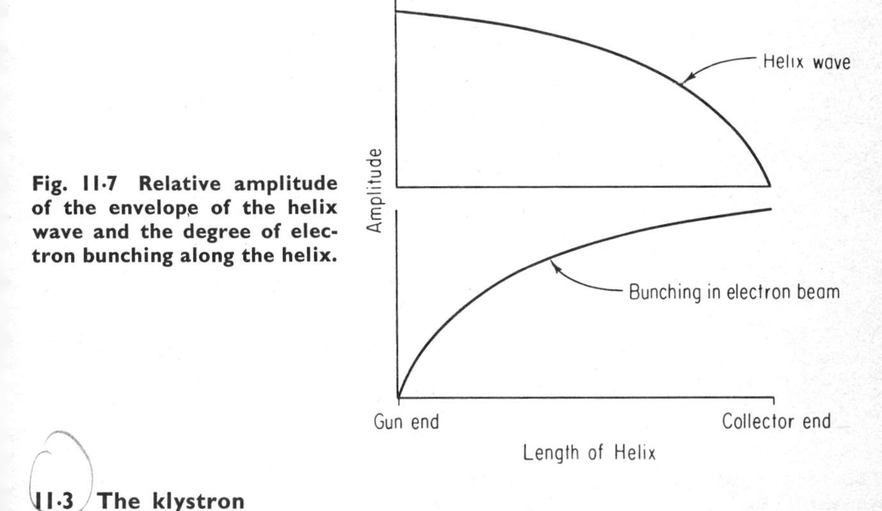

Fig. 11·7 Relative amplitude of the envelope of the helix wave and the degree of electron bunching along the helix.

11·3 The klystron

The klystron is a velocity-modulated tube in which the velocity modulation process produces a density-modulated stream of electrons. The oldest form of velocity variation device is the *two-cavity klystron amplifier*. The basic structure of the amplifier is illustrated in Fig. 11·8. The first grid controls the number of electrons in the beam and serves to focus the beam. The potential between the cathode and the cavity resonators is the accelerating potential and is commonly referred to as the *beam voltage*. The cavity is at ground potential, and the beam voltage is connected between the cathode and the collector. The velocity of the electron in the well-defined beam is determined by the beam accelerating potential. Upon leaving the region of the first grid, the electrons pass through the grids of the first cavity, which is referred to as the *buncher* cavity. The grids of the cavity enable the electrons to pass through but confine the magnetic fields within the cavity. The space between the grids is referred to as the *interaction space* or *gap*. When the electrons traverse this space, they are subjected to r-f potentials at a frequency determined by the cavity resonant frequency and/or the input signal frequency. The amplitude of this r-f potential between the grids is determined by the amplitude of the incoming signal in the case of an amplifier or by the amplitude of the feedback signal from the second cavity if used as an oscillator. Electrons which traverse the interaction space when the r-f potential on grid 3 is positive with respect to grid 2 are accelerated by the field, while electrons traversing the gap one-half cycle later are decelerated. In this process, essentially no energy is taken from the cavity since the average number of electrons slowed down is equal to the average number of electrons speeded up. The electrons

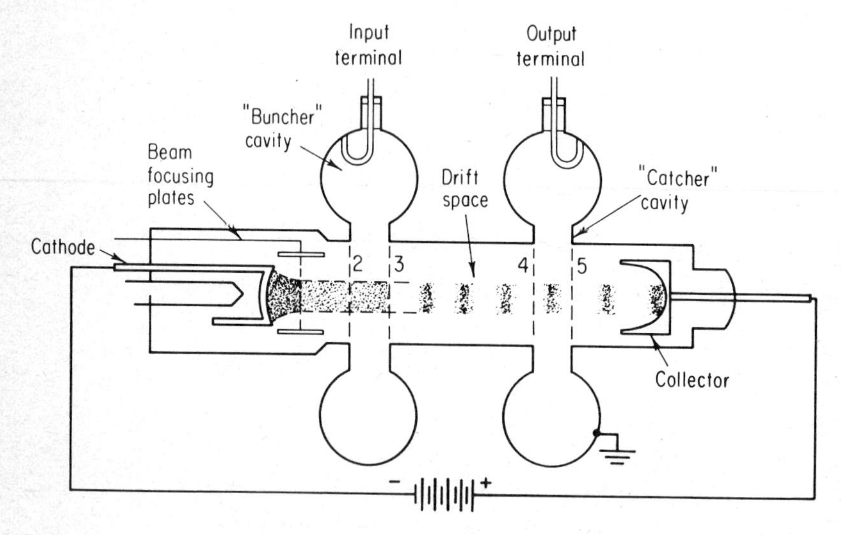

Fig. 11·8 Two-cavity klystron amplifier.

which are slowed down give up energy to the fields inside the cavity. Electrons that are speeded up absorb energy from the fields in the cavity.

Upon leaving the interaction space, the electrons enter a space called the *drift* or *bunching* space in which the electrons that were speeded up overtake the slower ones, thus forming *bunches*. This process results in the formation of electrons into bunches in space and produces the density modulation. A second cavity resonator called the *catcher* is located at the point where the bunching is maximum. This cavity is tuned to the same frequency as the input cavity resonator. The power output at the second cavity is obtained by slowing down the electron bunches. If an alternating field exists at the output cavity resonator, and grid 4 is positive with respect to grid 5, then the bunches of electrons passing through the grids are slowed down and deliver energy to the output cavity. The oscillation in the output cavity builds up in proper phase to retard the electron bunches because the oscillation amplitude cannot build up unless the electrons deliver energy to the cavity resonator. The power output is equal to the difference in kinetic energy of the electrons averaged before and after passing the interaction gap. The positive collector electrode beyond the cavity collects electrons and prevents secondary emission. Secondary emission is possible because of the impact of electrons which reach the end wall. The amplifier can be made into an oscillator by employing feedback from the output cavity to the input cavity in proper phase and of sufficient amplitude to overcome the losses in the system. The

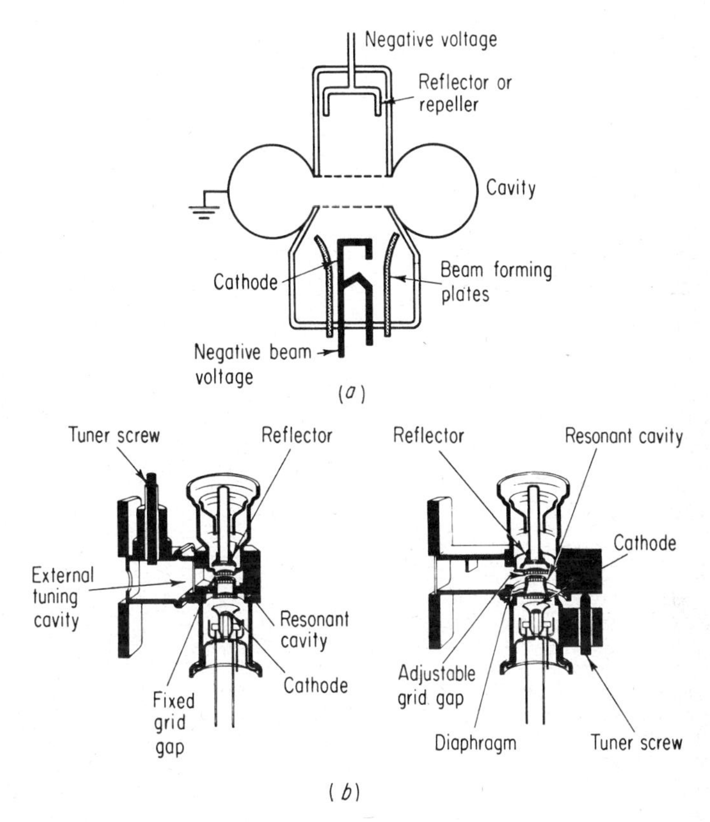

Fig. 11·9 Reflex klystron. (*a*) Schematic representation. (*b*) Internal construction. (*Varian Associates.*)

initial start of oscillations is caused by random fluctuations of the electron beam current and by noise voltages which generate alternating fields in the cavity. for low power applications

Reflex Klystron. The reflex klystron has been the most used source of microwave power in laboratory applications; it is shown in Fig. 11·9. This tube differs from the tube of Fig. 11·8 in that the output cavity is omitted and the *repeller* or *reflector electrode*, placed a short distance from the single cavity, replaces the collector electrode. The single cavity resonator performs the bunching and catching process, and feedback is developed by reflecting the electron beam so that it passes through the resonator a second time. The negative potential of the reflector produces a strong decelerating electric field. Therefore, electrons which are accelerated toward the reflector experience this strong decelerating field. The electrons come to a complete stop and experience acceleration in the reverse direction back through the cavity grids. The electrons in the bunched beam return through the cavity toward the cathode, where they are collected by the cavity grid near the cathode, by the

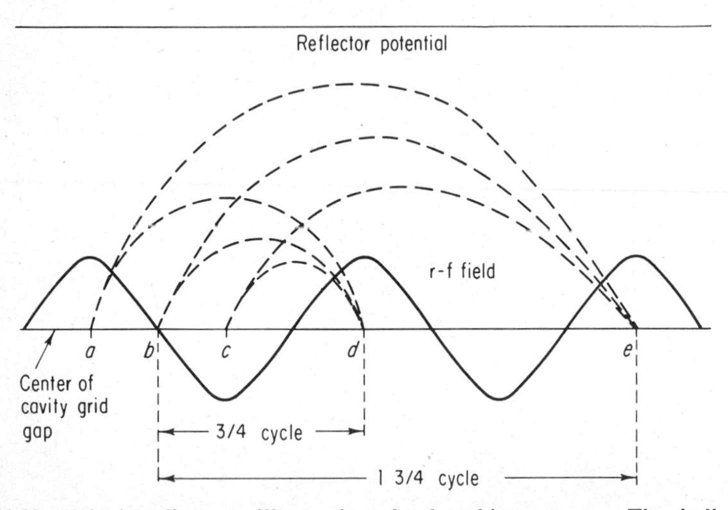

Fig. 11·10 Velocity diagram illustrating the bunching process. The indicated positive r-f field accelerates electrons traveling into the reflector region and retards electron bunches returning from the reflector region. The negative r-f field retards electrons going toward the reflector and accelerates electrons returning from the reflector field.

cavity walls, or by other accelerating grids that may be present. A few electrons also continue on toward the cathode. A fairly complete picture of what happens inside the tube can be obtained without being exact or including the many phenomena which, in practice, must be included in the operational theory. The simplified theory does not include multiple transit of electrons, space charge effects in the reflector region, different drift times for different electron paths, secondary emission, or the loss of electrons on grids and the edges of apertures. In the simplified theory, it is assumed that the electrons enter the interaction gap with the same initial velocity and that the retarding field between the gap and the reflector is uniform.

The velocity diagram of Fig. 11·10 is instructive in the study of the reflex klystron. The beam voltage connected between the cavity and the cathode accelerates the electrons through the cavity grids. The initial start of oscillations is caused by random fluctuations of the beam current. The beam current fluctuations produce r-f fields across the interaction space or gap between the grids. The frequency of these oscillations is determined by the physical dimensions of the cavity. The diagram shows the displacement of electrons relative to the center of the interaction space for electrons crossing the gap at different times of the r-f cycle. Electrons crossing the gap at *a* experience acceleration because the r-f field is a maximum value in the direction shown. The electrons traversing the gap gain maximum kinetic energy and penetrate farthest into the region of the reflector field. These electrons return to the cavity as indicated by the maximum height of the electron orbit. Electrons traversing the grid gap at *b* do not experience any additional force, either acceleration or deceleration, because the r-f field is zero. However, electrons traversing the grid gap at *c* experience maximum deceleration, penetrate a shorter distance into the reflector voltage region, and return to the cavity grid gap at essentially the same instant *d* as electrons leaving the grids at *a* and *b*. The bunching action is a function of grid spacing, the alternating gap voltage, and the retarding reflector field. In order to deliver maximum power to the cavity, the returning bunches of electrons must encounter the peak value of r-f field in the direction to decelerate the electrons. The maximum delivery of kinetic energy to the r-f field is indicated at *d* on the diagram. Analysis of the bunching process and maximum power delivery, as shown in Fig. 11·10, indicates that those electrons which pass through the cavity grids when the field is zero at *b* become the center of the bunch. Also, it is noted that the unaccelerated electrons at *b* must remain in the reflecting field for three-quarters (¾) of a cycle of the r-f cycle or an integral number of cycles plus three-quarters of a cycle. This condition is indicated by the dotted lines and the designated 1¾ cycle distance on the diagram. The term *mode* is again used to describe a certain set of conditions. In this case the term refers to the repeller or reflector voltage settings which are required to cause the electrons to remain in the reflector field the designated $(n - \frac{1}{4})$ cycles. The term *mode* therefore defines the number of cycles of electron transit time N and is given by

$$N = n - \frac{1}{4}$$

where n is a positive integer greater than zero. Klystron reflector voltage modes are designated ¾ mode, 1¾ mode, etc.

The basic frequency of oscillation of the klystron is determined by the cavity dimensions. The cavity dimensions are varied by flexing a portion of the cavity wall and/or changing the spacing of the cavity grids. In addition, the oscillation frequency can be adjusted by the reflector voltage. The maximum power output point is referred to as the "top" of the mode. If the reflector voltage is set to any other voltage, the transit time is changed, and the klystron oscillates at a different frequency with decreased power output. Therefore, in order to obtain maximum power output at a specific frequency, it is necessary to adjust the cavity resonant frequency and the reflector voltage to obtain maximum power output for any fixed setting of other electrode potentials.

Power and Frequency Characteristics. The variations of frequency and power output are plotted in Fig. 11·11 as a function of reflector voltage. These curves can be obtained by plotting the power output as the reflector voltage is varied to successive values or by sweeping the klystron reflector voltage with a sawtooth waveform and observing the rectified power output

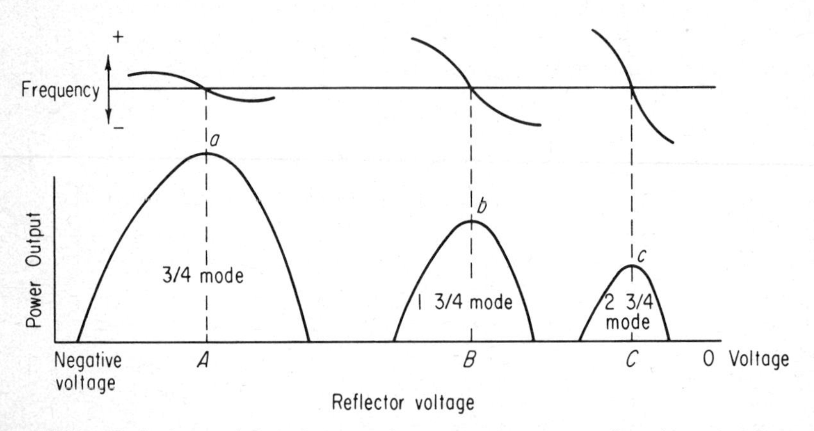

Fig. 11·11 Variations of frequency and power output as a function of reflector voltage.

as displayed on an oscilloscope. Oscillations occur over the range of reflector voltage values which cause the electron bunches to return through the cavity grids in the proper phase to deliver energy to the cavity.

The points a, b, and c at the top of each mode represent the maximum power output point where the transit time is $(n - \frac{1}{4})$ cycles. *Therefore, each of these points also represents the same frequency.*

Consider a klystron oscillating with a reflector voltage indicated at A in Fig. 11·11. If the reflector voltage is made more negative, the electrons are retarded more and the transit time is made shorter. Therefore, the bunches are not properly formed and return through the cavity too soon. Since bunching is not complete, the power delivered to the cavity is less, the resulting r-f signal has less amplitude, and the power output is lowered. The klystron now oscillates at a higher frequency with less output power. Alternately, if the reflector voltage is made less negative from A, the electrons travel farther into the reflector region and become debunched by the time they arrive back at the cavity grids. The debunching again causes the r-f amplitude to decrease, the power output is decreased, and in this case the frequency decreases. The power output and frequency stabilize at this new value. The power supplied to the cavity in the bunching process increases as the oscillation grows and continues to increase until the r-f current reaches its maximum possible value. This is the optimum bunching point where maximum power is supplied to the cavity. If the r-f voltage increases to a larger value, the electron bunches are formed too soon, less power is delivered to the cavity, the r-f current decreases, and the amplitude of the r-f voltage is automatically decreased to the value where maximum power is again supplied to the cavity.

It is now necessary to examine the power output and *electronic tuning range*. The electronic tuning range is defined as the total change in frequency from one end of the mode to the other. There is also an *electronic tuning sensitivity* associated with changes in reflector voltage. The electronic tuning sensitivity is usually measured in megacycles per volt of reflector voltage change.

The important characteristics of the klystron modes are represented in Fig. 11·11. The lower output power of the higher order modes can be accounted for in terms of the simplified theory. At lower reflector voltages labeled B and C on the diagram, the electrons penetrate farther into the reflector region. There is less bunching effect on the return through the cavity (the electron becomes debunched), and less power is returned to the cavity. Also, optimum bunching requires lower r-f cavity voltage due to the longer drift time. The overall result is decreased feedback and lower output power for the higher order modes.

The frequency sensitivity characteristics observed in Fig. 11·11 indicate that the tuning sensitivity and tuning range are higher for the higher order modes. Referring to Fig. 11·10, it is noted that for a given small change in reflector voltage, the bunched electrons return to a point in the vicinity of d before there has been a significant change in phase of the r-f voltage. Therefore, the klystron frequency change is not great. If the same relative change in reflector voltage is considered for the higher order mode, it is found that the longer transit time causes the bunches to return to a point in the vicinity of e. In this case, however, there is a greater change in the phase of the r-f voltage due to the longer transit time (the bunched electrons are farther displaced from the point e where maximum power delivery would occur). Since the klystron adjusts to this new set of conditions, the shift in frequency from the resonant frequency of the cavity is far greater than the shift which was experienced for the corresponding change in reflector voltage considered for the lower mode.

The power output from the cavity is obtained by loop-coupling for klystrons having coaxial outputs. Klystrons which have coaxial outputs are usually limited to frequencies below 8,200 Mc. At higher frequencies, the power output of the tube is obtained through an aperture in the cavity into a section of waveguide, as shown in Fig. 11·12.

Hysteresis. The term *hysteresis*, as applied to the klystron, refers to the multiple dependence of power output and frequency upon the direction from which the electrode potentials are made to approach a given voltage. In particular, the discontinuous behavior of klystron modes is observed when the reflector voltage is alternately increased and decreased over the range of a particular mode. In increasing the reflector voltage through a mode, the amplitude and frequency may exhibit smooth tuning characteristics up to a certain critical value of reflector voltage where a sudden change in frequency

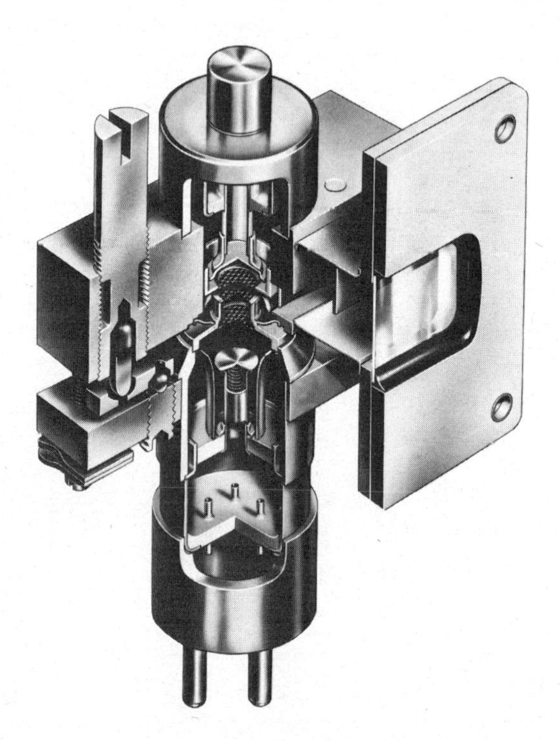

Fig. 11·12 Cutaway of a reflex klystron. (*Varian Associates.*)

or amplitude may be encountered. Several such changes may occur on one side of the mode or even on both sides. If the reflector voltage is now varied in a decreasing direction, some of the discontinuities may not occur at all, or amplitude and frequency changes may occur at different reflector voltages. Also, the maximum power output may occur at a different reflector voltage.

The various phenomena which can cause hysteresis were neglected in the simplified operational theory. Multiple transit effects may occur if electrons pass through the cavity more than once. The additional transits of the electrons, along with additional bunching action in the space between the resonator and cathode, result in additional varying components of admittance. Hysteresis effects can be caused by the nature of the load and by variation of the phase of the electronic impedance. The electronic impedance is the ratio of the r-f voltage to the r-f current associated with the bunched electron beam.

11·4 Magnetrons

The magnetron is employed as a generator of microwave frequencies in the centimeter and millimeter wavelength ranges. Magnetrons are available for use as pulsed or c-w generators but find greatest application as radar transmitters since they are capable of furnishing high peak power (in the order of megawatts in the centimeter range).

The main component parts of all magnetron oscillators consist of a cathode, anode, tank circuit, and a magnet. The cylindrical cathode emits electrons by thermionic and/or secondary emission. Except for the parallel-plane magnetron, the anode structure is cylindrical, coaxial with the cathode, and is usually provided with slots parallel to the axis of the cylinder. The tank circuit is connected to, or is a part of, the anode structure and is tuned to the desired frequency of oscillation. The tank circuit incorporates a means by which the output power is coupled to a transmission line. The magnet provides a constant magnetic field which is oriented parallel to the axis of the cathode and anode.

The three basic types of magnetron oscillators are identified on the basis of the mechanism by means of which energy is transferred to the r-f field.

1. The *cyclotron-frequency magnetron* operates by virtue of resonance between the period of r-f oscillation and the rotational motion of the electron.
2. The *negative-resistance* or *split-anode* magnetron operates on the principle of a static negative resistance characteristic between two anode sections.
3. The *traveling-wave* or *resonant-cavity magnetron* depends upon the mean translational velocity of the electron being in synchronism with the velocity of a traveling-wave component of the r-f interaction field in the space between the anode and cathode.

Cyclotron-frequency Magnetron.[3] The simplest form of cyclotron-frequency magnetron is illustrated in Fig. 11·13. The diode, formed by the anode-cathode coaxial cylinders, is positioned between the poles of a permanent magnet or electromagnet so that the magnetic lines of force are parallel to the axis of the anode and cathode.

The representative electron orbits corresponding to three distinct portions of the current versus magnetic field characteristics of a Hull magnetron are shown in Fig. 11·14. Electrons emitted from the cathode are accelerated along radii toward the anode by the electric field set up by the anode voltage. An electron traversing a magnetic field is acted upon by a force which depends on the magnitude of the electron velocity and on the strength of the magnetic field. The force exerted upon the electron is directed perpendicular to the direction of motion and also perpendicular to the direction of the magnetic field. The force is proportional to the magnetic field, the electron velocity, and the sine of the angle between the direction of the magnetic field and the

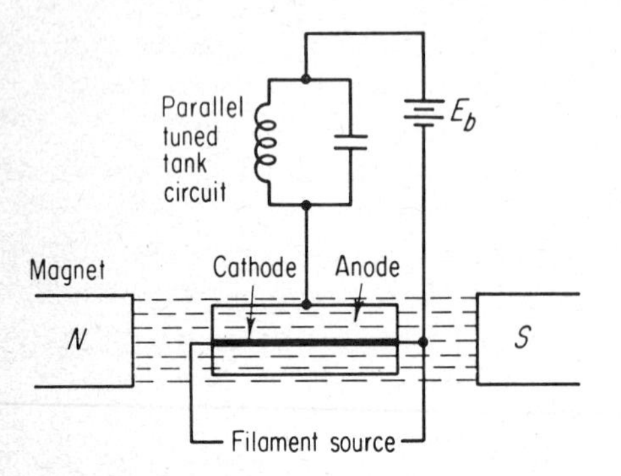

Fig. 11·13 Circuit arrangement of cylindrical anode magnetron.

direction of the electron velocity. The low value of magnetic field causes the electron to follow the curved path as shown at *a*. A critical value of magnetic field is noted at *b* where the electron, moving in a circular orbit, just grazes the anode. Any increase in magnetic field for this particular electric field

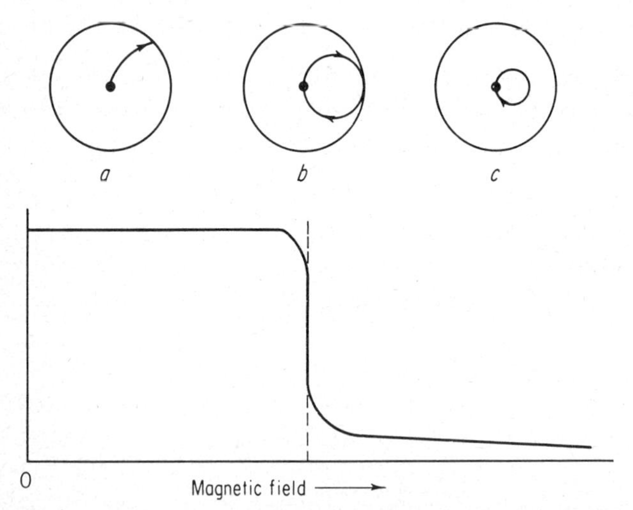

Fig. 11·14 Current versus magnetic field characteristics of a cylindrical magnetron The anode potential is held constant, and the magnetic field is varied.

strength results in electrons completely missing the anode as shown at *c* At the *critical* or *cutoff* value of magnetic field, there is an abrupt decrease in anode current. This same effect can be produced by holding the magnetic field constant and varying the cathode potential.

Electrons leaving the cathode are accelerated by the anode voltage and receive energy from the electric field. The magnetic field causes the electrons to follow a curved path which takes them in a circular orbit away from the cathode and then back again. In traveling back to the cathode, the electrons are moving against the d-c field and are slowed down. The kinetic energy initially acquired by the electrons is now transferred via the electric field to the anode, causing a variation in the anode voltage which causes an alternating voltage to be superimposed on the anode voltage. A parallel-tuned circuit or a quarter-wave parallel-line resonator connected between the electrodes can be used to build up and transfer the energy to a load. When the anode voltage contains an alternating component, the energy transfer

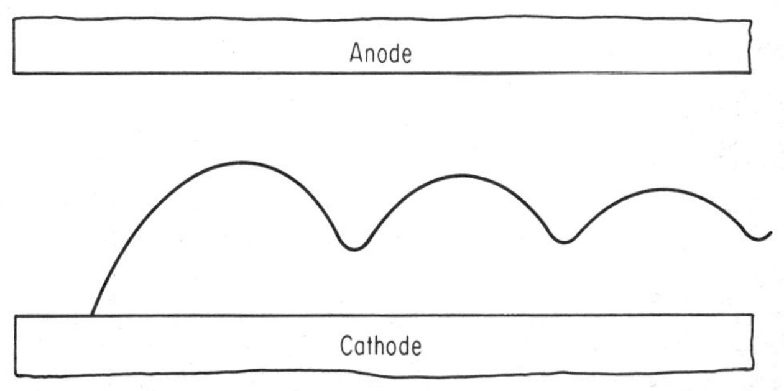

Fig. 11·15 Cycloidal paths of an electron in a parallel-plane magnetron.

to the d-c field must be greater than the energy loss to the electrons. If an electron leaves the cathode at an instant in the cycle when the r-f field aids the acceleration toward the anode, the electron gains energy from the alternating r-f field. On its return path toward the cathode, the r-f field has now changed direction and accelerates the electron toward the cathode. Electrons performing this orbit receive energy from the r-f field traveling out and back. The electron strikes the cathode, giving up its kinetic energy in the form of heat, and does not continue to absorb energy from the r-f field. If an electron leaves the cathode when the r-f field is in the opposite phase, the electron gives up energy when moving toward the anode and again when moving back toward the cathode. Electrons in this type of orbit do not reach the cathode after the first trip. Instead, they reverse direction and start a second orbit or loop of smaller amplitude. The electrons in this phase relationship with the r-f field continue in this manner until the electrons give up all their energy to the r-f field. The electrons are removed by conducting end plates at the ends of the electrodes or by special tilting arrangements of the electrodes and magnet.

An approximate orbit of an electron which loses energy to the r-f field, as explained above, is shown for the parallel-plane magnetron in Fig. 11·15. After a number of oscillations, the electron starts regaining energy from the r-f field, and its amplitude will again increase. It is very difficult to remove the electrons at the proper stage of their orbits. Also, the adjustments of magnetic field orientation and electrode potentials are critical.

Negative-resistance Magnetron. The *negative-resistance magnetron* has the anode split parallel to the axis in two halves, and for this reason it is sometimes referred to as the *split-anode magnetron.* Magnetrons of this type operate at frequencies up to 1,000 Mc. The paths of electrons in the interelectrode space were plotted by Kilgore.[4] A circuit arrangement of the negative-resistance magnetron, shown in Fig. 11·16, illustrates the modification of the normal radial field by the r-f alternating field and the resulting

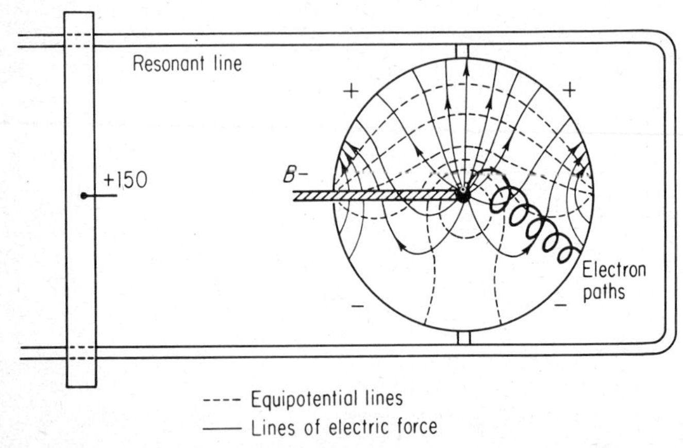

Fig. 11·16 Circuit arrangement of the negative-resistance (split-anode) magnetron. Electrons terminate on the segment which has lowest potential. R-f is generated in the external circuit between segments.

electron path. During the time that the electrons complete the several complete loops, the alternating field has not changed significantly. The electrons terminate on the anode segment which is at the lowest potential. This produces the negative resistance characteristic. There is a general drift of electrons along the equipotential surfaces, and since this drift is toward the negative region, the electrons terminate on the anode segment which has the lowest potential. It is noted that each loop in the electron path is in the direction opposing the alternating field across the gap and that the electrons give up energy due to retardation by this field.

Traveling-wave Magnetron. The *traveling-wave* or *multicavity magnetron* consists of a cylindrical anode structure containing a number of equally spaced cavity resonators with slots along the anode surface adjacent to the cylindrical cathode. A permanent magnet is usually used to provide the necessary magnetic field. The power output can be coupled out through a slot in the cavity or by means of a coupling loop.

As in other types of oscillators, the start of oscillations is due to random phenomena in the electron space charge and in the cavity resonators. The cavity oscillations produce electric fields which fringe out into the interaction space from the slots in the anode structure, as shown in Fig. 11·17. Energy is transferred from the radial d-c field to the r-f field by interaction of the electrons with the fringing r-f field. The first orbit of electron *A* in Fig. 11·17 occurs when the r-f field across the gap is in a direction to retard electron *A*. The transfer of energy is from the electron to the tangential component of the r-f field. The electron comes to a stop and is again accelerated by the radial d-c field. This orbit of the electron is adjacent to the next cavity slot

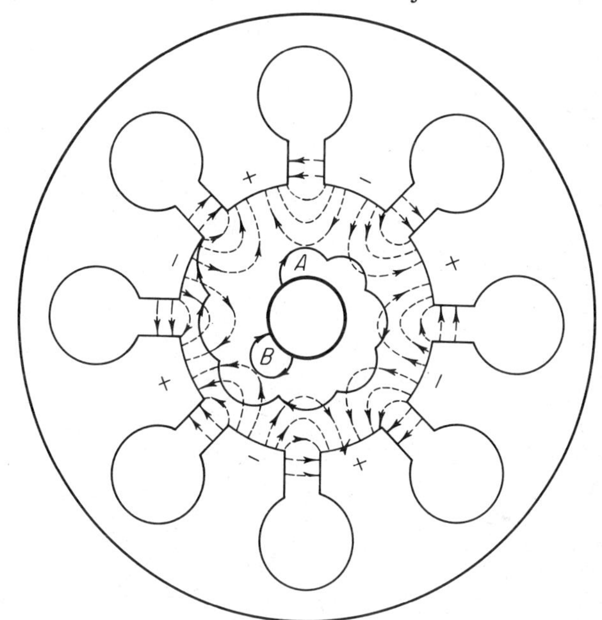

Fig. 11·17 Possible trajectory of an electron from cathode to anode in an eight-cavity magnetron operating in the π mode.

as shown in the diagram. If the r-f field across the cavity slot has changed phase by 180°, the direction of the r-f field is opposite the direction shown, and the electron is again slowed down and gives up energy to the r-f field. The dotted line indicates that energy transfer from the electron to the tangential component of the r-f field will occur at each cavity if the electron moves in synchronism with the r-f field. The electron gives up most of its energy before it finally terminates on the anode surface. The electron at *B* absorbs energy from the field and is accelerated and quickly returned to the cathode. There is a net delivery of energy to the cavity resonators because electrons that absorb energy from the r-f field are quickly returned to the cathode, whereas the energy in the rotational component of motion of the electron in the retarding r-f fields remains practically unaffected, and the electron may orbit around the cathode many times.

The definite phase relationship of the fields appearing across adjacent slots is used as a means of defining the different *modes* of oscillation. The total phase shift in such a system of resonators must be a multiple of 2π rad. The phase difference between adjacent resonators for a system of N resonators may assume values of $2\pi n/N$, where $n = 0, 1, 2, \ldots, N/2$. The phase angle is π rad for the $N/2$ mode shown in Fig. 11·17. This mode is called the *pi mode*.

The traveling-wave representation of the multicavity magnetron is shown in Fig. 11·18. An electron in motion in opposition to the r-f field is indicated

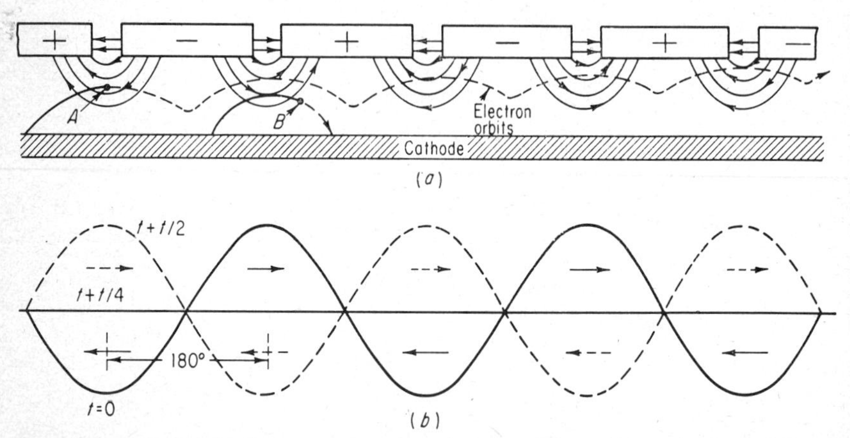

Fig. 11·18 (a) A plot of the π mode cavity anode potential at $t = 0$. The dotted line represents the trajectory of the electron. (b) The standing-wave diagram indicates the anode potential at three instants of time. The solid line represents the instantaneous r-f field distribution across the slots at time $t = 0$. The solid arrows represent the direction of force exerted on electrons at $t = 0$. At $t + t/4$, the forward and reverse traveling waves cancel, and the r-f field is zero between all slots (solid center line). At $t + t/2$ (180°), the instantaneous maximum field distribution is shown by the dotted waveform. The dotted arrows represent the direction of force exerted on electrons at $t + t/2$.

at the first slot. The continued orbits of the electron in step with the r-f field are indicated by the dotted line.

The traveling wave is represented by the standing-wave diagram. The solid line is used to represent the instantaneous r-f fields across the slots, as shown, and the solid arrows indicate the direction of force upon the electrons moving into the r-f fields from the cathode. The negative maximum represents the time t when maximum retardation occurs for the electrons crossing the first slot. The positive cycle and the solid arrow indicate that electrons crossing the second slot are accelerated and returned to the cathode as previously discussed. The solid center line represents zero instantaneous r-f field across the gaps at $(t + t/4)$ when the forward and backward traveling waves cancel (previously explained for incident and reflected waves in Chap. 3). The dotted line on the standing-wave diagram represents the maximum instantaneous r-f field at $(t + t/2)$ when the r-f fields have reversed for each cavity slot. It is noted that if the electron is traversing the second slot at this instant, it experiences retardation by the r-f field. This condition is indicated on the diagram. By the same reasoning, the additional electron orbits can be followed at succeeding intervals of time.

Phase Focusing. The mechanism by which electron bunches are formed and by which electrons are kept in synchronism with the r-f field is called *phase focusing*. The phase-focusing effect is illustrated by the simplified diagram in Fig. 11·19. The translational velocity of the electron is increased

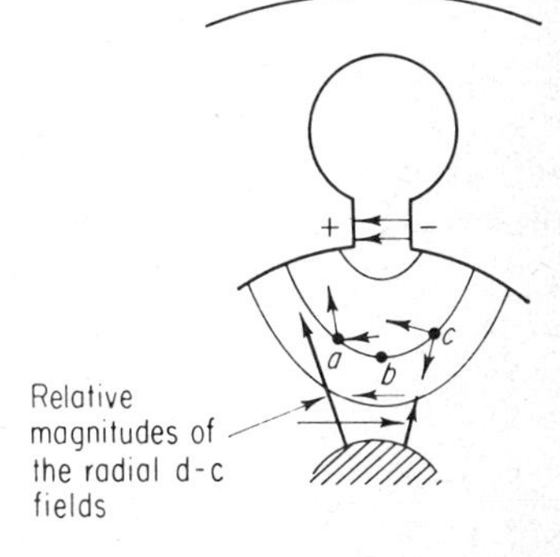

Fig. 11·19 Phase-focusing effect in the magnetron.

or decreased depending upon whether the r-f field is aiding or opposing the radial d-c field. An electron at *a* is in the vicinity of the positive anode, and the component of the r-f field aids the radial d-c field. The stronger resultant field increases the velocity of the electron. The electron at *b* experiences the maximum retarding effect of the r-f field which is about equal to the increase in velocity due to the radial d-c field. An electron at *c* is nearer the negative anode and experiences the lower relative magnitude of radial d-c field as shown. The velocity of the electron at *c* is thus decreased and tends to fall back. This causes a bunching action around the electron whose relative position is indicated at *b*. This action is applied to all of the electrons making up the space-charge in the interaction space. The selective grouping of electrons results in a spoke-shaped space-charge cloud of electrons spinning around the cathode in synchronism with the traveling wave on the anode. Each spoke of the space-charge wheel is in a region of maximum-retarding field, as shown in Fig. 11·20.

Mode Separation. Two identical resonant cavities will resonate at two frequencies when they are tightly coupled together. The two resonant frequencies lie above and below the resonant frequency of the individual resonators. As in the case of two coupled resonators, *N* resonators have *N* modes of oscillation because of the effect of mutual coupling. A common means of separating the *pi mode* from adjacent modes is by a method called *strapping*. The straps consist of wires of either circular or rectangular cross

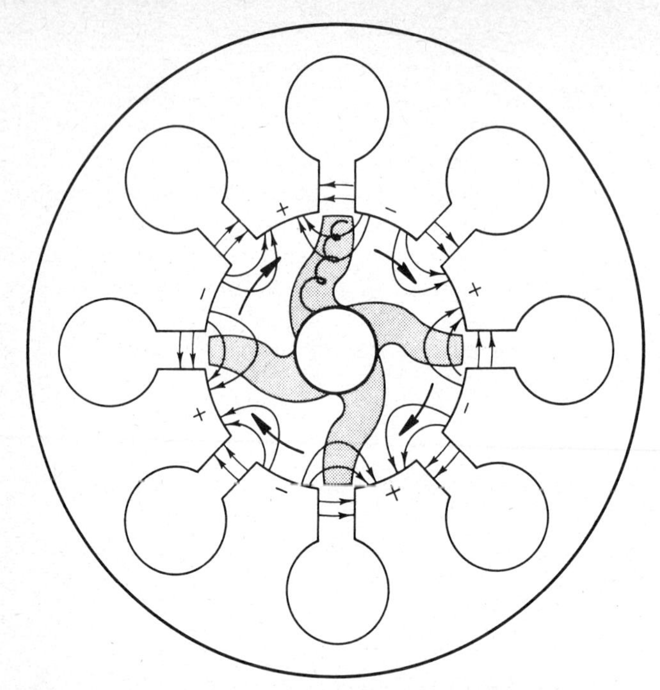

Fig. 11·20 Rotating space charge in the multicavity magnetron.

section connected to alternate segments of the anode block, as shown in Fig. 11·21. The straps are usually recessed into the resonator structure. In the pi mode, each strap joins anode segments that are at the same voltage. Each strap is capacitively coupled to the segment that it passes over. This capacitance, effectively in parallel with the resonator, reduces the frequency of the pi mode. For lower modes, the strap causes a decrease in the capacitance because of the different voltages which exist at the end of the straps.

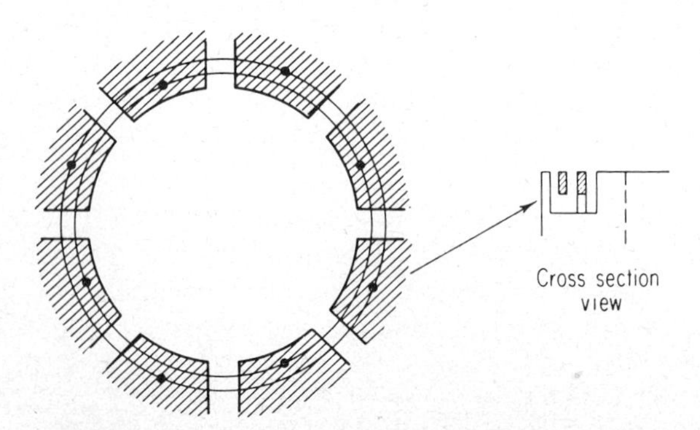

Fig. 11·21 Schematic illustration of strapping.

There is also inductance associated with the strap. The combined effects increase the resonant frequency of the lower modes, thus increasing the separation between modes.

Fig. 11·22 Rising-sun type of resonator system.

The *rising-sun* type resonator of Fig. 11·22 obtains division of modes into two groups by employing an equal number of small and large cavities.

11·5 Parametric amplifiers

The *parametric amplifier* is a low-noise amplifier capable of amplifying microwave frequencies. The name is descriptive and represents the fact that the operation of the amplifier depends upon the variation of one or more parameters with time. The basis of the parametric amplifier is a voltage-variable reactance (capacitive) diode. The diode is located in a microwave structure appropriate for the frequency of operation. Input- and output-matching networks and a diode-biasing circuit complete the circuit requirements.

The basic principle of operation can be illustrated using the parallel LC circuit in Fig. 11·23. The circuit has a resonant frequency f as indicated by the capacitor voltage as a function of time. Imagine that the capacitor plates can be physically separated and returned to the original separation. If the capacitor plates are separated at a when the voltage is a maximum, there is a decrease in capacitance and a corresponding increase in voltage since $E = q/C$ and the total charge q on the capacitor must remain what it was when the voltage reached its peak value. Work is done when the plates are pulled apart to separate the charge. The total energy in the system has been increased by the source which separated the capacitor plates. The energy transfer is referred to as *pumping*, and the source supplying the energy is referred to as the *pump*. From point *a* the capacitor begins to discharge, and the energy which was stored in the electric field between the capacitor plates is converted into energy in the magnetic field surrounding the inductance L. If the plates of the capacitor are returned to their original separation at b, when the voltage goes through zero and there is no charge on the plates, no work is done and the voltage is unchanged. If the capacitor plates are separated when the voltage

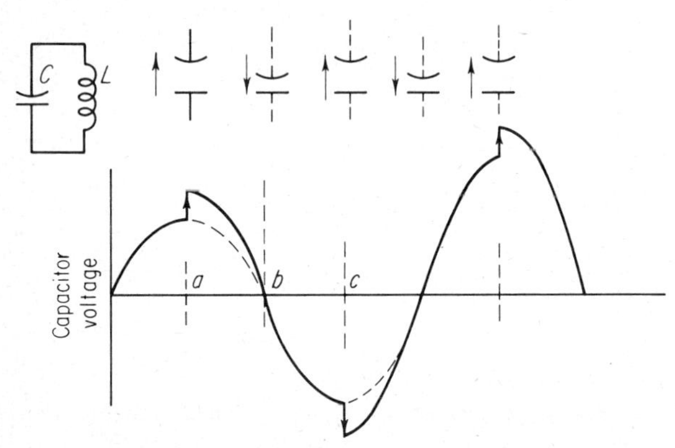

Fig. 11·23 Principle of parametric amplification.

is at the maximum negative point c, additional energy is supplied to the system. In order to obtain amplification, the plates must be separated at the precise maximum positive and negative points (a and c), and they must be returned to their original position b at the precise instant when the voltage is zero between the capacitor plates. The frequency of the pumping source is twice the resonant frequency of the tank circuit. If the phase of the pump input and the resonant frequency of the tank circuit are not correct, there will be attenuation of the tank circuit oscillations.

The device most frequently used to obtain the variable capacitance at microwave frequencies is the varactor diode. The varactor is a low-loss diode which has a highly nonlinear capacitance-voltage relationship. The equivalent circuit and capacitance variation as a function of bias voltage are shown in Fig. 11·24. The voltage-sensitive barrier capacitance C is in series with the small spreading resistance r. The figure of merit for the diode is referred to as the Q and is defined as

$$Q = \frac{1}{\omega C r}$$

The frequency at which the Q is unity is referred to as the *cutoff* frequency.

Fig. 11·24 Varactor capacitance variation as a function of bias voltage.

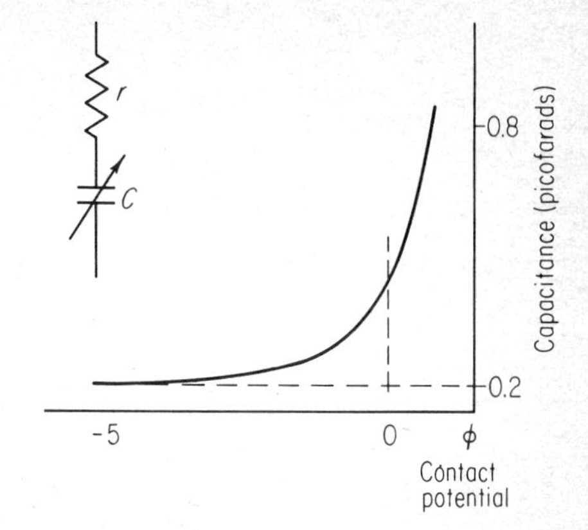

Diodes have been made with cutoff frequencies necessary to extend the parametric amplifier principle into the millimeter wavelength range.

The phase sensitivity of the circuit previously discussed can be overcome by adding another resonant circuit termed the *idler*. It is termed the idler circuit because there is no energy supplied directly to it from an external source. The idler frequency f_1 automatically adjusts to an in-phase condition so that work is done by the capacitance variation causing both signal and idler voltages to increase. Figure 11·25 illustrates the two resonant circuits coupled together by the capacitance varying at $f_p = f + f_1$. The tuned circuits are resonant at the signal frequency f and the idler frequency f_1. In the case where the pump frequency is equal to the signal frequency plus the idler frequency, the varying capacitance and idler look like a negative resistance to the signal, and the signal is amplified. If the pump frequency is

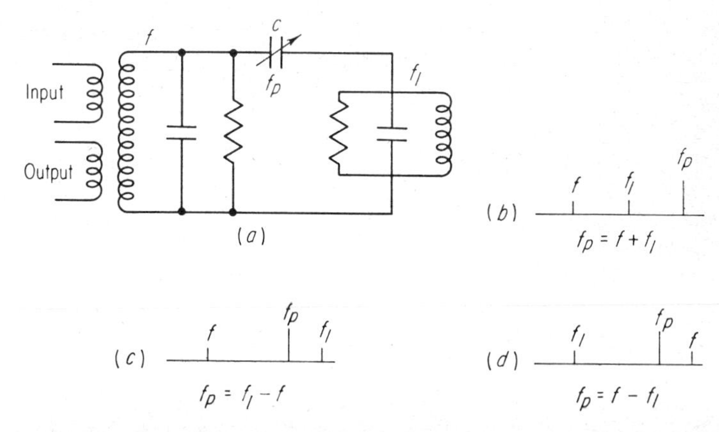

Fig. 11·25 Schematic representation of the three-frequency parametric amplifier (*a*) and the types of operation. (*b*) Sum-pumped parametric amplifier. (*c*) Up-converter. (*d*) Down-converter.

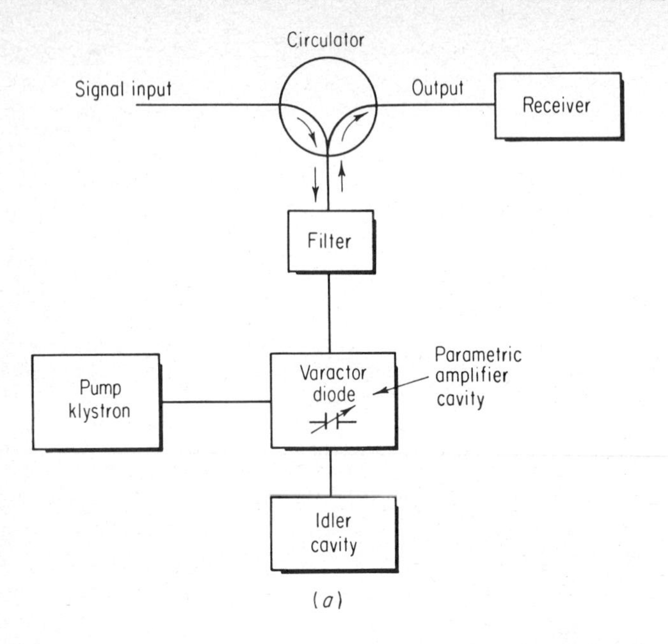

(a)

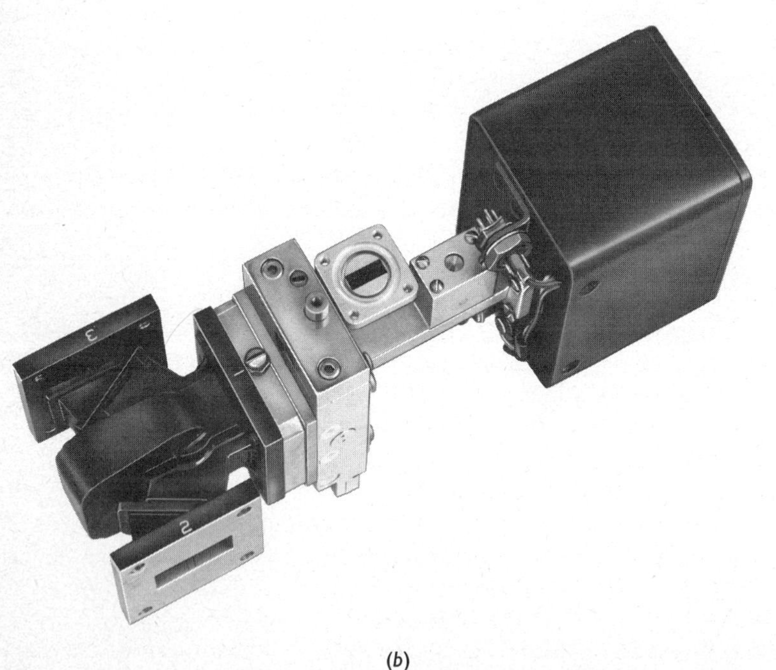

(b)

Fig. 11·26 (a) Block diagram of a typical parametric amplifier. (b) X-band nondegenerate parametric amplifier. (*Hughes Aircraft Company, Microwave Components Department, Aerospace Group.*)

equal to the difference between the signal frequency and the idler, then the capacitance and idler look like a positive resistance, and the input signal cannot be amplified.

The type of resonant circuit amplifier requiring the lowest pump frequency and power to operate, and the least complex to design, is the *degenerate* amplifier. This amplifier has its input circuit tuned to one half the pump frequency so that the signal and idler frequency are identical.

The *nondegenerate* amplifier is characterized by a pump frequency considerably higher than twice the signal frequency. Input and output are at the same frequency.

The *up-converter* parametric amplifier also has pump frequencies considerably higher than the signal frequency. For the case of the up-converter, $f_p = f_1 - f$ and, as previously indicated, the variable reactance and idler present an effective positive resistance to the signal so that the input signal cannot be amplified. However, the output can be taken at the idler frequency. The output frequency is higher than the input frequency. The conversion gain can be as high as the ratio of output to input frequencies. The opposite frequency ratio holds for the *down-converter* which experiences conversion loss.

A typical parametric amplifier is schematically illustrated in Fig. 11·26. The variable capacitance diode is placed in a resonant cavity. The idler is a resonant cavity coupled to the diode cavity. The filter is used to isolate the pump frequency from the output circuit. The circulator is used to isolate the diode from the input and output circuits.

In addition to the single cavity amplifiers, there are traveling-wave or multiple-cavity parametric amplifiers. They are comprised of a number of varactor diodes mounted in a series transmission line form.

11·6 The maser

The name *maser*[5] stands for *Microwave Amplification by Stimulated Emission of Radiation*. The magnetic moments of bound electrons in a gas or solid form the basis for the operation of the maser. The magnetic moment and precessional frequency of the electron were illustrated in the discussion of ferrites in Chap. 10. The motions of electrons in atoms and molecules located in a magnetic field are such that an electron can only exist in certain energy states or at certain energy levels. Atoms are capable of existing at any of several energy levels for brief periods of time. The vibrational energies may be illustrated by the energy level diagram of Fig. 11·27. The energy that is required to raise the atom from a lower to a higher energy level is directly related to the r-f frequency. If the r-f energy is applied at the frequency which corresponds to the precise difference in energy between two levels, then transitions of atoms from one energy state to the other will occur. If an atom in the lower energy state is placed in an r-f field of this frequency, it will absorb energy and hop to the upper state. On the other hand, an atom in the upper state will give up energy to the r-f field and drop to a lower level. Whether a system exhibits a net absorption or emission of energy depends on whether more molecules are in the lower or upper energy state. At thermal equilibrium there are more atoms in the lower states. The difference in populations increases as the energy spacings increase and also as the temperature is decreased. A population inversion can be obtained by application of a strong r-f signal (called the pump) at a frequency appropriate to the

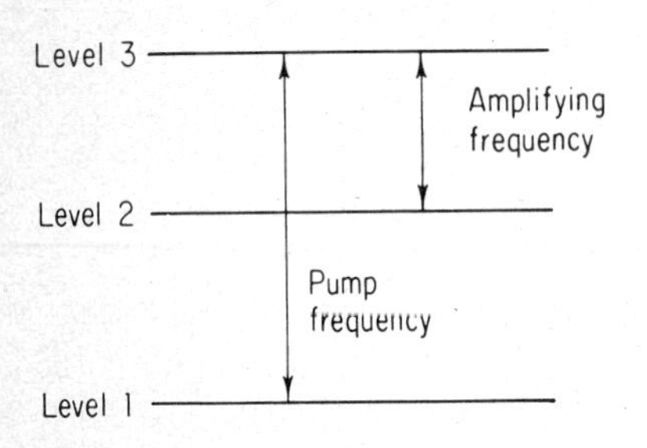

Fig. 11·27 Energy levels in a three-level maser material. The r-f frequency corresponds to the precise difference in energy between two levels.

energy difference between the lowest and highest levels. With this population distribution, the atoms which have been elevated to level 3 are induced to give up their energy by application of r-f energy at a frequency corresponding to the energy difference between levels 3 and 2. The induced emission is in phase with the applied signal and gives up energy to this field. There is no applied field between levels 1 and 2 so the atoms drop back to level 1 at a slower rate. This action is illustrated in Fig. 11·27.

The maser consists of the following:

1. A material which has the proper magnetic energy levels when placed in a d-c magnetic field
2. A source (pump) to provide r-f fields necessary for population inversion (pumping from level 1 to level 3, etc.)
3. Signal frequency to be amplified (in case of maser amplifiers)
4. Liquid helium or liquid nitrogen to cool the maser element down to a point where thermal processes will not mask the maser effect

Maser oscillators provide excellent frequency standards. The original ammonia maser constructed by Townes at Columbia University in 1955 had deviation less than one part in 100 billion, and the deviation for the hydrogen maser developed at Harvard in 1960 was less than one part in a million billion,[6] which amounts to an error of one second in 33 million years when used as a timing device.

An ammonia-type maser is shown in Fig. 11·28. The maser operation is based on the controlled flow of ammonia molecules, within a low pressure system, into a region of an electrostatic field. The electrostatic field is surrounded by a chamber which contains liquid nitrogen and provides a surface for condensation of all the ammonia molecules which impinge upon the surface of this chamber. The electrostatic field, commonly called the focuser, acts upon the ammonia molecules so that the molecules in the upper energy state of the ammonia inversion level are focused into a narrow beam which permits passage into a cavity operating in the TM_{01} mode. The molecules

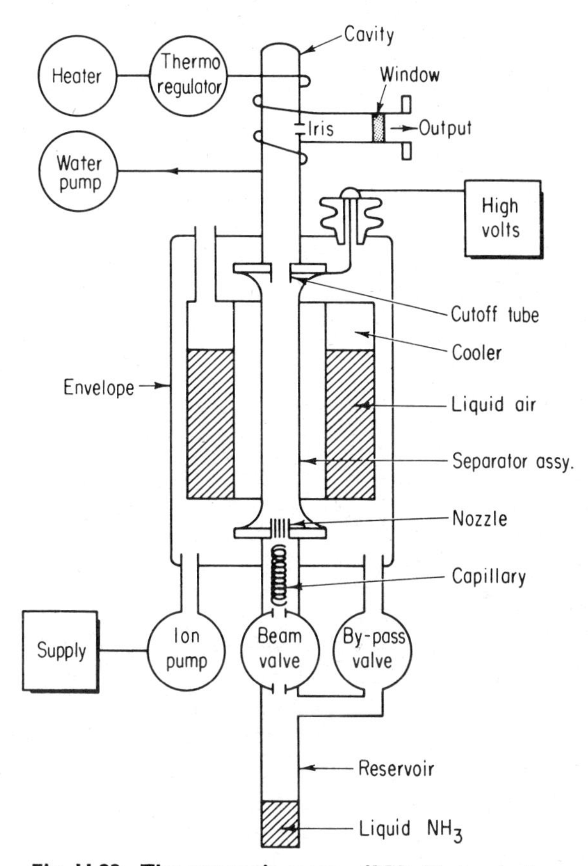

Fig. 11·28 The ammonia maser. (*PRD Electronics.*)

in the lower energy states are deflected away from the beam and condense on the walls of the cooling chamber.

When ammonia molecules enter the high Q cavity, transitions of the molecules to their lower energy states result in stimulated emission at a stable frequency. When a sufficient number of such molecules are present in the molecular beam, the resultant electromagnetic oscillations are sufficiently strong to become sustained as long as the molecular beam strength exceeds a critical minimum value. The ammonia may be recirculated to the reservoir after approximately 50 hours of operation.

The high vacuum is maintained within the system because of the incorporation of a getter-ion tube which continuously removes the rare gases and which is periodically flashed to remove remaining gases.

Stability of the cavity is assured by using Invar and by utilizing a temperature-control system which controls the temperature to within 0.02°C.

Liquid nitrogen must be added to the cooling condenser at the rate of ¼ pint every hour in order to maintain proper temperature conditions.

11·7 The laser

The name *laser* stands for *Light Amplification by Stimulated Emission of Radiation*. Lasers operate on the same physical principles as those already

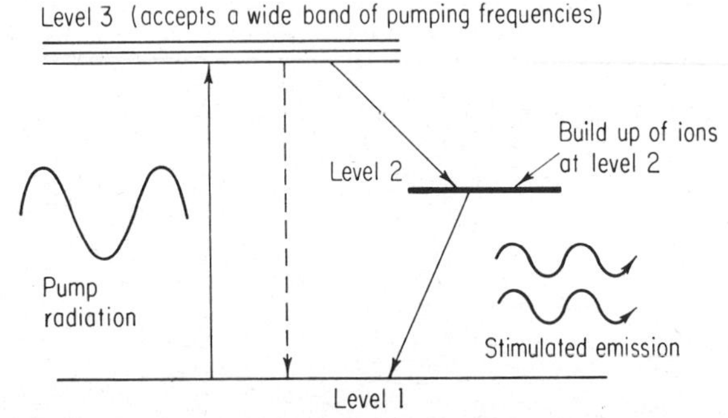

Fig. 11·29 Energy level diagram for laser action in ruby.

described for masers, but the range of operation extends to the infrared and visible light spectrum. In 1958, A. L. Schawlow and C. H. Townes[7] suggested that the maser principles could be extended to permit operation in the light spectrum. The first successfully operated laser was built by T. H. Maiman at the Hughes Aircraft Company in 1960. The active material used in this type of device is ruby, an aluminum oxide in which a few of the aluminum atoms have been replaced by chromium atoms. The single ruby crystal is in the shape of a cylinder. The ends are optically flat and silvered. One end is thinly coated and acts as an opaque mirror, while the other end has a non-transmitting coat of silver and acts as a reflecting mirror. A spiral flash tube, capable of high intensity light output, is placed around the ruby cylinder and emits an intense flash of white or green light when pulsed from a charged capacity bank. The light in the green region of the spectrum is absorbed by the ruby and raises the chromium atoms to an excited state from which two steps are required to carry them back to the ground state, as shown in Fig. 11·29. The use of light to raise the energy levels is referred to as *optical pumping*. After a finite elapsed time, some of the chromium ions at level 3 drop to level 2, and some drop directly back to level 1. There is a greater rate at which ions drop to level 2 and, in addition, they hold their energy at level 2 for a short time before dropping to level 1. In this way, the necessary population inversion, which is essential for producing stimulated emission, is obtained.

When the chromium ions drop from level 2 to level 1, they emit *photons* (considered as being bundles of light possessing wavelike properties) in all directions. A radiated photon tends to stimulate radiation of the same wavelength from other ions in its path. Upon release, the emitted photon falls

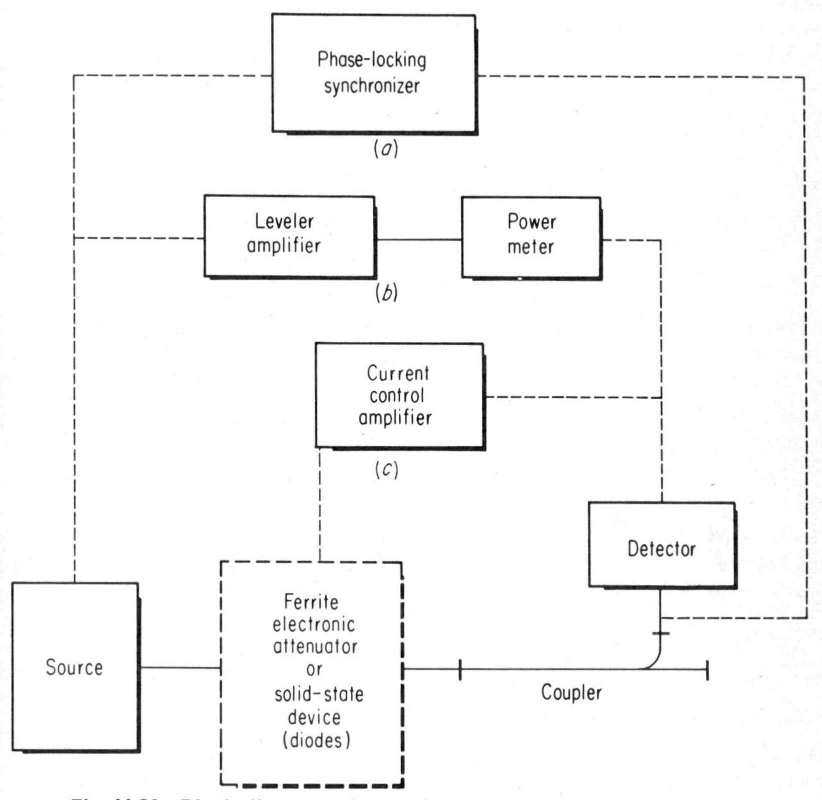

Fig. 11·30 Block diagram of representative stabilization techniques.

precisely in phase with the photon that triggered its release. Most of the emitted photons escape through the side wall of the ruby and are lost. The photons which travel parallel to the longitudinal axis of the cylinder are reflected by the silvered ends and reflect back and forth thousands of times. The photons continue to stimulate radiation from other ions during each passage, and eventually the light wave acquires enough energy to break through the escape port end of the ruby cylinder and emerges as a hot pulse of coherent light formed in an extremely narrow beam. The brightness of the beam may be millions of times that of the sun.

11·8 Stabilization of microwave power sources

The block diagram in Fig. 11·30 illustrates techniques which are representative of the many ways in which power and frequency stabilization of microwave sources may be obtained. A phase-locking system is used to obtain high frequency stability and spectral purity of microwave signal sources. The Dymec DY 5650A oscillator synchronizer is used to phase-lock a reflex klystron to the desired frequency between 1 and 12.4 Gc. A sample of the signal source is compared to a harmonic of a stable frequency which is generated in the synchronizer. The voltage output from the comparator circuit controls the electrode potentials of the source and maintains the necessary frequency stability. In this system, the source is operated c-w, and an electronically controlled modulating device must be used if the source is used in an audio-detection-type attenuation-measurement system.

The power meter and leveler amplifier shown at *b* can be used to stabilize the output of klystrons or backward-wave oscillators. Square-wave modulation may be applied to the leveler amplifier.

The technique shown at *c* can be used when the source is operated c-w or square-wave modulated. A current-control amplifier is used to control the output level with an electronically controlled ferrite attenuator or solid-state diodes.

Stabilization can also be obtained by using a high *Q* resonant cavity tightly coupled externally to the klystron. A major part of the source power is lost because of the high insertion loss of the cavity.

REFERENCES

1. Courtesy of the Hewlett-Packard Company, *Application Note 14*, Appendix I, 1955.
2. Courtesy of the Hewlett-Packard Company, *Application Note 12.*
3. A. W. Hull, *Phys. Rev.*, vol. 18, pp. 31–57, 1921.
4. G. R. Kilgore, Magnetron Oscillators for the Generation of Frequencies between 300 and 600 Megacycles, *Proc. IRE*, vol. 24, pp. 1140–1157, August, 1936.
5. J. P. Gordon, H. J. Zeiger, and C. H. Townes, The Maser—New Type of Microwave Amplifier, Frequency Standard, and Spectrometer, *Phys. Rev.*, vol. 99, pp. 1264–1274, August, 1955.
6. J. P. Gordon, The Maser, *Sci. Am.*, December, 1958.
7. A. L. Schawlow and C. H. Townes, *Phys. Rev.*, vol. 112, 1958.

CHAPTER

12

ATTENUATION MEASUREMENTS

Introduction. The purpose of this chapter is to discuss the structure, characteristics, and applications of a microwave system used in performing precision microwave attenuation measurements. The stringent equipment requirements for performing precision attenuation measurements are not always necessary when performing rough laboratory measurements, but a thorough understanding of the precision measurement system enables one to establish the system conditions necessary in order to perform measurements within a specified accuracy.

The determination of attenuation depends upon the *power ratio*, which can be found conveniently and accurately by several methods. A detailed explanation of all the available methods used in the measurement of attenuation requires more space than is available in this text. The methods discussed here are limited to audio and microwave (r-f) substitution, a combination of the two, and the i-f substitution method.

The commonest and most convenient system for measuring r-f attenuation is the audio-detection system, which consists of a square-wave-modulated r-f signal source, a detecting element, and a means to measure the resulting audio voltage. The source used in the present system is the reflex klystron, but the principles set forth concerning the modulation characteristics and proper tuning hold for other r-f sources. The detecting element is a barretter or crystal, depending upon the particular method of measurement. The audio voltage indication is obtained by using a linear amplifier and voltmeter.

The procedure for performing attenuation-type measurements using the audio and r-f substitution techniques, as well as a combination of r-f and audio-substitution techniques, will be discussed.

The measurement of directional coupler directivity characteristics and the use of high directivity couplers in precision reflectometer systems is covered in detail. This is basically a study of reflections; a thorough understanding of the principles and practical applications set forth is extremely useful to the person involved in the evaluation of microwave systems.

12·1 Attenuation

Attenuation is defined as the ratio of power absorbed by the load without the network in the line (P_1) to the power absorbed by the load with the network in the line (P_2) *when the signal source and load are perfectly matched.*

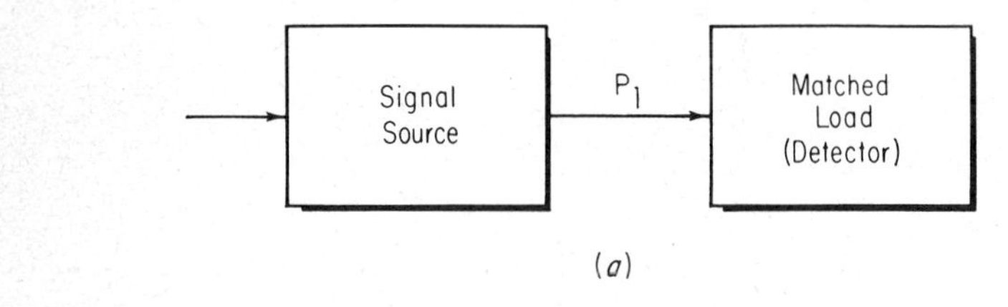

(a)

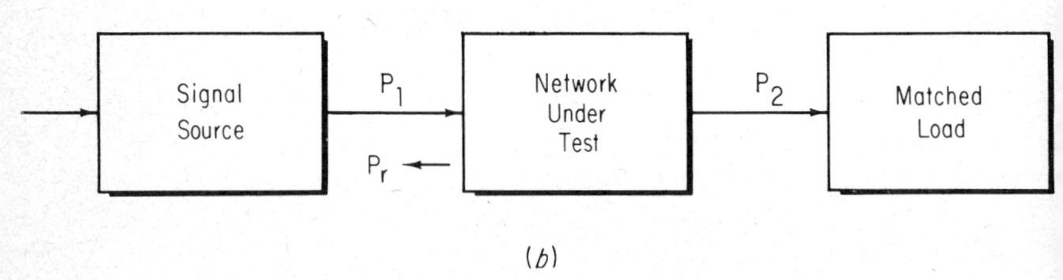

(b)

Fig. 12·1 Attenuation and insertion loss variables.

Attenuation is expressed as

$$A = 10 \log \frac{P_1}{P_2} \tag{12·1}$$

Under the conditions of a matched source and load, the attenuation measurement includes the dissipation losses in the network and the reflection losses caused by the mismatch between the network and the source. This is illustrated in Fig. 12·1*a*, where the matched load is a matched detector when performing an actual measurement.

Another term used to describe the dissipation and reflection losses caused by a network is *insertion loss*. Insertion loss is defined as the ratio of power absorbed by a load before and after the network is inserted in the line. It is noted that the definitions of attenuation and insertion loss are the same with the exception of the specific condition of a matched source and load for the definition of attenuation. The insertion loss measurement therefore includes losses caused by mismatched source and load. When measuring networks which have low dissipation losses in a mismatched system, it is possible to obtain a positive or negative insertion loss value depending upon the phase relationship of the reflections from the mismatched source and load. There seems to be a general exception to this definition of insertion loss. The minimum attenuation of variable attenuators, the forward loss of isolators, the main line loss of couplers, and the loss of phase shifters are often referred to as insertion loss.

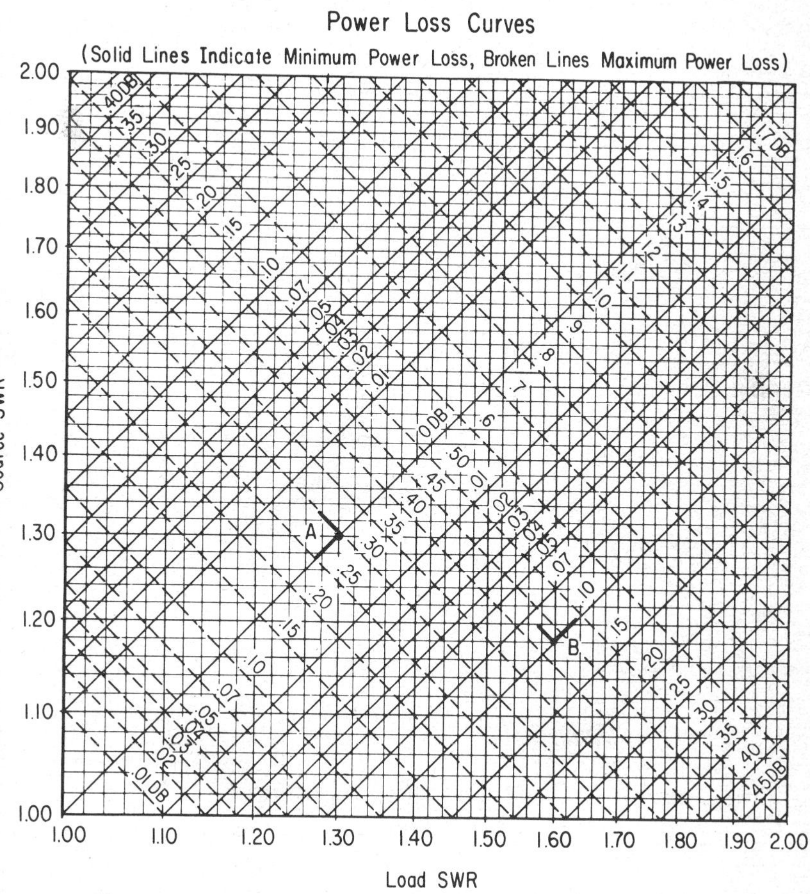

Fig. 12·2 Mismatch loss curves. (*Hewlett-Packard Company.*)

The *mismatch loss* is defined as the difference between the power absorbed by the network if it were perfectly matched to the power absorbed by the network with its existing mismatch. The mismatch loss shown in Fig. 12·1*b* is expressed in decibels

$$A_m = 10 \log \frac{P_1}{P_1 - P_r} \tag{12·2}$$

Mismatch loss curves are shown in Fig. 12·2. If the relative phase of the two mismatches is such that they cancel, a minimum standing-wave ratio occurs and there is a minimum loss. Correspondingly, a large standing-wave

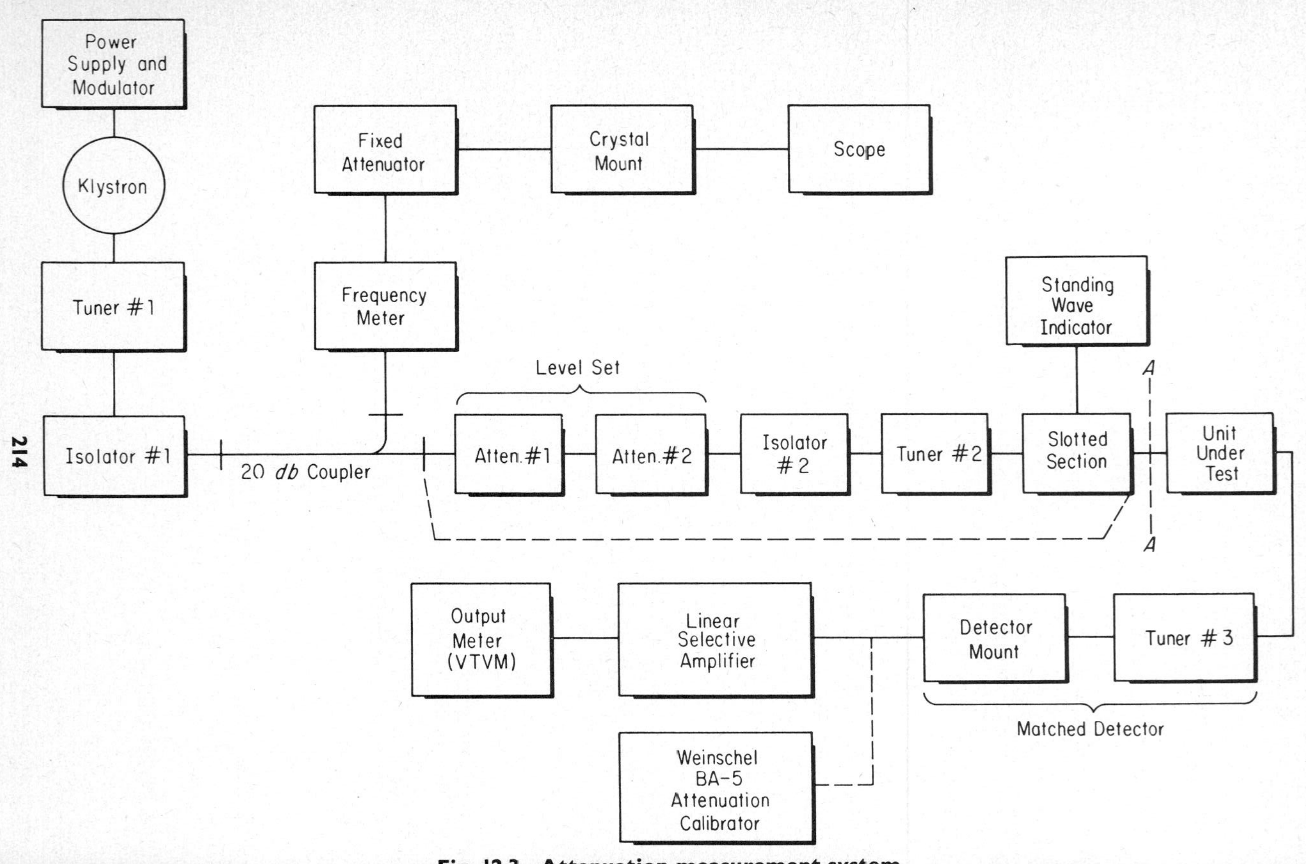

Fig. 12.3 Attenuation measurement system.

ratio exists and there is a maximum loss when the source and load mismatches add in phase. Examples of the range of loss due to source and load VSWR are shown in Fig. 12·2. At *A*, a source and load VSWR of 1.3 results in a minimum error of 0 and a maximum error of approximately 0.295 db. A source VSWR of 1.18 and a load VSWR of 1.6 result in a minimum error of 0.10 db and a maximum error of 0.43 db, as indicated at *B*.

12·2 The attenuation measurement system

A block diagram of an attenuation measurement system is shown in Fig. 12·3. *This basic system can be used at all frequency bands where circuit components of the desired characteristics are available.* The same basic system is used when performing attenuation measurements in a coaxial system *at any frequency* where the necessary components are available.

The power supply provides the necessary signal source potentials and modulation signals.

Tuner No. 1 is used to tune the source for maximum output power. This tuner is not necessary if the source output power is sufficient for the desired range of measurement.

Isolator No. 1 prevents source pulling when signal levels are varied with the level set attenuators. An isolator which provides a minimum reverse attenuation of 20 db is usually used here, although 10 db of isolation is sufficient when the level set attenuators are well matched.

The 20-db coupler, frequency meter, pad, crystal detector, and scope form the signal-monitoring portion of the system. It provides a means of properly tuning the source to a proper frequency and provides an all-important means by which any changes in signal source operation are easily observed. The pad or isolator placed between the frequency meter and crystal detector is necessary in cases where the mismatch of the crystal detector causes frequency pulling of the frequency meter and is only necessary in cases where a precise frequency setting is required.

The two variable attenuators are used to change power levels in the system. These attenuators should have low input and output VSWR characteristics over the complete range of variation. A precision-calibrated attenuator is recommended as one of these level set attenuators.

Isolator No. 2 prevents any change in the source VSWR due to variations of the level set attenuators. The source VSWR is the VSWR looking back toward the generator from the point in the circuit where the unit under test is to be connected.

Tuner No. 2 is used to tune out the source VSWR, which is the VSWR looking back into isolator No. 2. This tuner can be eliminated if the output VSWR of isolator No. 2 is such that it does not affect the required accuracy of measurement. The possible errors in measurement due to source and load mismatch are shown in Fig. 12·2.

Tuner No. 3 and the detector mount with its element form the matched detector.

The linear amplifier provides bias for the detecting element in the case of the barretter, and amplifies the audio signal from the detecting element (barretter or crystal). The amplified signal is then applied to the output voltmeter (VTVM) or other indicating device. In addition, there are many laboratory applications where the standing-wave indicator can be used as the amplifier, bias source, and output-indicating device.

12·3 Oscillator tuning

The reflex klystron is used in the following explanation of oscillator tuning, but the tuning principles set forth apply to other modulated or c-w sources.

In order to perform accurate attenuation measurements, a highly stable oscillator is needed as a signal source. This places stringent requirements on the power supplies used with high-frequency oscillators. Water cooling, oil baths, and special stabilization circuits have been successful in obtaining stable operation with the reflex klystrons.

Considerable care must be exercised in tuning the klystron so that it will operate at the peak of its mode curve where any shift in the operating voltages will cause least variation in the output power. The klystron is square-wave modulated in order to prevent frequency modulation. The modulation frequency is variable so that it can be adjusted to the fixed frequency required by the bolometer amplifier.

The applied potentials and modulation circuitry for a reflex klystron are shown in the simplified schematic representation of Fig. 12·4*b*. The repeller voltage characteristics of the klystron are shown in Fig. 12·4*a*. In this example, as the d-c repeller voltage (without modulation) is varied from -600 volts to -100 volts as indicated by the sloped line, the power output and frequency characteristics of the klystron modes are plotted as indicated. The dip in each mode represents the reaction of the frequency meter shown in the monitoring portion of the attenuation measurement system.

The present measurement system requires that the klystron be modulated at 1,000 cps. This modulation is obtained by turning the klystron on and off at this 1,000 cycle rate. In the ideal case, the klystron oscillates at a *single frequency* when turned on. This places specific requirements upon the characteristics of the square-wave modulation signal.

1. The square wave must have a fast rise time. If the rise time is not fast enough, the klystron will oscillate at other frequencies before the total modulation voltage is applied. A practical example of this is shown by the spikes on the square-wave output indication of Fig. 12·4*a* just to the right of the $3\frac{3}{4}$ mode curve. The finite rise time causes oscillations coming to the top of the mode and again when leaving the mode.

2. The square wave must have a constant amplitude. The constant amplitude is obtained by clamping the modulation voltage to the beam voltage, as shown in Fig. 12·4*b*, or by an appropriate clipping circuit. The author has incorporated this particular circuit in several klystron power supplies.

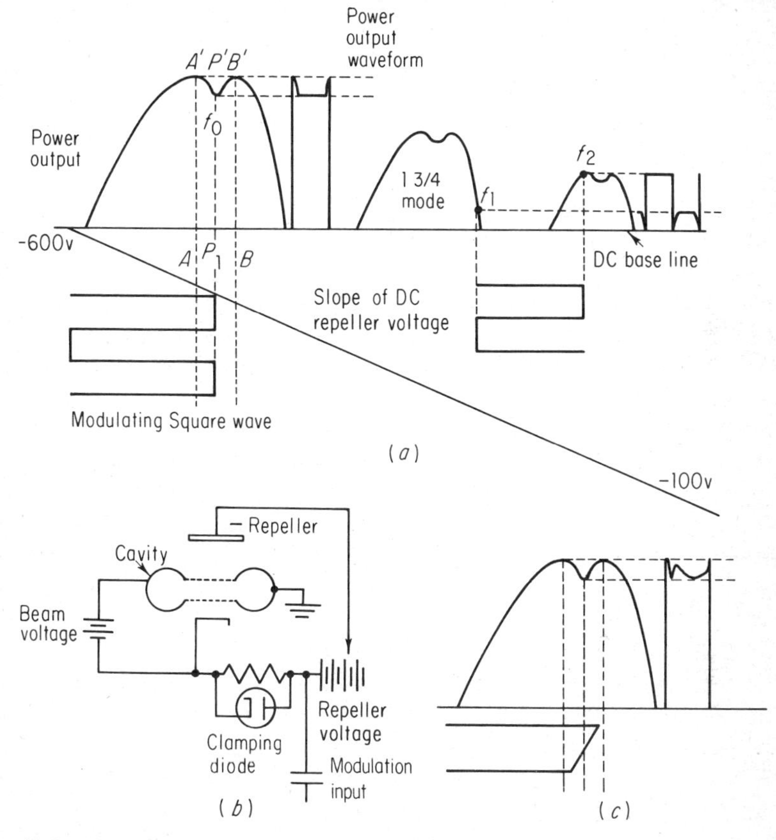

Fig. 12·4 Klystron power output versus reflector voltage is illustrated at (*a*). The proper setting and indication for square-wave modulation is shown on the ¾ mode, and modulation in two modes is shown for the case of modulation waveform being too large. A schematic of the basic operating potentials for the klystron is shown at (*b*). Frequency modulation due to poor square-wave-modulating waveform is indicated at (c).

Since only one side of the square wave is clamped, the repeller voltage is adjusted to obtain modulation with the clamped side of the square wave. When the modulation is turned off, the klystron oscillates c-w, and only a slight adjustment of the repeller voltage is required to obtain maximum power output (oscillation at the top of the mode). This circuit also provides a measure of protection in cases where a high beam voltage is suddenly applied to the klystron. Sudden application of the beam voltage causes a high current flow in charging the coupling capacitor in the modulation circuit. A large current flow through the coupling resistor could cause the repeller to go positive if the repeller voltage was set at a low value, resulting in possible klystron damage. Figure 12·4*c* shows the case where the top of the square wave is not flat. The resulting frequency modulation is evident by observing the square-wave modulation on the scope. This is avoided in some commercial power supplies by clipping both sides of the modulating square wave.

3. The square-wave amplitude must be sufficient to obtain oscillation at the top of the mode and to completely cut the tube off, as shown in Fig. 12·4*a*. If the square wave is too low in amplitude, the klystron could be oscillating at two frequencies in the same mode. This problem is easily recognized by using a scope with a d-c base line. When the tube is off, the lower level is at the base line, but when the square-wave amplitude is too small, the square wave will move up from the base line, indicating a power output. Many possible waveforms can be obtained with this type of problem.

   If the square wave is too large in amplitude when using a lower mode, it is possible to obtain a signal such as the one illustrated in Fig. 12·4*a*, in which a frequency f_1 is obtained in the 1¾ mode and another frequency f_2 is obtained in the 2¾ mode. If the base line on the drawing represented the base line on the scope, the frequency f_2 is noted as being up from the base line, thus indicating that the tube is oscillating at two frequencies.

In general, microwave sources are quite expensive and in many cases may be easily damaged. Therefore, the user is required to know the characteristics of the source, the value of the necessary operating potentials, and the procedure to be followed when applying and adjusting the operating potentials.

In the case of the klystron considered, the procedure for applying the potentials may depend upon the particular switching arrangement of the power supply. It is always necessary to make sure that the repeller voltage is available. The repeller voltage must be applied to the tube before the beam voltage is applied. Usually the repeller voltage is applied and set to the approximate required value during the warm-up; then the beam voltage is turned on and set to the required value. The repeller voltage is then adjusted for proper oscillation of the klystron. In the case of square-wave modulation, the repeller voltage and square-wave amplitude are adjusted to obtain maximum power output. If the klystron is to be set to a particular frequency indicated by the frequency meter, the klystron cavity is adjusted to bring the oscillation frequency to the required value.

The proper point of oscillation is adjusted as indicated on the ¾ mode in Fig. 12·4*a*. The klystron mode is shown as it would appear if a sawtooth voltage were applied. The square-wave modulation signal is shown on top of the d-c repeller voltage at P_1. A dip in the output square-wave amplitude, as indicated on the scope, is noted when the frequency of the klystron is tuned to the frequency-meter resonant frequency. When the d-c repeller voltage is adjusted from A to B through the point P_1, the output square-wave amplitude varies from A' to P' to B' as shown. If the frequency meter reaction is at the top of the mode, the square-wave output amplitudes will be equal at A' and B'. This is the method by which the klystron frequency is adjusted to the proper value indicated by the frequency meter reaction. If the amplitude of the output square wave is not equal on each side of the reaction (dip), then the klystron cavity must be adjusted until equal amplitudes are obtained as the repeller voltage is adjusted back and forth through the reaction (dip).

12·4 Source and load tuning

The basic representation of the unit under test placed between the source and load is shown in Fig. 12·1*b*, and the requirement of matched source and load was discussed in Sec. 12·1. In the attenuation measuring system, the source for the unit under test is the output flange of the slotted section, and the load is the matched detector. The source VSWR is the output end VSWR of isolator No. 2, assuming that the residual VSWR of the slotted section is negligible. Tuner No. 2 is used to tune out this source VSWR when the required accuracy of measurement requires a matched source. In order to tune out this VSWR, the signal power must be fed back in the direction of the source. This is sometimes accomplished by placing a waveguide switch in the system and using the necessary coaxial cable and adapter to send power back into the system output, as indicated by the dashed line in Fig. 12·4. When this technique is used, the level set attenuation must be set to a high value to make sure that no signal flows in the line from the direction of the source. The high isolation of the waveguide switch is not always sufficient, in which case tuner No. 2 would be used to tune out a leakage signal instead of the actual isolator VSWR. This could result in a very high source VSWR and errors in system-measurement accuracy.

When the waveguide switch and cable assembly is not incorporated in the system, the following method is recommended. Disconnect the section of the system to the right of the first level set attenuator, turn the section around, and feed power into the normal output end of the slotted section. Tuner No. 2 is then adjusted to tune out the VSWR looking into the output end of isolator No. 2. There should be no change in VSWR as the level set attenuator No. 2 is varied over its complete range.

After the source tuner is adjusted, the tuned section is turned around and reconnected for normal system operation.

Coaxial Slotted Section. When the previous source tuning is performed in a coaxial system using a coaxial slotted section, the source VSWR should not be interpreted as being the VSWR as read by the slotted section when tuner No. 2 is tuned. The slotted-section probe examines the reflection *beyond* the probe in the direction of power flow. When the slotted section is again connected into the system to perform measurements, it is noted that the actual source presented to the unit under test is the output connector of the slotted section. The VSWR of the connector has not been tuned out, and for some coaxial slotted sections it may be considerable. The output connector VSWR must be established in order to quote the actual source mismatch.

12·5 Detector tuning

The bias current for the detecting element (barretter) is set to the proper value with the current adjustment on the amplifier. Tuner No. 3 is adjusted so that it does not protrude into the waveguide. The power in the main line is adjusted to obtain an indication on the output meter, and the detector, if adjustable, is adjusted for maximum output as indicated on the output indicating meter. The adjustment is secured in position, and tuner No. 3 is used to tune out the detector mismatch. The detector and tuner adjustments are secured, and this combination now forms the *matched detector*. The former tuning procedures must be repeated each time the source frequency is changed. As previously indicated, these tuning procedures can be eliminated if isolator No. 2 has a low output VSWR. Tuner No. 3 and the tuning adjustment can be eliminated by placing a *well-matched* isolator in place of tuner No. 3. In some frequency bands, the detector mismatch is high enough to require an isolator or fixed attenuator between the tuner and detector.

12·6 Bolometer amplifier and output meter

In the discussion of barretter operation it was pointed out that even though the barretter has a long time constant it can respond to low audio frequencies. The r-f signal is audio modulated, and since the barretter is used as a square-law detector, the audio voltage developed across the barretter because of the incremental resistance change is directly proportional to the r-f input power. The audio voltage is amplified and read on the output indicating device.

The amplifier which supplies the bias for the barretter and amplifies the audio voltage is usually referred to as a *bolometer amplifier*. The basic bias source was shown in Chap. 8, Fig. 8·13. The high-impedance bias circuit insures constant current operation, and the audio-bypass capacitor is necessary in order to obtain maximum audio signal across the transformer input of the amplifier. The amplifier usually incorporates a selective filter (twin tee) in a feedback circuit to obtain narrow bandwidth and low noise characteristics. The center frequency of the tuned amplifier is usually 1,000 cps, which requires that the source must be modulated at the same rate. The amplifier has high gain and provides linear amplification.

The amplifier output is applied to the output meter. Ballantine Model-320 voltmeters (precision calibrated) are frequently used in this application. The meter scales are calibrated for linear detection (direct voltage measurements) which gives 20 log V_1/V_2, and since the barretter is a square-law detector, the Ballantine decibel reading must be divided by 2. The Model 320 full-scale reading is 5 db instead of the indicated 10 db. Meters such as the standing-wave indicator are calibrated for use with square-law detectors and provide direct decibel readings.

A significant measurement error occurs if the amplifier input resistance is equal to the barretter resistance. These resistances are in series in the circuit, and the audio voltage developed is divided between the resistances. The average resistance of the barretter varies from one reading to the next because of the change in power level, and the voltage to the amplifier is not proportional to the voltage generated by the barretter. If the input resistance is much greater than the barretter resistance, it can be seen that the average resistance change of the barretter is small compared to the total series resistance, and the voltage change across the amplifier input is not significant.

The gain control of the bolometer amplifier is usually kept at a maximum setting, and output levels are adjusted using the r-f attenuators. The current adjustment is used to set the barretter current to the proper value. The particular instructions concerning connecting and disconnecting the detecting element must be followed in order to prevent damage to the detecting element.

12·7 Attenuation measurements using the bolometer amplifier-voltmeter method

The basic techniques of measuring attenuation by the power ratio method are set forth in this section. A block diagram of the attenuation measuring system is shown in Fig. 12·3 and is to be referred to in the present discussion.

The basic measurement is performed by setting up a level on the output voltmeter without the unknown in the system and then measuring the change in output when the unknown is placed in the system. The attenuation of the unknown is the difference between the initial and final voltmeter readings.

Since the barretter is subject to burn out at low r-f power levels, the r-f level set attenuation should be set to a high value of attenuation prior to turning on the r-f source. With the equipment set up as shown in Fig. 12·3, the proper potentials are applied to the source, and the necessary adjustments are performed in order to obtain the required 1,000 cps modulated source signal (usually 1,000 cps square-wave modulation).

The amplifier gain is set to a maximum, the barretter bias is adjusted to the proper value, and the source and load mismatches are tuned out as outlined previously.

The voltmeter chosen for this discussion is the Ballantine Model 320 VTVM. The voltmeter range that may be used in practice is limited by nonlinearity of the barretter at the higher power level and by noise at the low power level. For power ratio measurements up to 20 db the voltmeter can be used over the range from the top of the 32-volt scale. This range is convenient when measuring fixed devices which have attenuation values up to 20 db. A safe range usually extends downward from the top of the 10-volt scale on the voltmeter. The range of voltmeter operation is therefore chosen as extending from the top of the 10-volt scale to the top of the 0.32-volt scale covering three scale ranges, 10 volt, 3.2 volt, and 1 volt. The 0.32-volt range is used for measurements between 45 and 50 db as explained later. In order to obtain greatest accuracy, a precise calibration should be established for each dial division on the voltmeter.

WARNING: The noise level at the output should be kept below 3 mv. In an actual measurement system it is often necessary to reverse the power cords on various system instrumentation equipment in order to obtain this low noise level with no r-f power input to the barretter. It is sometimes necessary to isolate the separate ground lead on equipment which is provided with a third lead on the line cord. Also, excessive noise due to a waveguide-system ground loop can be eliminated by placing a thin dielectric insulator such as mica somewhere in the system to break the ground loop.

12·8 Fixed attenuator measurements

The actual measurement of fixed attenuation devices (less than 15 db) is performed as follows:

1. Connect the matched detector at *A-A*. Adjust the level set r-f attenuation to obtain a full-scale reading on the 10-volt range of the voltmeter.
2. Place the component under test at *A-A* as shown in Fig. 12·3 and record the change in power level. One-half the difference between the reference level and the final voltmeter reading gives the power attenuation of the component under test.

Measurements of attenuation values between 15 and 30 db are performed by a partial r-f substitution technique as follows:

1. Set level attenuator No. 1 to zero.
2. Connect the matched detector at *A-A* and adjust level set attenuator No. 2 to obtain a full-scale reading on the 10-volt range of the voltmeter.
3. Level set attenuator No. 1 is set to 15 db by decreasing the power level to obtain a full-scale reading on the 0·32-volt scale. Use level set attenuator No. 2 to return the output meter reading to full scale on the 10-volt range. (15 db of attenuation is now stored in attenuator No. 1.)
4. Place the component under test at *A-A* as shown in Fig. 12·3.
5. Remove the 15-db attenuation by returning level set attenuator No. 1 to zero dial reading. The value of attenuation of the component under test is computed by adding this 15-db value to the additional drop in output

as indicated on the output voltmeter. When the value of attenuation is greater than 30 db, then it is necessary to store 30 db in attenuator No. 1 using the same procedure.

12·9 Variable attenuator measurements

When measuring variable attenuation devices, the loss at zero dial reading (insertion loss) is obtained by the procedure set forth for fixed attenuation devices. The variable attenuator can be calibrated using the following procedure:

1. Set the dial of the unit under test to zero dial reading.
2. Adjust level set attenuators to obtain full-scale indication on the 10-volt scale of the voltmeter.

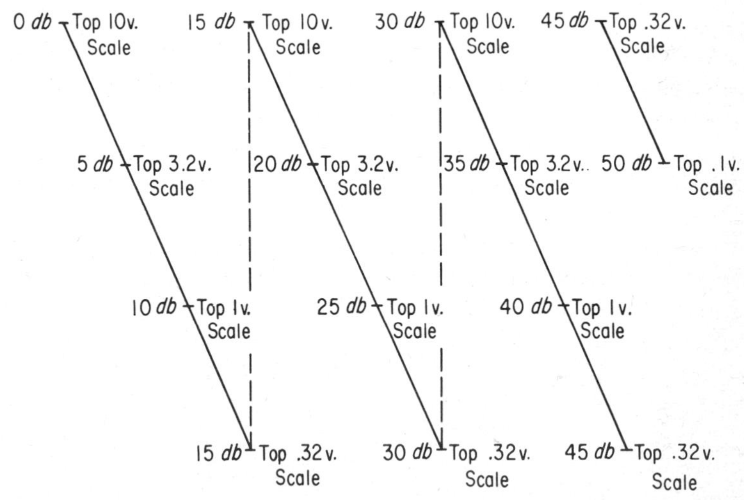

Fig. 12·5 Illustration showing voltmeter scale uses in measuring attenuation up to 50 db. The dotted lines indicate insertion of microwave power to return the power level to the top of the 10-volt scale.

3. Adjust the attenuator under test to obtain the desired incremental change in output power level as indicated on the output voltmeter and record the corresponding dial reading of the attenuator.
4. Repeat this procedure until a decrease of 15 db is reached (top of 0·32-volt scale); then adjust the level set attenuators for full-scale reading on the 10-volt range.
5. Adjust the attenuator under test in the desired increments for another 15 db; then adjust the level set attenuators to return to the top of the 10-volt scale. The attenuator under test can now be varied to obtain calibration values up to 50 db by decreasing the output level to a reading at the top of the 0.1-volt range of the voltmeter. The procedure set forth above is illustrated in Fig. 12·5, and the technique can be applied to other

voltmeters with different ranges and scale lengths. This particular technique is used in order to operate the barretter over a small range in power level, thus preventing nonlinear effects due to non-square-law characteristics.

12·10 Audio substitution measurement technique

In the *audio substitution* method of measurement a calibrated audio-attenuation standard is used to maintain a constant level at the output indicator with the unknown in and out of the line. The output level is obtained at the indicator with the unknown out of the system; then the unknown is placed in the system, and the necessary attenuation is removed from the audio standard attenuator to obtain the original reference level. Since the input power to the barretter is changed when the unknown is placed in the line, the accuracy of measurement depends upon the response of the barretter to the audio-modulated r-f signal. The square-law response characteristics of the barretter were discussed in the previous sections where it was pointed out that the decibel change on a linear output voltmeter is divided by 2 in order to obtain the r-f power level change in decibels when using a square-law detector. In the present measurement the value of the unknown (r-f attenuator) is obtained by taking one-half the value of attenuation change of the standard attenuator. This is necessary because it is a-c voltage input to the standard attenuator which is proportional to the r-f power level (the same as in the bolometer-amplifier voltmeter method).

Several commercial attenuators can be used as calibrated standards. In order for the calibration of the attenuator to hold, the input impedance to the amplifier following the standard attenuator should be adjusted to the characteristic impedance of the attenuator, and the impedance looking back from the standard attenuator must be matched. In case of mismatches, appropriate corrections must be made.

Since the audio attenuator is inserted between the barretter and the amplifier, it is independent of the linearity of the amplifier.

The Weinschel BA-5 attenuation calibrator shown in Fig. 12·6 incorporates the required precision audio attenuator, bias for the barretter, output amplifier, and indicating output meter.

The barretter bias-current adjust and current meter are located on the front panel. The precision attenuator readings are in terms of the square-law detector. Therefore, the decibel readings correspond to the decibel changes in microwave power (provided the detector is operating square law).

The partial block diagram of the original attenuation measurement system is shown in Fig. 12·7 and will be considered in explaining the single channel method of series substitution. The general procedure for performing attenuation measurements is as follows:

1. The source is adjusted to the desired frequency, the modulation frequency

Fig. 12·6 Weinschel BA-5 attenuation calibrator. (*Weinschel Engineering.*)

is adjusted to the proper value, and the necessary source and load tuning is accomplished.

2. The r-f power level at the barretter is set to 200 μw with the unit under test out of the system. This value is obtained by setting all attenuators on the BA-5 calibrator to zero, setting the gain controls for minimum gain, and then adjusting the r-f power level with the level set attenuator to obtain full-scale reading on the BA-5 output meter.
3. The precision attenuators are then set to a value *greater than* the anticipated value of the unknown to be measured. The gain on the BA-5 attenuation calibrator is then increased to obtain a convenient reference level (specific instructions are given in the BA-5 instruction book).
4. The unit under test is now placed in the system, causing a decrease in the output reference level. The precision attenuators of the BA-5 are adjusted

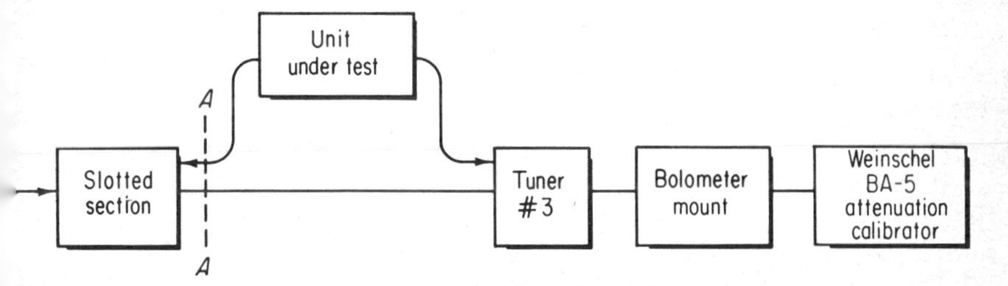

Fig. 12·7 Measurement section of a single-channel attenuation measurement system using the Weinschel attenuation calibrator.

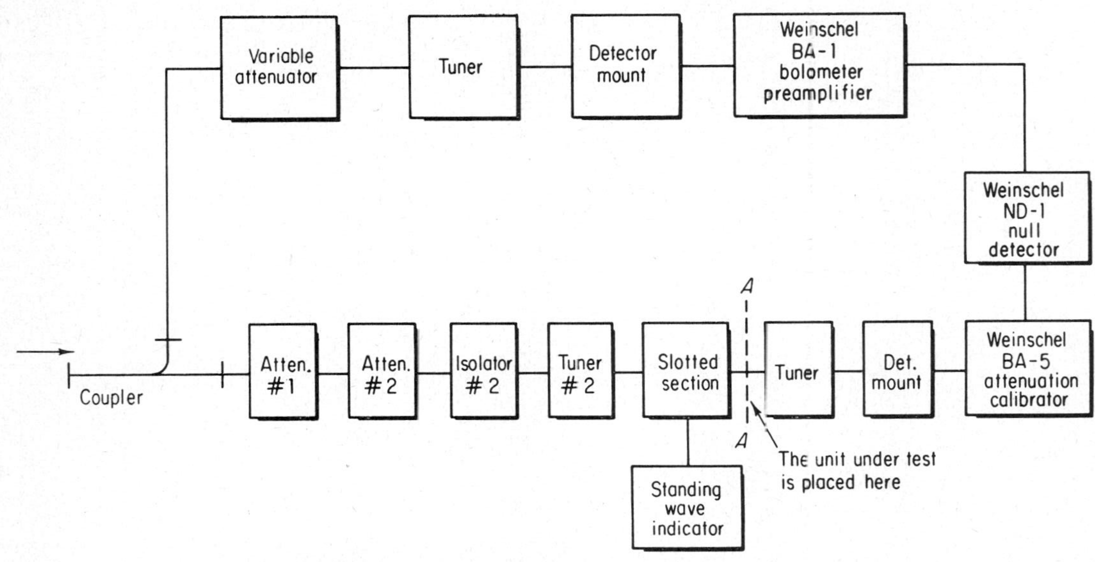

Fig. 12·8 Dual-channel attenuation measurement system.

to return the output level to the same reference point, and the value of attenuation is recorded.

The BA-5 attenuation calibrator can be used over a one-step range of 30 db. The 30-db range limitation is determined by the square-law response of the barretter used and the internal noise of the instrument. The BA-5 is usually used over a one-step range of 20 db. When measuring fixed attenuation values greater than 20 db, the same technique of partial r-f substitution as outlined under the bolometer amplifier-voltmeter method can be used. Also, variable attenuators can be calibrated using the same technique set forth using the Ballantine 320 voltmeter, but in this case the range of operation is 20 db at a time instead of 15 db, i.e., the variable attenuator is calibrated to 20 db, then r-f power is adjusted to return to the original reference level when the 20 db is set back in on the level set attenuators (200 μw is again applied to the barretter).

12·11 Dual channel system

Weinschel Engineering developed a dual channel system which eliminates inaccuracy of measurements caused by source instability. A block diagram of the system is shown in Fig. 12·8.

The lower channel is the original single channel system. The proper levels can be obtained as explained for the single channel case with the unit under test out of the system. The reference channel is then adjusted to obtain a null on the null detector. The unknown is placed in the system, and the BA-5 attenuation calibrator is then adjusted to obtain the null. The value of attenuation is recorded from the BA-5 precision attenuator readings. Since both channels operate at the same power level, any change in the source affects both channels the same.

12·12 Microwave substitution method

In the microwave substitution method, the microwave power at the detector is maintained at a constant level with the aid of a calibrated precision microwave attenuator. The calibrated standard attenuator is placed in series with the unknown and can be located in the generator section or the detector section. It is preferable to locate the standard attenuator at a point where a good match is observed in both directions. The unit under test is placed in the detector section at *A-A* as indicated on the system block diagram. Since any mismatch existing between the attenuator under test and the line results in an attenuation, it is not advisable to place the attenuator under test and the standard attenuator adjacent to each other in the detector section because of the interaction of the two possible mismatches.

The measurement of a fixed attenuation device can be performed by either of the following techniques:

1. *a.* The unknown is placed at *A-A* in the system, and the standard microwave attenuator is set to zero.
   b. The output reference level at the detector, as indicated on the output voltmeter, is set using the level set attenuator.
   c. The unknown is removed from the system. Caution: The microwave attenuator is set to a high value of attenuation in order to prevent damage to the detecting element.
   d. The detector is connected at *A-A*, and the output reference level is obtained by adjustment of the standard attenuator. The attenuation value of the unknown is read on the standard attenuator.
2. *a.* The microwave standard attenuator is set to a value greater than the anticipated value of the attenuator under test.
   b. With the detector connected at *A-A*, an output reference level is obtained by adjusting the level set attenuators.
   c. The unknown is placed at *A-A*, and the output reference is again obtained by adjustment of the standard attenuator. The difference between the initial and final readings of the standard attenuator gives the value of the unknown attenuation device.

Calibration of variable attenuation devices is obtained by setting the standard attenuator to a high value and obtaining an output reference level with the variable device set to zero. Attenuation is inserted by the variable device under test and alternately removed with the standard attenuator. The changes in the standard attenuation values are recorded as the calibration values for the variable device under test.

The response law of the detector does not have to be known and there are no limitations due to considerations of linearity of the detector when using the microwave (r-f) substitution technique because the power level into the detector remains constant. Crystals are used in this method of measurement especially when high values of attenuation are being measured. This is due to the higher sensitivity of the crystal detector compared to the barretter.

12·13 I-f substitution method

In the *i-f substitution* or *heterodyne* method, the r-f signal is applied to a frequency converter, and the resulting i-f frequency is channeled through an i-f standard attenuator into an i-f receiver which is used as the detector. In this system, the change in detected output when the unknown microwave attenuator is inserted in the system is compensated for by adjusting the i-f standard attenuator. The measurement procedure is essentially the same as that described in the previous section.

12·14 Directivity measurements

The directivity is a measure of the discrimination between waves traveling in two directions in the main line of a directional coupler as defined in Chap.

8. Directivity is measured as the ratio of two power outputs from the auxiliary line when a given amount of power is alternately applied in the forward and reverse directions in the main line of the coupler.

For the ideal case of a *perfect* matched termination and no reflections from the flange connection, the directivity measurement would consist of the simple measurement illustrated in Fig. 12·9 and outlined below.

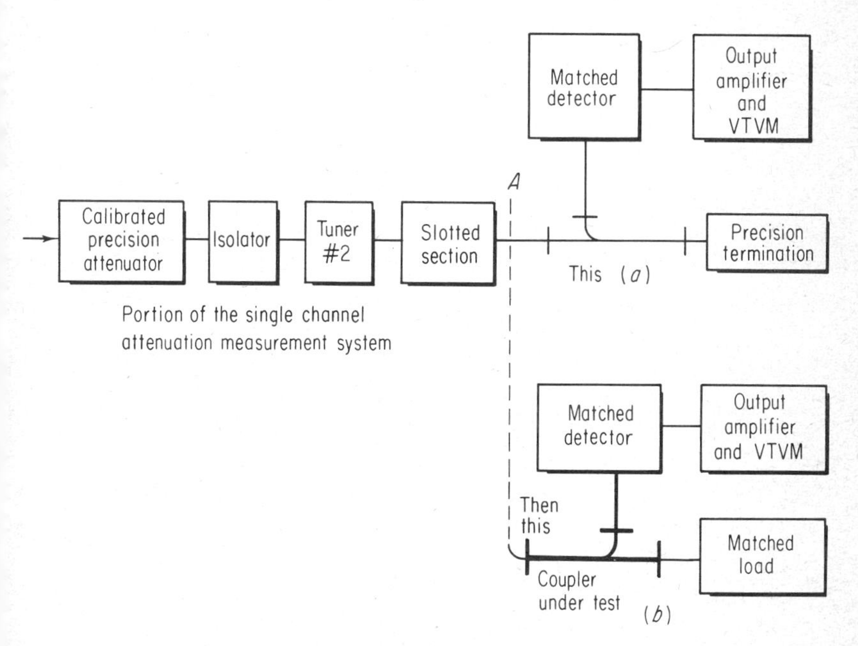

Fig. 12·9 Directivity measurement.

1. Set the calibrated precision standard attenuator to zero attenuation reading and adjust the level set attenuators to obtain a reference level at the output meter.
2. Set the standard attenuator to a high value of attenuation and reverse the coupler as indicated in the diagram.
3. Adjust the standard attenuator to obtain the original reference level at the output meter. The change in decibel readings of the standard attenuator is the directivity value.

Actually, it is not necessary to reverse the coupler as indicated in step 2 of the procedure. Instead, the standard attenuator is set to a high value, and the matched termination is replaced by a short circuit. Thc standard attenuator is adjusted to obtain the original reference level and the resulting value of directivity. The main line loss of the coupler must be subtracted from this value if a short circuit is used instead of reversing the coupler.

In practice, the measurement is not this simple because the matched termination is not perfect, and reflections from the flange mismatch and from the termination combine with the inherent directivity signal at some arbitrary phase angle. When the reflections are out of phase with the directivity signal, there is a cancellation of signals, and the resulting measured value of directivity is greater than the actual directivity of the coupler. Likewise, in-phase signals result in measured directivity values less than the actual directivity.

A sliding load permits the reflection from the load to be identified from the directivity signal and the fixed phase signal from the flange mismatch. The contributions due to flange reflections are minimized by precision alignment and mating of the coupler and termination flanges. The technique which is used to perform accurate directivity measurements using a sliding load (termination) is referred to as the *sliding-load* technique. There are two similar methods used to perform the measurement; the particular method chosen depends upon the magnitude of the directivity to be measured and the return loss or equivalent VSWR of the sliding termination. Also, the value of coupling can determine the particular method and will be explained later. If the directivity to be measured is less than the return loss of the sliding load, the directivity signal is larger than the signal due to the reflection from the load. *A large return loss corresponds to a low VSWR or well-matched load.* When the directivity value is greater than the return loss of the load, then the signal from the sliding load is greater than the directivity signal. The technique of directivity measurement for each of the above cases will be considered in detail.

Certain relationships have been derived which are general for the phasor combination of any two signals; mathematical calculations have been performed for a wide range of relative signal levels, and the resulting quantities have been tabulated. These relationships as applied to the measurement of directivity are illustrated in Fig. 12·10. The physical significance of the various parameters and functions illustrated in the diagram can be shown by considering the measurement technique. First it is necessary to define the parameters and functions shown in Fig. 12·10.

D = directivity

V_d = the r-f voltage equivalent of the leakage voltage for which directivity is defined

V_r = the r-f voltage equivalent of the sliding-load return loss signal (the signal reflected from the load)

m or n = the relative attenuation, in db, of the 180° out-of-phase sum of V_d and V_r with respect to the in-phase sum of V_d and V_r. These are measured quantities which are obtained as decibel variations on the output meter when the sliding load is varied to change the phase of the load reflection signal by 360°

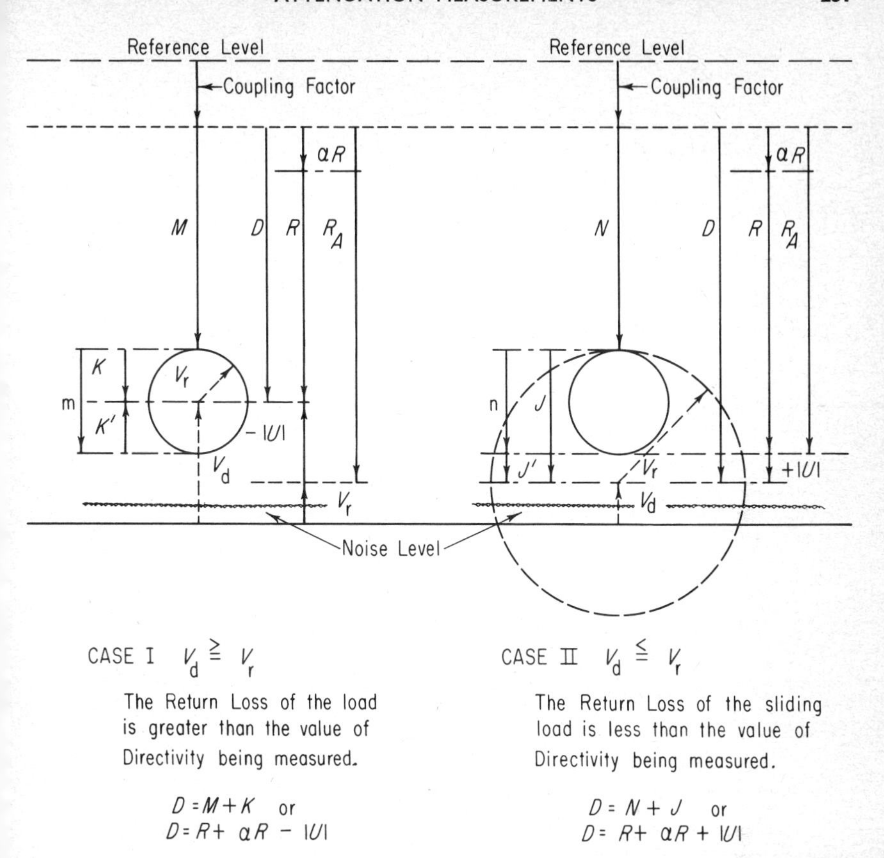

Fig. 12·10 General relationships for the phasor combination of two signals

K = the relative attenuation, in db, of V_d with respect to the in-phase sum of V_d and V_r when $V_d \geqq V_r$. It is the correction factor which must be added to M in order to obtain the directivity D

$$K = 20 \log \frac{V_d + V_r}{V_d} = 20 \log \frac{2 \times 10^{m/20}}{10^{m/20} + 1} \tag{12·3}$$

K' = relative attenuation, in db, of V_d with respect to the 180° out-of-phase sum of V_d and V_r when $V_d \geqq V_r$

J = relative attenuation, in db, of V_d with respect to the in-phase sum of V_d and V_r when $V_d \leqq V_r$. J is a correction factor which must be added to N in order to obtain the directivity (D)

$$J = 20 \log \frac{V_d + V_r}{V_d} = 20 \log \frac{2 \times 10^{n/20}}{10^{n/20} - 1} \tag{12·4}$$

J' = relative attenuation, in db, of V_d with respect to the 180° out-of-phase sum of V_d and V_r when $V_d \leqq V_r$

U = universal ratio function, the relative attenuation, in db, of V_r with respect to V_d. For case I where $V_d \geqq V_r$, $|U|$ is negative. For case II where $V_d \leqq V_r$, $|U|$ is positive

Case I:

$$U = 20 \log \frac{V_r}{V_d}$$

$$= 20 \log \frac{10^{m/20} + 1}{10^{m/20} - 1}$$

Case II:

$$U = 20 \log \frac{V_r}{V_d}$$

$$= 20 \log \frac{10^{n/20} + 1}{10^{n/20} - 1}$$

αR = insertion loss of the coupler for the reverse power flow direction

R_A = apparent return loss of the sliding load; it is equal to the return loss of the load plus the main line attenuation of the coupler under test

R = true return loss of the sliding load

M and N = the minimum measured directivity values for cases I and II, respectively

The designations M, N, J, K, m, and n were set forth in Wind and Rapaport, "Handbook of Microwave Measurements," vol. I, Edwards Brothers, Inc., Ann Arbor, Michigan, 1954.

It should be noted that Eq. (12·3) *can be obtained by considering* V_d and V_r *as incident and reflected voltages* E_i *and* E_r, *respectively*. Equation (12·3) becomes $20 \log 2\rho/(\rho + 1)$ which shows that $10^{m/20}$ is the VSWR. Also, the U function of case I becomes

$$U = 20 \log \frac{E_i}{E_r} = 20 \log \frac{\rho + 1}{\rho - 1} = 20 \log \frac{1}{\Gamma}$$

which is *precisely* the *return loss* of Eq. (5·1) in Chap. 5. *The U function table is therefore a table of the return loss where W is the SWR in decibels.*

Matched-sliding-load Method. The matched-sliding-load method of measuring directivity is used when the return loss of the sliding load is greater than the directivity value to be measured. This condition is illustrated as case I in the diagram of Fig. 12·10. This method is usually used when measuring directivity values less than 35 db with matched sliding loads which have return loss values of 50 db or greater.

Directivity measurements are performed using the attenuation measurement system. Part of the attenuation measurement system, including a calibrated standard attenuator, is shown in Fig. 12·9. The coupler is connected as shown in (*a*) with a matched sliding load connected to the main line input of the coupler. The measurement procedure is as follows:

1. Connect the coupler in the reverse direction as shown in Fig. 12·9*a* with a matched sliding load connected to the normal main line input of the coupler.
2. Set the standard attenuator to zero dial reading and use the system level set attenuator to obtain an indication on the output meter.
3. Vary the sliding load to obtain a maximum output indication on the output meter and adjust the level set attenuator to obtain a convenient reference output indication (preferably at the top of the scale on a calibrated range of the meter).
4. Vary the sliding load and measure the decibel variation from maximum to minimum output meter indications. Call this value m.
5. Vary the sliding load to obtain the original maximum output reference level of step 3.
6. Set the calibrated standard attenuator to a high value. This is necessary in order to prevent possible damage to the detecting element when step 7 is performed.
7. Reverse the coupler as shown in Fig. 12·9*b* or replace the sliding load with a short-circuit termination.
8. Adjust the standard attenuator to obtain the original reference level of step 5. Record the standard attenuator reading (change in attenuation from the setting in step 2) and call this value M. *It is noted that in Fig.* 12·10 *the value M is the minimum measured directivity value.*
9. Obtain K from the curve of m versus K in Fig. 12·11.
10. Calculate the directivity.

$$D = M + K$$

Example: The value of m in step 4 is 1.0 db, and the value of M in step 8 is 28 db. From Fig. 12·11, the value for K is found to be 0.49 db.

$$D = M + K = 28 + 0.49 = 28.49 \text{ db}$$

If the value of J is obtained from Fig. 12·12 using the same value of m for n on the curve, it is found that for $n = 1.0$ db, J is 25.25 and $M + J = 28 + 25.25 = 53.25$ db, which is the return loss of the load. It is noted that the return loss is *greater than* the value of directivity being measured. The possible error in the measurement can be found on the universal error curve as explained in a subsequent section.

Mismatched-sliding-load Method. The mismatched-sliding-load method of measuring directivity is necessary when measuring high directivity of couplers which have high coupling values regardless of the directivity value. A mismatched sliding load is obtained by placing a piece of metal foil on a matched sliding load. Usually a broken polyiron load is used and can be accurately calibrated in terms of return loss.

In the case of high directivity values, the directivity signal is the same order of magnitude as the signal reflected from a matched sliding load. This condition results in large (db) variations at the output when the sliding load is varied. In cases where the two signals are equal in magnitude, the output signal goes down to the noise level and cannot actually be determined. In this case, the value of M is determined and the actual directivity is greater than the value M by some unknown value.

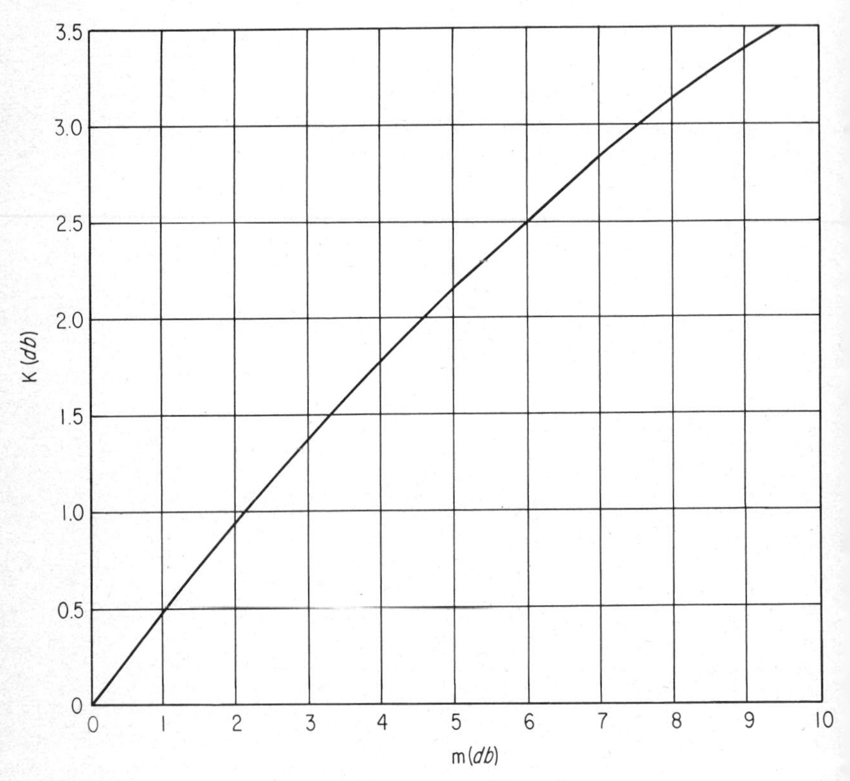

Fig. 12·11 ***m* versus *K* curve.**

In the case of large coupling values, a large power input would be required in order to set up the original reference level when the coupler is connected in the reverse direction, as shown in Fig. 12·9*a*. As an example, a 40-db coupling value and 35-db directivity result in a 75-db attenuation of the signal. Since the total range of attenuation measurements in the audio detection system is usually less than 60 db, the signal at the output meter is still below the noise level. Therefore, in order to obtain an output signal for a reference level, it is necessary to connect a mismatched load on the main line of the couplers in Fig. 12·9*a*. This is represented on the diagram of Fig. 12·10 as case II, in which the return loss signal from the sliding load is greater than the magnitude of the directivity signal, i.e., *the return loss is less than the value of directivity*.

The measurement procedure is the same as for the matched-sliding-load method except that a mismatched sliding load is used in step I, and the designations m, M, and K are replaced with n, N, and J, respectively. The value of J is read from the curve of n versus J in Fig. 12·12. The directivity is then equal to $N + J$.

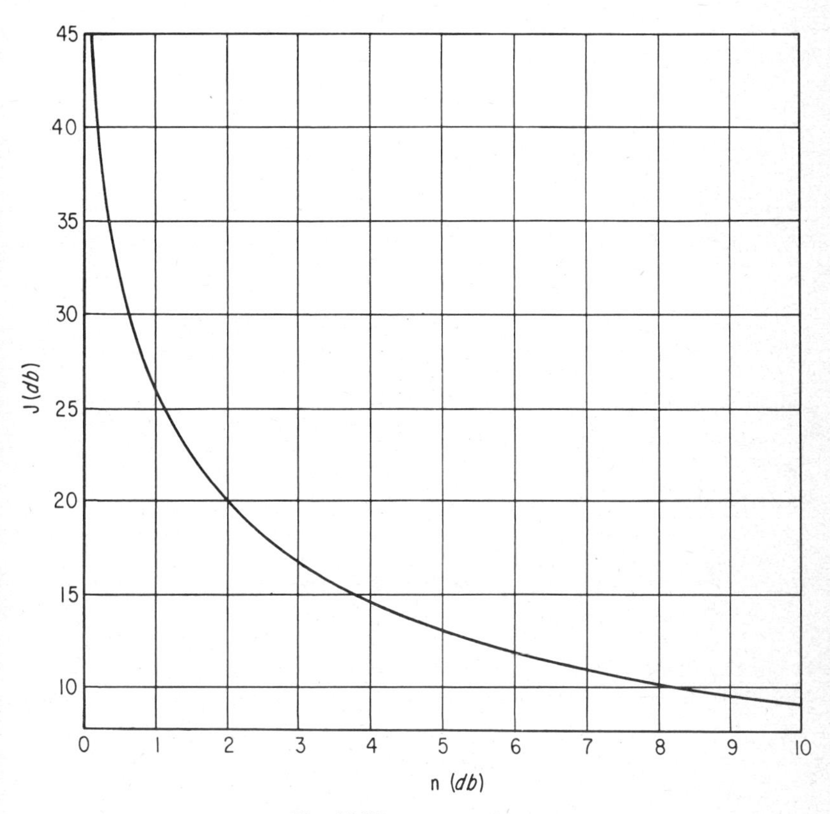

Fig. 12·12 ***n* versus *J* curve.**

Example

1. The coupler is arranged in the system as shown in Fig. 12·9*a* with a mismatched sliding load terminating the normal main line input of the coupler.
2. The calibrated standard attenuator is set to zero, and the r-f level set attenuators are adjusted to obtain a reading on the 10-volt scale of the output meter.
3. The sliding load is varied for a maximum output reading, and the

output indication is set to the top of the 10-volt scale by adjusting the r-f level set attenuators.

4. The sliding load is varied for a minimum reading on the output meter. The change from the top of the scale is 0.65 db ($n = 0.65$ db).
5. The sliding load is varied to set the output reference back to the top of the 10-volt scale.
6. The standard attenuator is set to 50 db.
7. The coupler is reversed as shown in Fig. 12·9*b*, and a matched termination is placed on the main line output as shown.
8. The standard attenuator is adjusted downward to obtain the original reference level at the top of the 10-volt scale. The attenuator reads $N = 17.5$ db.
9. The value of J from the curve in Fig. 12·12 is 28.8 db.
10. $D = N + J = 17.5 + 28.8 = 46.3$ db.

The apparent return loss of the load is equal to $N + K$, where K is found from the curve of m versus K in Fig. 12·11. For this example K is found to be 0.32 db. $N + K = 17.82$ db $=$ the apparent return loss of the load. If the coupler being measured has a main line attenuation value of 0.5 db (10-db coupler), then the true return loss of the load is $17.82 - 0.5 = 17.32$ db.

The universal ratio function (U) is the result of an error analysis by James U. Clark (former colleague of the author in the Microwave Standards Laboratory of Hughes Aircraft Company). The error curve was derived in order to predict the maximum possible error in the measurement of a signal whose approximate attenuation is known but cannot be isolated from the effect of an unwanted signal whose approximate attenuation is also known or may be estimated. That is, a particular value of attenuation *including* the effect of the unwanted signal is measured, and the maximum possible error in the *measured* value was to be determined.

There is such an implicit relationship between U and the correction factor that any discrete value of U corresponds to a unique value of K, K', and J', depending upon whether U is negative or positive.

The universal error function curve of Fig. 12·13 is a plot of the relationships of the correction factor K, K', J, and J' to the independent variable U. The curve is interesting in that it shows the functions K and J to be in reality two parts of a single continuous function and the functions of K' and J' to be two parts of a single discontinuous function. The separation in decibels between the K and K' functions on the curve, for any discrete value of U, is exactly the value m of case I. The separation in decibels between J and J' functions, for any discrete value of U, is exactly the value of n of case II.

The curve is therefore a composite diagram of *all* the interrelationships among the functions. If a particular value of any one of the functions is known, the corresponding value of each of the other functions may be taken

Table 12·1 Table of the universal ratio function U *

$$U = 20 \log_{10} \frac{10^{W/20} + 1}{10^{W/20} - 1}$$

W	0	1	2	3	4	5	6	7	8	9
0.00	∞	84.80	78.78	75.25	72.76	70.82	69.23	67.89	66.74	65.71
0.01	64.80	63.97	63.21	62.52	61.87	61.28	60.71	60.19	59.69	59.22
0.02	58.78	58.35	$57.9\bar{5}$	57.56	57.19	56.84	56.50	56.17	55.85	$55.5\bar{5}$
0.03	55.25	54.97	54.69	54.43	54.17	53.92	53.67	53.43	53.20	52.98
0.04	52.76	52.54	52.33	52.13	51.93	51.73	51.54	51.35	51.17	50.99
0.05	50.82	$50.6\bar{5}$	50.48	50.31	$50.1\bar{5}$	49.99	49.83	49.68	49.53	49.38
0.06	49.23	49.09	$48.9\bar{5}$	48.81	48.67	48.54	48.41	48.28	$48.1\bar{5}$	48.02
0.07	47.90	47.77	47.65	47.53	47.41	47.30	47.18	47.07	46.96	46.84
0.08	46.74	46.63	46.52	46.42	46.31	46.21	46.11	46.01	45.91	45.81
0.09	45.71	45.62	45.52	45.43	45.33	45.24	45.15	45.06	44.97	44.88
0.1†	44.80	43.97	43.21	42.52	41.87	41.28	40.71	40.19	39.69	39.22
0.2	38.78	38.35	$37.9\bar{5}$	37.56	37.19	36.84	36.50	36.17	35.85	$35.5\bar{5}$
0.3	35.26	34.97	34.69	34.43	34.17	33.92	33.67	33.43	33.20	32.98
0.4	32.76	32.54	32.33	32.13	31.93	31.73	31.54	31.36	31.17	31.00
0.5	30.82	$30.6\bar{5}$	30.48	30.31	30.15	29.99	29.84	29.68	29.53	29.38
0.6	29.24	29.09	28.95	28.81	28.68	28.54	28.41	28.28	28.15	28.02
0.7	27.90	27.78	27.66	27.54	27.42	27.30	27.19	27.07	26.96	26.85
0.8	26.74	26.63	26.53	26.42	26.32	26.22	26.11	26.01	25.91	25.82
0.9	25.72	25.62	25.53	25.44	25.34	25.25	25.16	25.07	24.98	24.89
1.0	24.81	24.72	24.63	24.55	24.47	24.38	24.30	24.22	24.14	24.06
1.1	23.98	23.90	23.82	$23.7\bar{5}$	23.67	23.60	23.52	$23.4\bar{5}$	23.37	23.30
1.2	23.23	23.16	23.08	23.01	22.94	22.87	22.80	22.74	22.67	22.60
1.3	22.53	22.47	22.40	22.34	22.27	22.21	22.14	22.08	22.02	21.96
1.4	21.89	21.83	21.77	21.71	$21.6\bar{5}$	21.59	21.53	21.47	21.41	21.35
1.5	21.30	21.24	21.18	21.13	21.07	21.01	20.96	20.90	$20.8\bar{5}$	20.79
1.6	20.74	20.69	20.63	20.58	20.53	20.47	20.42	20.37	20.32	20.27
1.7	20.22	20.17	20.11	20.06	20.01	19.97	19.92	19.87	19.82	19.77
1.8	19.72	19.67	19.63	19.58	19.53	19.49	19.44	19.39	$19.3\bar{5}$	19.30
1.9	19.26	19.21	19.17	19.12	19.08	19.03	18.99	18.94	18.90	18.86
2.0	18.81	18.77	18.73	18.69	18.64	18.60	18.56	18.52	18.48	18.44
2.1	18.39	18.35	18.31	18.27	18.23	18.19	18.15	18.11	18.07	18.03
2.2	17.99	17.96	17.92	17.88	17.84	17.80	17.76	17.73	17.69	17.65
2.3	17.61	17.58	17.54	17.50	17.46	17.43	17.39	17.36	17.32	17.28
2.4	$17.2\bar{5}$	17.21	17.18	17.14	17.11	17.07	17.04	17.00	16.97	16.93
2.5	16.90	16.86	16.83	16.80	16.76	16.73	16.69	16.66	16.63	16.59
2.6	16.56	16.53	16.50	16.46	16.43	16.40	16.37	16.33	16.30	16.27
2.7	16.24	16.21	16.18	16.14	16.11	16.08	16.05	16.02	15.99	15.96
2.8	15.93	15.90	15.87	15.84	15.81	15.78	$15.7\bar{5}$	15.72	15.69	15.66
2.9	15.63	15.60	15.57	15.54	15.51	15.48	15.45	15.43	15.40	15.37
3.0	15.34	15.31	15.28	15.26	15.23	15.20	15.17	15.14	15.12	15.09
3.1	15.06	15.03	15.01	14.98	14.95	14.93	14.90	14.87	14.84	14.82
3.2	14.79	14.76	14.74	14.71	14.69	14.66	14.63	14.61	14.58	14.56
3.3	14.53	14.50	14.48	14.45	14.43	14.40	14.38	14.35	14.33	14.30
3.4	14.28	14.25	14.23	14.20	14.18	14.15	14.13	14.10	14.08	14.06
3.5	14.03	14.01	13.98	13.96	13.94	13.91	13.89	13.86	13.84	13.82

* T. Mukaihata, M. F. Bottjer, and H. J. Tondreau, Rapid Broadband Directional Coupler irectivity Measurements, *IRE Trans. Instr.* vol. I-9, no. 2, pp. 196–202, September, 1960.

† Increment of argument changes to 0.01.

Table 12·1 *(continued)*

W	0	1	2	3	4	5	6	7	8	9
3.6	13.79	13.77	13.7$\bar{5}$	13.72	13.70	13.68	13.65	13.63	13.61	13.59
3.7	13.56	13.54	13.52	13.49	13.47	13.4$\bar{5}$	13.43	13.40	13.38	13.36
3.8	13.34	13.32	13.29	13.27	13.25	13.23	13.21	13.18	13.16	13.14
3.9	13.12	13.10	13.08	13.06	13.03	13.01	12.99	12.97	12.9$\bar{5}$	12.93
4.0	12.91	12.89	12.87	12.84	12.82	12.80	12.78	12.76	12.74	12.72
4.1	12.70	12.68	12.66	12.64	12.62	12.60	12.58	12.56	12.54	12.52
4.2	12.50	12.48	12.46	12.44	12.42	12.40	12.38	12.36	12.34	12.32
4.3	12.30	12.28	12.26	12.24	12.23	12.21	12.19	12.17	12.1$\bar{5}$	12.13
4.4	12.11	12.09	12.07	12.05	12.04	12.02	12.00	11.98	11.96	11.94
4.5	11.92	11.91	11.89	11.87	11.85	11.83	11.81	11.80	11.78	11.76
4.6	11.74	11.72	11.71	11.69	11.67	11.65	11.63	11.62	11.60	11.58
4.7	11.56	11.5$\bar{5}$	11.53	11.51	11.49	11.48	11.46	11.44	11.42	11.41
4.8	11.39	11.37	11.35	11.34	11.32	11.30	11.29	11.27	11.25	11.24
4.9	11.22	11.20	11.19	11.17	11.15	11.14	11.12	11.10	11.09	11.07
5.0	11.05	11.04	11.02	11.00	10.99	10.97	10.95	10.94	10.92	10.91
5.1	10.89	10.87	10.86	10.84	10.83	10.81	10.79	10.78	10.76	10.7$\bar{5}$
5.2	10.73	10.72	10.70	10.68	10.67	10.65	10.64	10.62	10.61	10.59
5.3	10.58	10.56	10.54	10.53	10.51	10.50	10.48	10.47	10.45	10.44
5.4	10.42	10.41	10.39	10.38	10.36	10.3$\bar{5}$	10.33	10.32	10.30	10.29
5.5	10.27	10.26	10.24	10.23	10.21	10.20	10.19	10.17	10.16	10.14
5.6	10.13	10.11	10.10	10.08	10.07	10.05	10.04	10.03	10.01	10.00
5.7	9.98	9.97	9.96	9.94	9.93	9.91	9.90	9.88	9.87	9.86
5.8	9.84	9.83	9.82	9.80	9.79	9.77	9.76	9.7$\bar{5}$	9.73	9.72
5.9	9.71	9.69	9.68	9.66	9.65	9.64	9.62	9.61	9.60	9.58
6.0	9.57	9.56	9.54	9.53	9.52	9.50	9.49	9.48	9.46	9.45
6.1	9.44	9.42	9.41	9.40	9.39	9.37	9.36	9.3$\bar{5}$	9.33	9.32
6.2	9.31	9.29	9.28	9.27	9.26	9.24	9.23	9.22	9.20	9.19
6.3	9.18	9.17	9.15	9.14	9.13	9.12	9.10	9.09	9.08	9.07
6.4	9.05	9.04	9.03	9.02	9.00	8.99	8.98	8.97	8.96	8.94
6.5	8.93	8.92	8.91	8.89	8.88	8.87	8.86	8.85	8.83	8.82
6.6	8.81	8.80	8.79	8.77	8.76	8.75	8.74	8.73	8.72	8.70
6.7	8.69	8.68	8.67	8.66	8.64	8.63	8.62	8.61	8.60	8.59
6.8	8.58	8.56	8.55	8.54	8.53	8.52	8.51	8.49	8.48	8.47
6.9	8.46	8.4$\bar{5}$	8.44	8.43	8.42	8.40	8.39	8.38	8.37	8.36
7.0	8.3$\bar{5}$	8.34	8.33	8.31	8.30	8.29	8.28	8.27	8.26	8.2$\bar{5}$
7.1	8.24	8.23	8.22	8.20	8.19	8.18	8.17	8.16	8.15	8.14
7.2	8.13	8.12	8.11	8.10	8.09	8.07	8.06	8.05	8.04	8.03
7.3	8.02	8.01	8.00	7.99	7.98	7.97	7.96	7.9$\bar{5}$	7.94	7.93
7.4	7.92	7.91	7.90	7.89	7.87	7.86	7.85	7.84	7.83	7.82
7.5	7.81	7.80	7.79	7.78	7.77	7.76	7.75	7.74	7.73	7.72
7.6	7.71	7.70	7.69	7.68	7.67	7.66	7.65	7.64	7.63	7.62
7.7	7.61	7.60	7.59	7.58	7.57	7.56	7.55	7.54	7.53	7.52
7.8	7.51	7.50	7.49	7.48	7.47	7.46	7.45	7.44	7.44	7.43
7.9	7.42	7.41	7.40	7.39	7.38	7.37	7.36	7.3$\bar{5}$	7.34	7.33
8.0	7.32	7.31	7.30	7.29	7.28	7.27	7.26	7.25	7.2$\bar{5}$	7.24

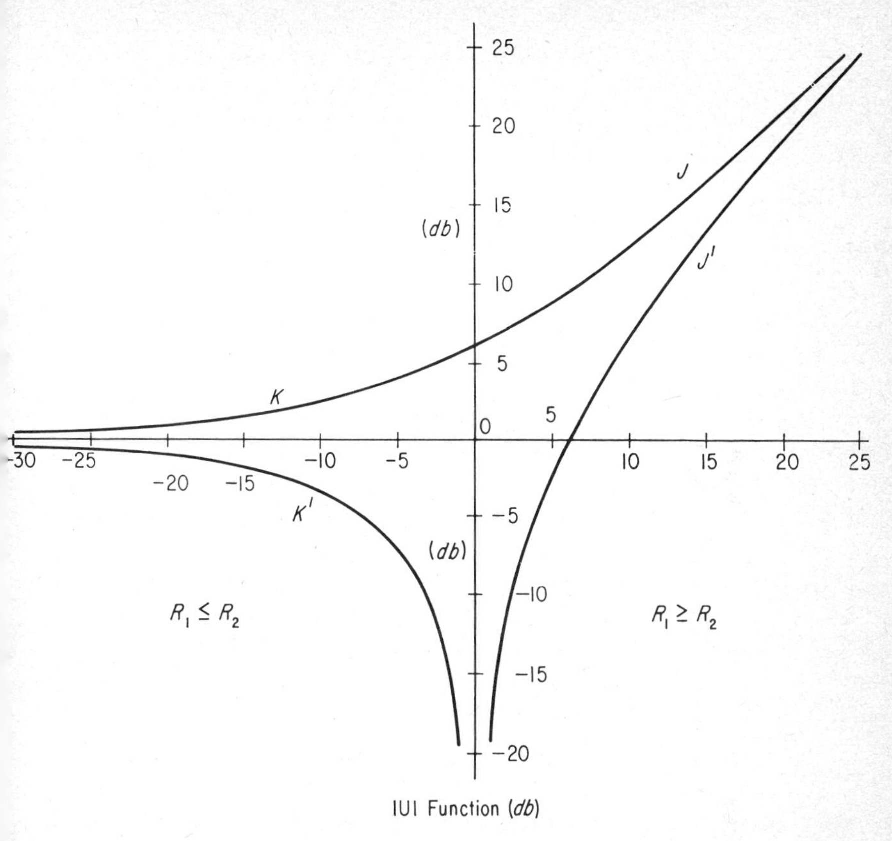

Fig. 12·13 Universal error function curve. (*Hughes Aircraft Company, Primary Standards Laboratory.*)

directly from the curve. For instance, if a particular value of m and n is known, one may use a pair of dividers or scale to find the point at which the separation between K and K' or J and J' curves is equal to m or n, respectively, and at that point read U, K, K', J, and J' as desired.

The curve is useful in predicting the *maximum* error in the measurement of the relative attenuation R_1 of a signal which is in combination at some random phase with an unwanted signal of relative attenuation R_2, which cannot be isolated from the wanted signal in the measurement. This can be shown by considering the possible errors due to flange effects when measuring directivity of a coupler. The measured value of directivity D includes the true directivity and the effect of reflections from the junction of the flanges of the couplers and the sliding-load casing. The error in the measured value D is therefore a function of the random phase reflection from this junction. The return loss of the flange can be determined independently.

Example 1: If the true directivity is 40 db and the flange return loss is 60 db. $U = 40 - 60 = -20$ db and the error from the error curve $K = +0.8$ db and $K' = -0.9$ db.

The measured value of directivity could therefore be in range of 39.1 to 40.8.

Example 2: The true directivity is 60 db, and the return loss of the load is 50 db. $U = 10$ db from the curve, $J = 12.3$ db, and $J' = 6.7$ db. The measured directivity could therefore be in the range of 47.7 to 53.3 db.

Universal Ratio Function Technique of Directivity Measurement. In the universal ratio function technique of directivity measurement, use is made of the fact that any discrete value of m or n corresponds to a unique value of U. Rapid wide-band directivity measurements are easily made using this technique. In order to assure a good match over the complete frequency band, the coupler is oriented in a system as shown in Fig. 12·14. The sliding load has a known (calibrated) return loss (terminations can be designed which have almost constant return loss over the complete waveguide range). It is only necessary to accurately measure the decibel variation on the output meter as the sliding load is varied. This decibel variation is labeled W on Table 12·1, the universal ratio function U table. W represents the value m for case I or n for case II as previously defined. The value of U is obtained and the directivity is calculated from

$$D = R + \alpha R + U \tag{12·5}$$

A convenient setup for performing wide-band directivity measurements is shown in Fig. 12·14 connected at AA of the attenuation system.

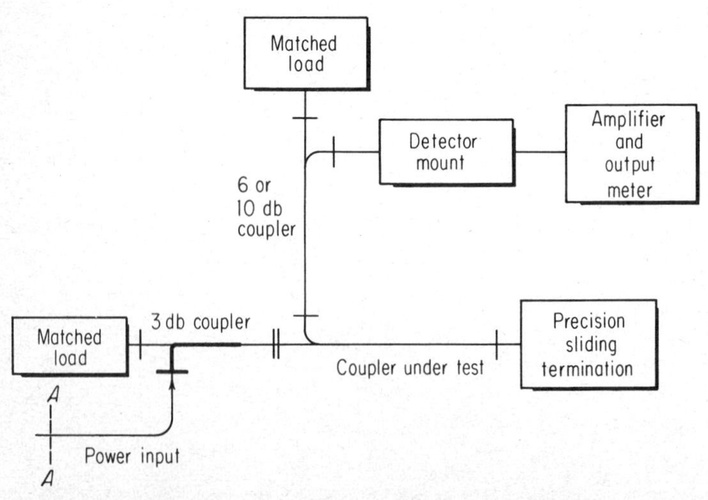

Fig. 12·14 Wideband coupler directivity measurement system.

The coupler and matched termination at the input provide a good match over the waveguide range. The coupler and matched load on the auxiliary arm of the coupler under test provide a match over the waveguide range. The calibrated sliding termination is located in a precision casing, and alignment pins are used when connecting this termination to the coupler under test.

Example: The coupler under test has a coupling value of 10 db. The return loss of the load R is 18.0 db, and the main line insertion loss is 0.53 db.

The measured decibel variations (W) at the output on the calibrated standing-wave indicator are 0.85 db. The value of $U = 46.21$ db is obtained from the universal ratio function table.

$$\text{Directivity} = R + \alpha R + U$$
$$= 18 + 0.53 + 46.21 = 64.74 \text{ db}$$

12·15 Return loss measurements

The return loss has been defined as the ratio, expressed in decibels, of the power incident upon a discontinuity to the power reflected from the discontinuity.

$$\text{Return loss} = -20 \log \Gamma = 20 \log \frac{1}{\Gamma} \tag{12·6}$$

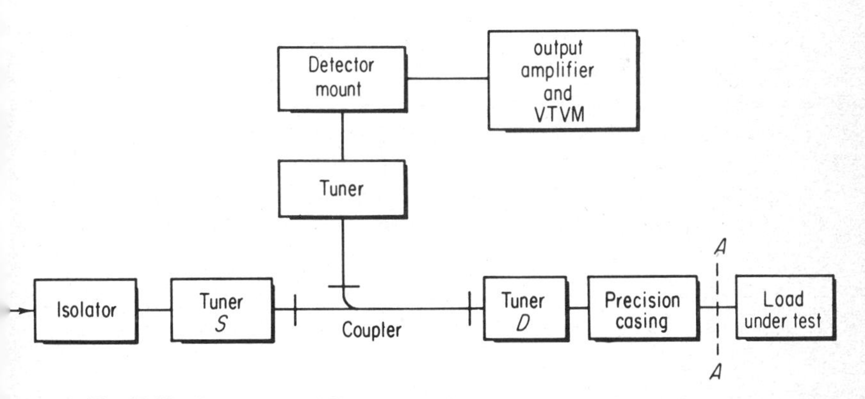

Fig. 12·15 System used for performing return loss measurement.

Return loss measurements can be performed using the setup illustrated in Fig. 12·9, but a more precise measurement system will be discussed. The system is illustrated in Fig. 12·15. The power input is from the attenuation measurement system previously discussed in detail. Actually, the tuner could be the source tuner (No. 2) of the attenuation measurement system. The return loss measurement can be extended to values greater than 20 db by using partial r-f substitution techniques, but will only be considered for return loss values less than 20 db at present.

The procedure for tuning the system is as follows:

1. A sliding short is placed at A and tuner S is adjusted for no variation on the output meter when the short is quickly varied over the range of one guide wavelength. No variation indicates a matched source.
2. Tuner D tunes the directivity of the coupler to a high value. A sliding mismatched load (35- to 40-db return loss) is placed at A and varied over a guide wavelength. The variation at the output is tuned to zero. This is a critical tuning adjustment. A 5-stub tuner is recommended for tuner D.

As an example, the termination and flange return loss values of a standard reflection are to be determined. The termination is labeled as having a 0.15 reflection coefficient.

Procedure

1. A precision short circuit is placed at A, and the calibration attenuator is adjusted for measuring an attenuation value of 20 db according to the manufacturer's instruction. (The amplifier-voltmeter technique could also be used here.)
2. After the reference level is obtained at the output meter, the short is removed and the standard reflection is placed at A.
3. The precision audio attenuator in the attenuation calibrator is adjusted to obtain an on-scale reading at the output, and the sliding termination in the standard reflection is varied to obtain *maximum* output reading. The audio attenuator is then adjusted to obtain the original reference level, and the decibel attenuation is recorded.
4. The sliding termination is varied for a *minimum* reading on the output meter, and the maximum to *minimum* decibel variation is recorded.
5. *a.* Label the decibel variation in step 3 as M and the decibel variation in step 4 as m. Find K from the curve of m versus K in Fig. 12·11.
   b. Label the decibel variation in step 3 as N and the decibel variation in step 4 as n. Find J from the curve of n versus J in Fig. 12·12.
   Calculate $n + J =$ return loss of the sliding termination.

Example: The decibel attenuation value obtained in step 3 is 16.3 db, and the decibel variation when the load is varied in step 4 is 0.5 db. M or $N = 16.3$ db; m or $n = 0.5$ db.

From the respective curves, $K = 0.25$ db, and $J = 35.4$ db. $M + K =$ 16.45 db, the flange return loss. The equivalent VSWR is 1.354.

The equivalent reflection coefficient is 0.1505. $N + J = 51.7$ db, the return loss of the termination, which has an equivalent VSWR of 1.0052.

It is noted that this measurement is similar to the directivity measurement, again pointing out the importance of obtaining a thorough understanding of the phasor combination of signals.

In the above problem it was assumed that the return loss of the short circuit was zero db. In practice the return loss is not zero, and a correction to the measurement is necessary.

PROBLEMS

12·1 If the source mismatch is 1.45:1 and the load mismatch is 1.2:1, determine the maximum and minimum values of reflected power which are possible if the input power is 200 mw.

12·2 The power input to a 6-db attenuator is 300 mw. The reflected power is 20 mw. Determine the standing-wave ratio and the mismatch loss.

12·3 The incident voltage is 100 volts and the reflected voltage is 50 volts. Determine the reflected power in per cent, the return loss, and the mismatch loss.

12·4 The directivity of a coupler is measured using the matched-sliding-load technique. If $M = 30.6$ db and $m = 1.7$ db, what is the value of directivity of the coupler and the value of the return loss of the sliding matched load?

12·5 A directivity measurement is performed using the mismatched-sliding-load technique. $N = 20.6$ and $n = 0.5$. What is the value of directivity of the coupler and the value of return loss of the mismatched load.

12·6 The main line loss of a coupler is 0.55 db, and the return loss of the sliding load is 16 db.

a. If the return loss of the load is less than the value of directivity being measured, and W is 0.55 db, what is the value of directivity?

b. If the coupler must meet a directivity specification of 40 db minimum, what is the maximum permissible decibel variation W when using the above sliding load?

c. As the sliding load is varied over its range, what is the maximum permissible decibel variation if the return loss of the sliding load is 20 db and the directivity specification is 38 db minimum.

12·7 Use the universal error function curve to determine the maximum possible errors in measurements of directivity in Probs. 12·4 to 12·6 if the flange return loss of the sliding termination is 55 db.

CHAPTER

13

MEASUREMENT OF MICROWAVE POWER

Introduction. Power is usually measured in terms of voltage, current, and power factor at low frequencies. The measuring instruments read the product of voltage, current, and power factor. These quantities tend to lose their meaning in the distributed circuits encountered in typical microwave applications. Power measurements are performed at microwave frequencies by means of various types of thermally sensitive elements. The methods for using these elements will be discussed in this chapter.

The various categories of power measurement usually depend on the power level, frequency, and the waveform of the applied power. The range of power measurements extends from the measurement of megawatts down to the thermal agitation level of resistors. This leads to arbitrary power ranges which, for convenience, can be referred to as low power (0 to 10 mw), medium power (10 mw to 10 watts), and high power (upwards of 10 watts). The waveform may be c-w or pulsed.

Most power-measuring devices respond to average power, and the power meter provides a measure of the average power over a number of cycles. Since the microwave sources are pulsed at an audio rate in many applications, it is necessary to measure the average power over several cycles or pulses of the modulation. In these cases it is usually necessary to know the peak power. The peak power P_p can be computed from the average power if the pulse length (τ) and the repetition frequency (f_r) are known. The relationship of the factors is $P_p = P_a \dfrac{1}{\tau f_r}$

where P_p = peak power, watts; refers to the average taken over a period of the pulse

P_a = average power, watts; refers to the average taken over a period between consecutive pulses

τ = pulse length, sec

f_r = pulse repetition frequency, pulses per sec

The quantity τf_r is called the *duty cycle*.

The methods available for power measurements at microwave frequencies are capable of good accuracy, but are fairly elaborate, and many sources of error are possible in each one. It is advisable to obtain comparison checks of the several types of measuring devices. By proper attention to details, one can obtain agreement to within a few per cent.

13·1 Power-sensitive elements

Three general types of thermal devices used in microwave power measurements are thermocouples, bolometers, and calorimeters.

The thermocouple is formed when two wires of different metals have one of their junctions at a higher temperature than the other. The difference in temperature between the junctions produces an emf proportional to the difference in temperature. The thermocouple is used to measure the temperature rise of a load which dissipates microwave power; by appropriate calibration this temperature rise is converted into an indication of the microwave power. Thermocouples have excellent sensitivity and are especially useful as power monitors.

The bolometer characteristics have been outlined in Chap. 8. The change in bolometer resistance due to application of microwave power can either be measured and calibrated in terms of power, or an equal resistance change can be effected by d-c or audio frequency power. The d-c or audio power is then used for low power measurements measured by conventional low frequency methods.

The calorimetric technique is the most fundamental method of measuring microwave power. Power is dissipated in a load, and the temperature rise of the load, or a cooling fluid flowing in or around the load, is measured. Calibration of the calorimeter includes correction for heat capacity, heat losses, operating frequency, and mismatch.

13·2 Bolometric method of power measurement

The bolometric method is most useful in the low and medium power ranges. This method employs a bolometer which operates in one arm of a Wheatstone bridge circuit, as illustrated in Fig. 13·1.

In the absence of microwave power, the bolometer resistance is set equal to the resistance of another arm of the bridge by adjusting the d-c power supplied by the bias source. Under balanced conditions, no current flows through the galvanometer. When microwave power is applied to the bolometer, the bridge circuit becomes unbalanced because of the change in bolometer resistance. The d-c bias power is adjusted to rebalance the bridge, and this change in d-c power to the bolometer is a measure of the microwave power. The validity of this substitution of d-c for r-f power was pointed out in the discussion of bolometers in Chap. 8.

The unbalanced current can also be calibrated against an audio frequency power so that a meter placed in the detector arm of the bridge can have its scale marked to read microwave power directly. The resistance-power curve of the bolometer (barretter or thermeter) enters into the power determination in this technique of measurement. The resistors R_1 and R_2 are temperature-compensating resistors which are selected so that their effective temperature coefficient equals the thermal coefficient of the bolometer.

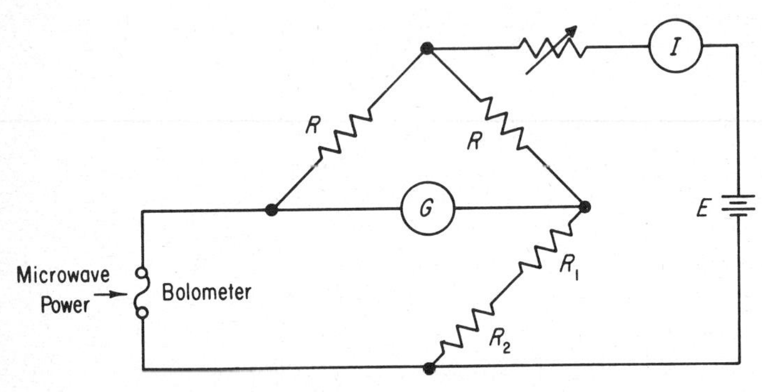

Fig. 13·1 Basic Wheatstone bridge.

The precision with which power measurements can be performed using this technique lies in the precision of the measuring instruments and components comprising the bridge circuit. This measurement requires measurement of the difference between two almost equal currents so that a small error in reading the current can cause a large error in the power measurement.

13·3 Balanced-bridge method

In the balanced-bridge method, the bridge is initially balanced with audio frequency and d-c bias power. When the bridge is unbalanced by the application of microwave power to the bolometer, sufficient bias power is removed to rebalance the bridge. The increment of power required to rebalance the bridge is taken as a measure of the microwave power. This method is more precise because the bolometer is always operated at a constant resistance. Numerous commercial bridges have been devised to facilitate the adjustment and use of the basic bridge circuit.

A self-balancing bridge circuit is illustrated by the block diagram in Fig. 13·2. In the absence of r-f power the bridge is brought to balance (zero meter reading) by supplying both d-c bias power and audio frequency power (approximately 10.8 kc) from an oscillator. Introduction of microwave power unbalances the bridge, and the audio power is automatically removed to rebalance the bridge. The amount of power removed is displayed on an output meter which is calibrated to show the *decrease* in audio power as an *increase* in r-f power. Commercial bridges of this type cover the range up to 10 mw. On the 10-mw range the balance is obtained with 12 mw of audio power. The amount of audio required for full-scale indication is less for the

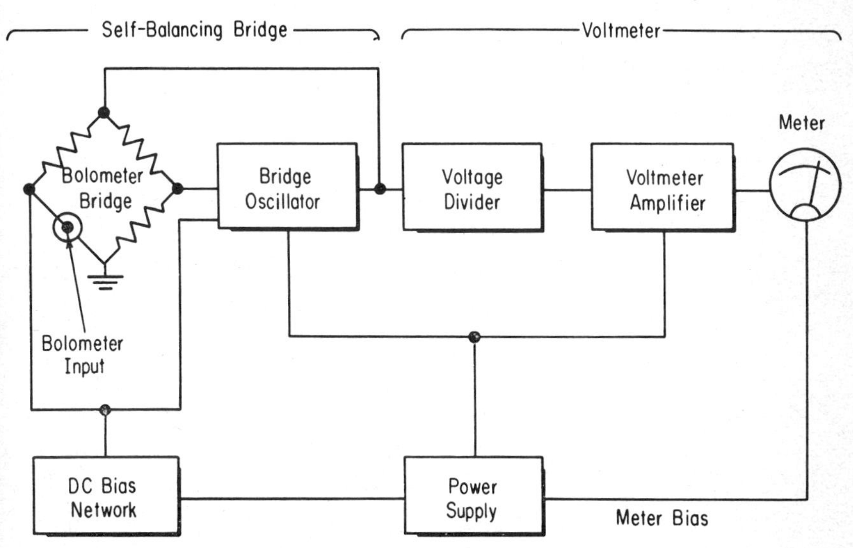

Fig. 13·2 Block diagram of the Model 430 power meter. (*Hewlett-Packard Company.*)

lower ranges, and more d-c is required to balance the bridge. In switching to a desired range the d-c bias is changed by approximately the required amount. A fine adjustment of d-c is then made, with no r-f applied, to balance the bridge.

13·4 Sources of error in power measurements

The sources of error encountered in microwave power measurements are due to the *mismatch loss*, *r-f loss*, *substitution error*, *and the error in substituted power measurement.*

Mismatch Loss. The dissipative element used in microwave power measurements is located in a transmission line mount as illustrated in Sec. 8·11. The power delivered to the mount will be a maximum when the r-f impedance of the mount is equal to the characteristic impedance of the transmission line. When the VSWR of the detector mount is not unity, power is reflected back toward the source. Also, power is reflected back from the source if the source is not matched. A chart of mismatch error (Fig. 10·2) was discussed in Chap. 10, and it was pointed out that the power delivered to the element of a mismatched mount and mismatched source lies between certain limits. A tuner can be used to tune out the mount VSWR, in which case a new source of error arises. This error is the r-f loss through the tuner and can easily be evaluated. A tuner can also be placed in the line to tune out the source VSWR, as explained in the discussion of the attenuation measurement system.

R-F Loss. The term *r-f loss* refers to the losses in the r-f components of the power measurement system. The r-f loss of a tuner used for tuning out the mismatch of a bolometer mount was mentioned in the previous section. Other losses are the resistive losses in the walls of the bolometer mount and associated transmission lines.

If the mount has adjustable elements, there are resistive losses due to poor contacts. The presence of dielectric materials can also cause r-f loss.

The resistive losses in the associated transmission lines can be evaluated by an attenuation measurement. The losses in the bolometer mount are usually described in terms of mount efficiency η.

$$\eta = \frac{\text{power absorbed by the bolometer}}{\text{power delivered to the bolometer mount}}$$

The r-f loss, mismatch loss, and substitution error are regarded as one loss called the *calibration factor*. The calibration factor of a good bolometer mount may be as high as 0.98. The National Bureau of Standards determines the calibration factor for bolometer mounts. This service has not been extended to cover all of the r-f frequency bands.

Substitution Error. The substitution error is due to the difference in heating effects of r-f and low frequency or d-c power. The error is due to the different current distribution along the wire caused by the presence of standing waves. If the bolometer element is small compared to the wavelength of the r-f, the error may be less than one per cent.

Error in Substituted Power Measurement. The accuracy of substituted power measurements depends upon the accuracy of the instrumentation equipment used. As an example, if a 200-ohm thermistor operating at 15 ma is used to measure 100 mw of power, it is necessary to measure the current with an accuracy of almost 0.03 per cent in order to obtain an accuracy of 10 per cent in the power measurement.

13·5 Extending the range of bolometric devices

Since thermistors and barretters have a limited power-handling capacity, it is necessary to reduce higher power levels to an acceptable level. The power to be measured can be reduced by the use of a calibrated-attenuator directional coupler, or power divider. There are many classes of power dividers and many forms of directional couplers which can be used. The directional coupler is frequently used, and for the purpose of illustration, a coupler and attenuator will be used in the following power measurement.

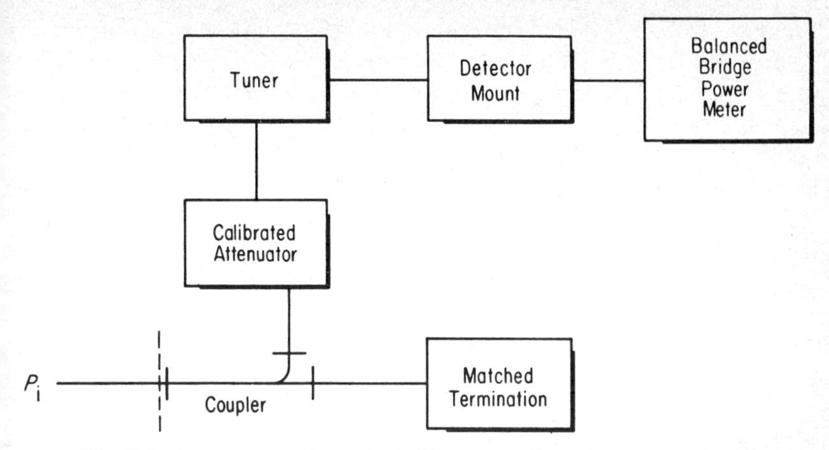

Fig. 13·3 System used to extend the range of power measurement.

Example: A system suitable for extending the range of power measurement is shown in Fig. 13·3. The calibrated attenuator is placed in the system for the convenience of setting the power meter reading to a convenient reference level in most cases, and is not an actual requirement in the extended range measurement system. A power measurement is obtained by balancing on the desired range scale of the power meter with no r-f applied. When r-f is applied, the attenuator can be used to set the power meter indication to a convenient reference such as 1 or 5 mw, etc. The power input to the coupler is calculated from the following formula.

$$P = \text{antilog} \frac{(C + \alpha R + AR)}{10} \qquad Pm\left(\frac{1}{1000}\right)$$

where P = power output, watts
αR = insertion loss of calibrated attenuator at zero dial reading
AR = attenuation reading on calibrated attenuator
C = coupling factor of directional coupler
Pm = power meter reading, mw

The factor 1/1,000 is necessary when the power meter reading is in milliwatts. This power calculation must also be corrected for the mismatch and r-f loss in the matched detector assembly. The actual source power can be obtained if the loss in the source tuner is known.

13·6 Calorimetric power meter

The calorimetric method of power measurement consists of dissipating all of the incoming power in some convenient medium and determining the resultant effect by calorimetry. Since the power absorbed is dissipated as heat, the temperature rise of the termination or cooling medium is a measure of the power absorbed. If a circulating cooling fluid is made to flow at a constant rate through or in contact with the load so that the power is dissipated in the circulating fluid, the temperature rise of the cooling fluid passing the load can be calibrated to provide a measure of the microwave power applied to the load. The calorimeter methods are most useful at high power. Water flow systems are usually unwieldy and are rarely used in field instruments.

A simplified block diagram of the Hewlett-Packard 434A calorimetric power meter is shown in Fig. 13·4. This instrument provides a means for direct measurement of power from 0.01 to 10 watts over the frequency range

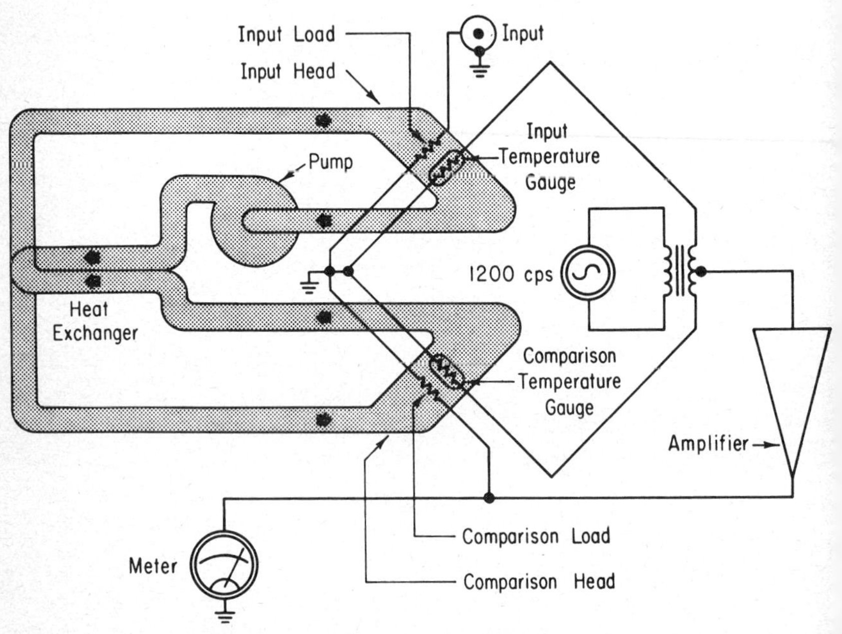

Fig. 13·4 Simplified diagram of the Model 434A calorimetric power meter. (*Hewlett-Packard Company.*)

from d-c to 12.4 Gc. The range can be extended as described in the previous section.

The model 434A consists of a self-balancing bridge which has identical temperature-sensitive resistors (gauges) in two legs, an indicating meter, and two load resistors—one for the unknown input power and one for the comparison power.

A low viscosity silicone oil of high stability is used in the circulating system. The flow rate must be equal in the two gauging regions but does not otherwise enter into the measurement. A high-performance parallel-flow heat exchanger ensures that the temperature of the oil is the same as it enters the measurement and comparison channels.

The input load resistor and one gauge are in close thermal proximity, so that heat generated in the input load resistor heats the gauge and unbalances the bridge. The unbalance signal is amplified and applied to the comparison load resistor, which is in close proximity to the other gauge, so that the heat generated in the comparison load resistor is transferred to the gauge and nearly balances the bridge.

The meter measures the power supplied to the comparison load to rebalance the bridge. The overall accuracy is specified as within 5 per cent or full scale. This is an overall figure which includes r-f loss and d-c calibration error. The accuracy can be increased by evaluating these losses.

13·7 Dry calorimeter

The PRD Model 666 dry calorimeter is illustrated in Fig. 13·5. The system consists of two identical calorimetric bodies. The active load receives the

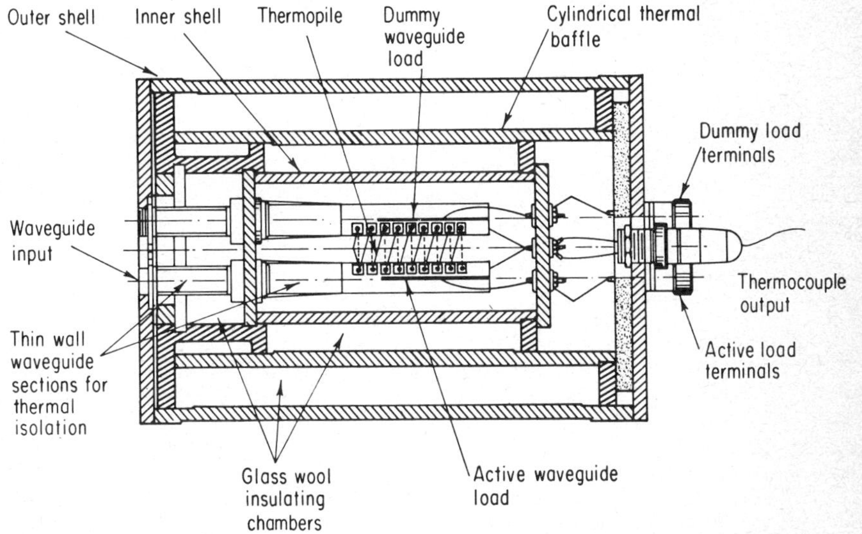

Fig. 13·5 The cross-sectional view shows the general construction of the calorimeter. (*PRD Electronics.*)

input power, and the other load is used as the thermal reference body. The temperature-sensitive detector is a thermopile. The thermopile junctions are electrically separated from the waveguide by an insulating layer. The thermopile output voltage is calibrated as a function of input power. The specified accuracy is ± 3 per cent.

13·8 Pulse power measurements using bolometers

When pulsed power is applied to the barretter, its resistance change tends to follow the pulse due to the short time constant. Therefore, errors in measurement can occur because the barretter resistance is not constant during the measurement. If the time constant is greater than the duration of the pulse and the repetition period is larger than the barretter time constant, the barretter temperature will rise linearly during the pulse and will decay exponentially to a constant value before application of the next pulse. The barretter resistance is not strictly proportional to temperature. Because of the above factors, the indicated average power in a measurement system would be in error. In the case of the bridge circuit, the bridge would read low. The change in resistance of the barretter during the pulse causes r-f impedance change looking into the mount and also causes fluctuations in the d-c bias in the bridge methods of measurement. The measurement error is directly proportional to the resistance change during the time of the applied pulse.

Thermistors and film-type bolometers are more suitable for application in pulse measurement because of the longer time constants.

13·9 Barretter integration-differentiation technique

The barretter integration-differentiation technique was developed at the Sperry Microwave Electronics Company and is incorporated in the Sperry *Microline* Model 31A2 peak power meter. This instrument operates on the principle of integrating the pulse or waveform and then differentiating the

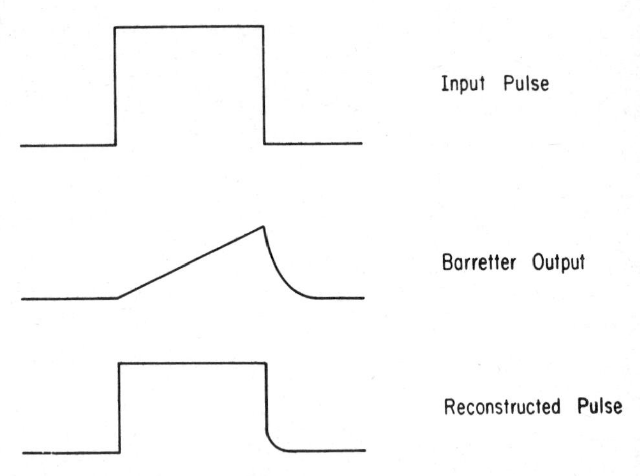

Fig. 13·6 Barretter integration-differentiation waveforms.

resulting waveform in order to restore the original waveform. This is conveniently illustrated in Fig. 13·6. In practice the pulse could be any shape since the integration-differentiation process is independent of waveshape. The integrator is the barretter which has a specific sensitivity (change in resistance per milliwatt per microsecond). The reconstructed pulse is measured with a peak-reading voltmeter which indicates peak power directly.

Direct peak power can be measured over the range from 5 to 300 mw, and the range can be extended using techniques previously discussed. The frequency range is from 1 to greater than 12.4 Gc with a dynamic range of 50 to 5,000 pulses per second, and the pulse width can be from 0.25 to 10 μsec. The accuracy is 10 per cent or better.

13·10 Notch wattmeter

The notch wattmeter system is used in applications when the average power in a pulsed r-f wave is too low to be measured directly by a power meter.

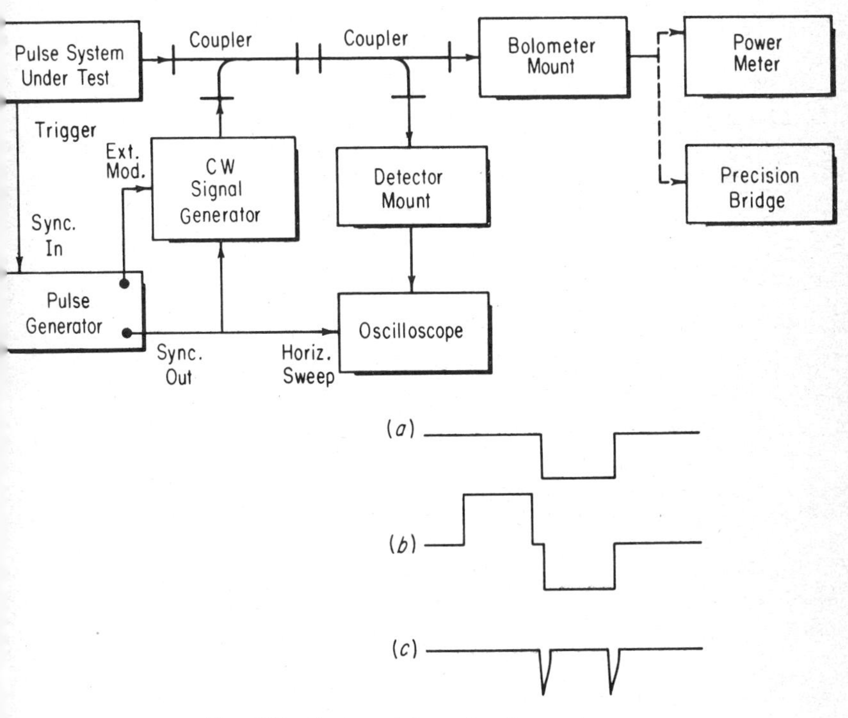

Fig. 13·7 Basic notch wattmeter system.

Various null or notch techniques are possible. A typical measurement system is shown in Fig. 13·7. The r-f pulse to be measured is compared with a c-w r-f wave which is pulsed off to form a notch. Waveform *a* represents the detected c-w signal from the signal generator which has been pulsed off to form the notch. Waveform *b* shows the detected r-f pulse and the detected notch when they are not coincident. At *c* the two waves are coincident, and the pulse to be measured is placed in the notch. Coincidence is obtained by adjusting the pulse width, delay, and repetition frequency at the pulse generator. The signal generator has been adjusted to be equal to the pulse

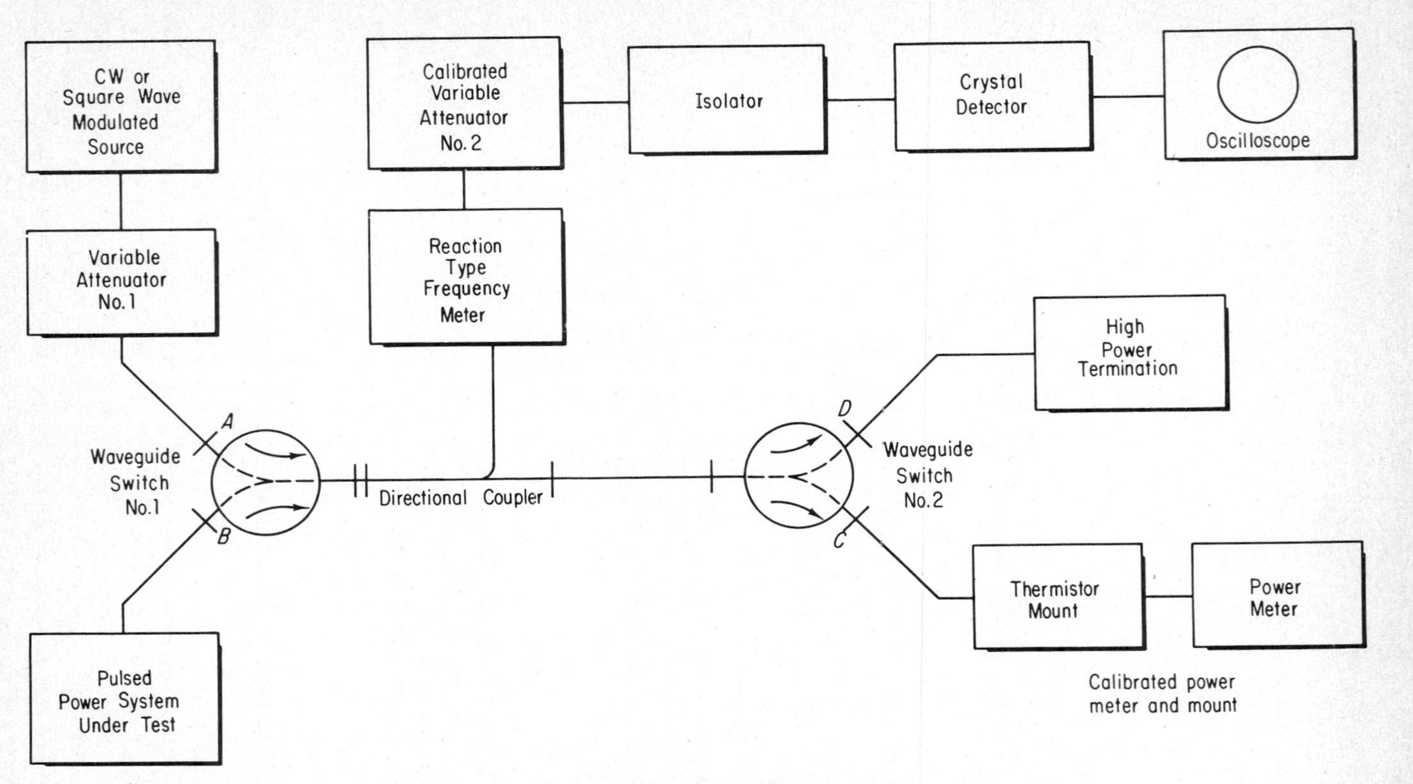

Fig. 13·8 A practical peak power measurement system.

power of the pulse under measurement. The peak power can be determined from the calibrated output of the signal generator or from the reading of the power meter to which the two signals are simultaneously applied.

13·11 A practical peak power measurement system

A system which can be used to measure peak pulse power is illustrated in Fig. 13·8. The pulse power to be measured is compared to a known power output at *C*. The input power to the calibrated detector mount and power meter combination can be square-wave modulated or c-w when the crystal is connected to the d-c input of the scope.

The detecting elements can be damaged by improper waveguide switch settings, therefore caution must be exercised when performing measurements.

A desired output reference is obtained on the output power meter by adjusting attenuator No. 1, and a reference output is obtained on the scope with the precision-calibrated attenuator (No. 2) set to zero-db reading. The pulse power is turned on and channeled to the high power termination, and the high precision attenuator (No. 2) is adjusted to obtain a pulse amplitude equal to the original reference amplitude as indicated on the scope. The decibel reading of the precision-calibrated attenuator (No. 2) plus the decibel loss from *B* to *C* indicates the power ratio of the two signals. The peak pulse power is now calculated from this decibel reading.

If square-wave modulation is used, an average power meter reading (reference) must be multiplied by 2 in order to obtain the peak reference power.

PROBLEMS

13·1 Calculate the peak power if the pulse length is one μsec, average power is 100 watts, and the interval between pulses is 1,000 μsec.

13·2 A c-w signal is supplied to the system in Fig. 13·3. The power meter reads 9 mw, the detector has a VSWR of 1.6 (not tuned out), the calibrated attenuator reads 17 db, the insertion loss of the attenuator is 0.5 db, and the coupling value is 21 db.

a. What is the value of input power P_i?

b. What is the peak power if the pulse width is two μsec and the pulse repetition rate is 1,500 pulses per second?

13·3 A square-wave modulated signal of 300 mw is applied to the input of the system in Fig. 13·3. The total attenuation to the matched detector is 21 db. Assume that there is no loss in the detector.

a. What is the output power meter reading if the detector is a barretter?

b. What is the power meter reading if the detector is a thermistor?

c. What would the total system attenuation be if the power meter read 10 mw?

CHAPTER

14

RAPID BROADBAND MEASUREMENT TECHNIQUES

Introduction. The application of precision microwave measurement techniques results in time-consuming point-by-point measurements. Even though the measurements may be performed with reduced accuracy, they are still laborious and incomplete. Production testing and engineering development of microwave components demand faster and more thorough methods of measurement.

The broadband reflectometer makes possible the direct and rapid measurement of standing-wave ratio and the magnitude of the reflection coefficient. The operating principles of the basic reflectometer system will be discussed prior to consideration of measurement applications and procedures incorporated in an improved reflectometer system. Also, a broadband attenuation measurement system will be discussed, and the procedures used to measure the characteristics of attenuators, isolators, filters, switches, couplers, amplifiers, and matched pairs of crystals or bolometers will be outlined.

14·1 Basic broadband reflectometer

The basic reflectometer system[1] is an assembly of components used to measure the magnitude of reflection coefficient, a complex vector quantity which is defined as the ratio of the reflected voltage wave to the incident voltage wave. The system samples the reflected and incident waves and calculates the ratio of their voltages. The separation of the incident and reflected waves is accomplished by directional couplers, and the calculation of their ratio is performed by a direct-reading ratio meter. The ratio meter provides a recorder output as indicated in Fig. 14·1. The sweep generator is modulated at a 1,000-cps rate since the ratio meter is designed for operation at this frequency.

The incident power is sampled by the 20-db coupler which is connected in the forward direction. The —20-db signal coupled into the auxiliary arm is detected and applied to the incident channel of the ratio meter. The sample of the incident wave which couples into the reversed coupler (−10 db) is absorbed in the internal termination of the coupler, and only a negligible portion of the power flows to the detector connected to the auxiliary arm. The remaining incident power impinges on the load termination of the reversed coupler. If the incident wave is not completely absorbed, there is a reflected wave which travels back toward the source. A sample of the reflected wave (−10 db) is coupled to the detector on the auxiliary arm and is applied

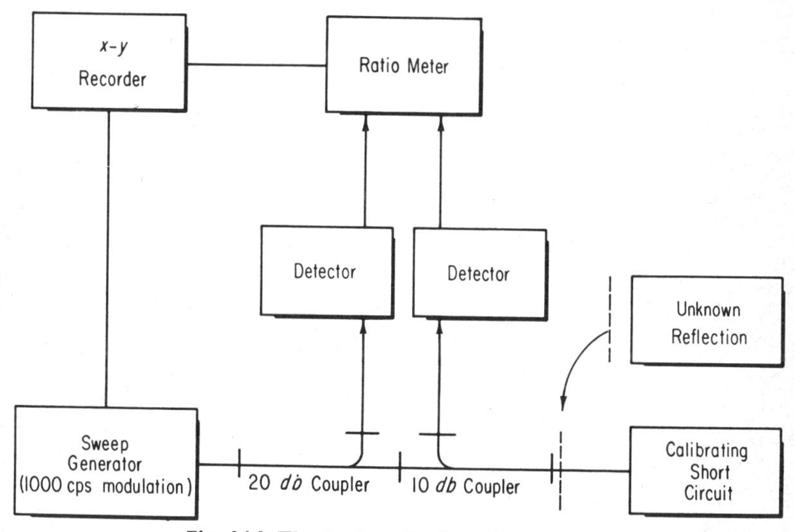

Fig. 14·1 The basic reflectometer system.

to the reflected channel of the ratio meter. As the reflected wave travels toward the generator, a 20-db signal is coupled into the internal termination in the auxiliary line of the forward-connected coupler. The remaining signal is absorbed in the generator impedance or is re-reflected down the line. The ratio meter measures and displays the magnitude of the ratio of the reflected to incident voltage, which is the reflection coefficient of the load. The detected voltage outputs from the square-law detectors are directly proportional to the sampled power. Therefore, the ratio meter is calibrated to indicate the square root of the detected power. The output from the ratio meter can be connected to an *x-y* recorder or a d-c oscilloscope which has a long persistence screen.

Various errors are associated with the basic reflectometer system because the system components do not exhibit ideal operational characteristics. The errors associated with the reflectometer system are due to *scalar quantities* and *vector quantities*.

Scalar Errors. Scalar errors result from changes in the operating characteristics of system components due to changes in signal amplitude and/or

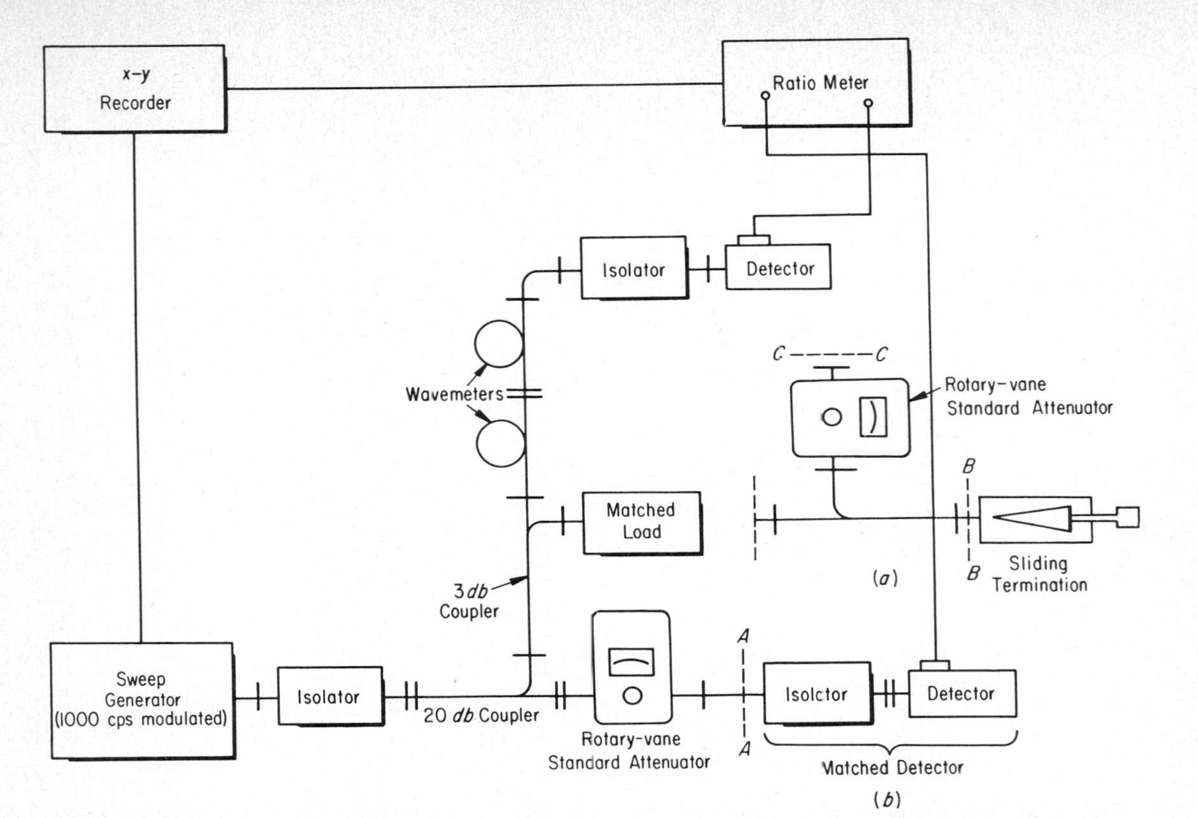

Fig. 14·2 Broadband reflectometer and attenuation measurement system. Directivity and reflectometer measurements are performed with the setup shown at (*a*). Gain characteristics of amplifiers and attenuation measurements are performed with the unknown placed at *A-A*, as indicated at (*b*).

signal frequency. These errors, usually referred to as sensitivity errors, are independent of the phase of various reflections in the system. There are frequency-sensitivity variations between the coupling coefficients of the two couplers and between the detection efficiencies of the detector mounts. The detection efficiencies of the two detectors can change as a function of signal amplitude, in which case there may be a departure from square-law response. This error is prominent when the system is calibrated at a high level and measurement is performed at a low level. For example, if the system shown in Fig. 14·1 is used to perform a measurement, a calibrating short circuit is used to obtain a known ratio of 1.0 at the ratio meter. The short circuit is replaced with the low-reflection unknown. A significant difference in signal level is encountered at the detector on the reversed coupler. In order to obtain accuracy with the basic reflectometer system, the scalar errors must be eliminated by calibrating and determining the magnitude of error across the frequency band. The error data is applied as a correction factor or is applied in the form of error-correction curves.

Vector Errors. Vector errors result from spurious signals composed of leakage signals and reflected signals which have relative phase relationships. These vector quantities may have any phase angle with respect to the reflected signal to be measured. The range of error is determined by the in-phase and 180° out-of-phase combination of these signals. The directivity of the forward coupler, reflections from the forward detector, and reflections from the source are usually reduced to low levels and can be neglected in a reflectometer system which employs a 20-db forward coupler. Therefore, the only spurious signals of consequence are the reflections from the reverse detector, the directivity leakage signal of the reverse coupler, and reflections from the source which cause re-reflections from the load. The directivity (in decibels) of the reversed coupler can be measured and the proportion of incident voltage, which will appear at the reverse detector, can be found by converting the decibel value of directivity to voltage ratio. This ratio defines the error which the directivity signal will introduce in a ratio reading. The spurious signals re-reflected from the load can be evaluated, and error curves can be constructed to form the area of ambiguity which is associated with the measurement of a specific reflection coefficient. The total scalar error plus the total spurious signal error defines the top limit of the error curve. The total scalar error minus the total spurious signal error defines the bottom limit of the error curve. Fortunately, thesc laborious and time-consuming error evaluations are not necessary. An improved reflectometer system[2] has solved the problem of scalar errors with a rather simple step.

14·2 Broadband attenuation measurements

The broadband attenuation measurement system, shown in Fig. 14·2, is used to obtain continuous calibration over the complete frequency range.

The system can be used to measure return loss, reflection coefficient, voltage-standing-wave ratio, coupler directivity, amplifier gain or loss characteristics, and attenuation characteristics of microwave structures and components.

The system is calibrated at the exact level at which measurements are made. The scalar errors, which are caused by frequency-sensitivity variations of the system components, are included in the calibration technique and can be disregarded. The scalar accuracy is limited by the accuracy of the rotary-vane standard attenuator. The overall system accuracy is limited by the directivity leakage in the reverse coupler.

The sweep generator provides the 1,000-cps-modulated variable frequency input signal. Power stabilization should be incorporated to maintain the signal level constant.

A 3-db directional coupler is placed in the auxiliary arm of the forward coupler. The main line attenuation characteristic compensates for the coupling characteristic of the 20-db coupler. The calibration lines are more level.

The two wavemeters are used to set accurate frequency limits. The wavemeters are also used to establish a calibrated frequency scale.

The ratio meter forms the ratio of the two crystal voltages which are proportional to the amplitudes of the two channel outputs. One crystal monitors the input signal to the unit under test, and the other crystal monitors the output signal from the unit under test.

The rotary-vane standard attenuator, located in the main line, is used to obtain calibration traces corresponding to the anticipated values of attenuation which are to be measured. The rotary-vane type attenuator is used because it has excellent frequency-sensitivity characteristics (usually within two-tenths of a decibel at 30- and 40-db settings). The errors due to the frequency sensitivity of the attenuator can be minimized by a point-by-point calibration over the frequency range.

General Measurement Procedures. Attenuation-type measurements are performed using the matched detector assembly, as indicated in Fig. 14·2*b*.

1. Adjust the sweep generator, recorder, and ratio meter to obtain a full-range sweep on the recorder.
2. Connect the matched detector at *A-A* and adjust the ratio meter and recorder to obtain an on-scale recorder reading.
3. Use the wavemeters to obtain a frequency scale calibration. Record the wavemeter reactions as the source sweeps slowly over the full frequency range. Actuate the recorder pin only in the area of the frequency meter reaction. Repeat the procedure to obtain the desired number of frequency calibration points.
4. Connect the matched detector at *A-A* and set the rotary-vane attenuator in the main line to the expected value of attenuation which is to be measured.

5. Record the calibration trace as the source is slowly swept over the frequency range. Obtain additional calibration traces around the expected calibration value.
6. Insert the unknown at *A-A* as shown in the diagram and plot the calibration characteristics as the source sweeps over the frequency range.

A typical calibration plot of cross-guide coupler characteristics is shown in Fig. 14·3.

When measuring the characteristics of an amplifier, the attenuator is used to attenuate the input signal to the amplifier in order to compensate for the anticipated gain of the amplifier. As an example, the nominal 20-db gain of an amplifier is to be measured. Prior to placing the amplifier in the line, calibration traces are plotted at several decibel values centered around the 5-db setting of the attenuator. The attenuator is set to 25 db, and the amplifier is inserted in the line. The final calibration sweep is plotted, and the amplifier frequency-gain response characteristics are obtained.

The isolator which is placed in front of the output crystal detector must be well matched. In cases where well-matched isolators are not available, the crystal detector can be connected to the auxiliary arm of a directional coupler which has a matched termination on the main line output.

14·3 Broadband reflectometer measurements

The broadband reflectometer system is formed by connecting the reversed coupler assembly (Fig. 14·2*a*) to the point *A-A* on the main line. *The main line rotary-vane attenuator is set to zero and is not used.* Attenuation values are set into the system with the rotary-vane standard attenuator mounted on the auxiliary line of the reversed coupler.

Reflection Coefficient and VSWR Measurements. The return loss is related to the reflection coefficient and the VSWR by

$$\text{Return loss} = 20 \log \frac{1}{\Gamma} = 20 \log \frac{\text{VSWR} + 1}{\text{VSWR} - 1}$$

The formulas are used to great advantage in the reflectometer system calibration procedure. The general procedure for measuring reflection coefficient and VSWR is as follows:

1. Terminate the reverse coupler at *B-B* with a short circuit.
2. Calculate the equivalent return loss value for the anticipated value of reflection coefficient or VSWR which is to be measured.
3. Set the rotary-vane standard attenuator, which is connected to the auxiliary arm of the reverse coupler, to the calculated value of return loss.
4. Obtain calibration traces using the same techniques outlined for attenuation measurements in Sec. 14·2.

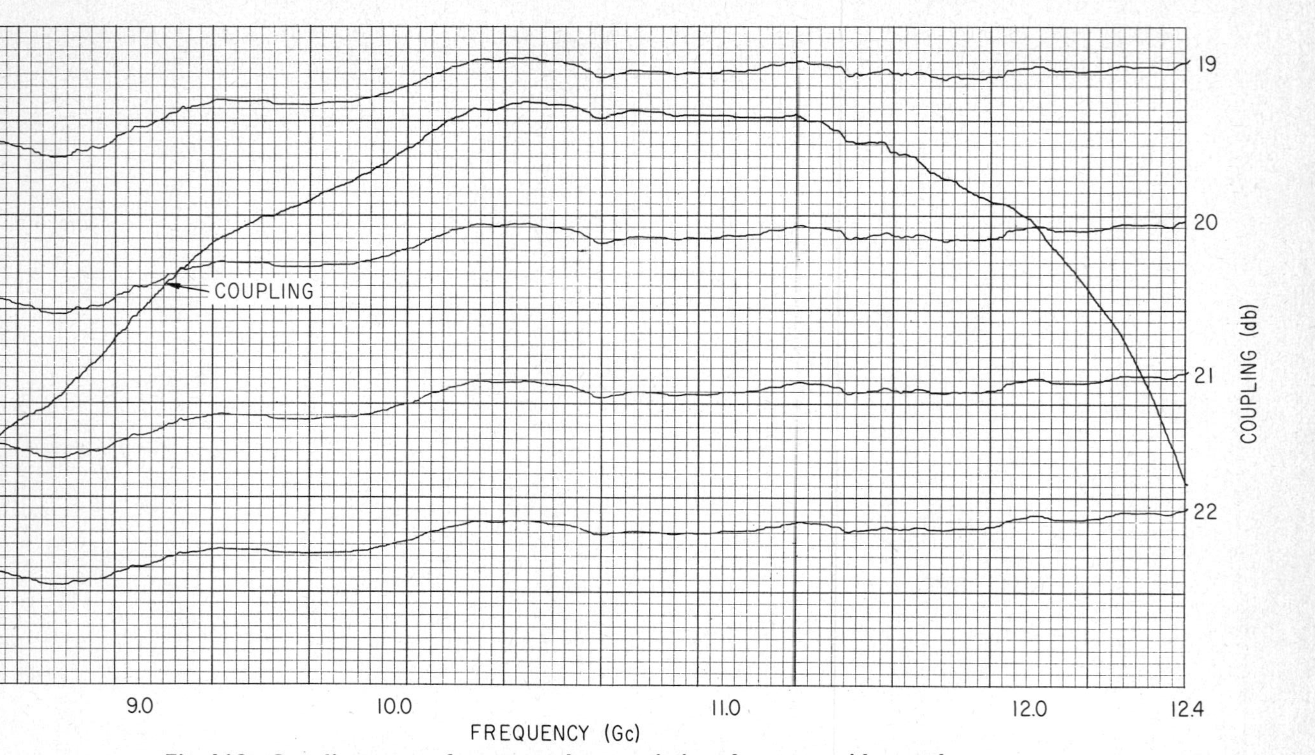

Fig. 14·3 Coupling versus frequency characteristics of a cross-guide coupler.

5. Record additional calibration traces around the anticipated value of reflection coefficient or VSWR using the same procedure.
6. Replace the short circuit at *B-B* with the unknown.
7. Set the standard attenuator to zero and obtain a recorded plot of the characteristics of the unknown termination.

Directivity Measurements. Precision directional-coupler directivity measurements were discussed in Sec. 12·14. A good approximation of the directivity value can be obtained using the present reflectometer system. The coupler under test is connected into the system as shown in Fig. 14·2*a*. A short circuit is connected at *B-B*, and the attenuator in the auxiliary arm is set to the value corresponding to the minimum specified directivity. A calibration trace is obtained, and additional traces, corresponding to higher and lower directivity values, are recorded as desired. The short circuit is replaced with a matched sliding termination. The attenuator is returned to zero, and the sweep generator is swept over the frequency range at a very slow rate. The sliding matched termination is varied back and forth over its range during the sweep. The sliding termination technique produces all possible phase combinations between the directivity signal and the signal reflected from the sliding termination. The maximum and minimum directivity values are obtained by the in-phase and out-of-phase variations of the two signals.

The idea of calibrating the reflectometer with standard reflections, or with the attenuator in the main line, may seem suited for obtaining calibration traces until the phase combinations of signals in the auxiliary arm of the reverse coupler are considered. When the short circuit is used, a large reflected signal couples into the auxiliary arm of the reverse coupler, and the small directivity signal combines with this large signal at some arbitrary phase. Both signals are attenuated by the standard attenuator in the auxiliary arm of the coupler and maintain their relative amplitudes. However, in the case of a low-reflection standard reflection connected at *B-B*, it is noted that a small reflected signal combines with the small directivity signal. It was shown in Sec. 12·14 that large errors are possible when measuring the phasor combination of two signals whose relative magnitudes approach the same value.

PROBLEMS

14·1 Calculate the values of attenuation which are introduced with the rotary-vane attenuator in the reflectometer system for the following values of reflection coefficient and VSWR.

$$\text{VSWR} = 1.10,\ 1.15,\ 1.20,\ 1.35,\ \text{and}\ 3.6$$

$$\Gamma = 0.05,\ 0.10,\ 0.20,\ 0.50,\ \text{and}\ 0.64$$

14·2 Draw a diagram to illustrate the use of a directional coupler to obtain a matched detector for the measurement system in Fig. 14·2.

REFERENCES

1. *Hewlett-Packard Journal*, vol. 6, no. 1–2, September–October, 1954, vol. 12, no. 4, December, 1960.
2. J. K. Hunton and Elmer Lorence, Improved Sweep Frequency Techniques for Broadband Microwave Testing, *Hewlett-Packard Journal*, vol. 12, no. 4, December, 1960.

CHAPTER

15

MICROWAVE MEASUREMENT AND CALIBRATION TECHNIQUES

15·1 Frequency measurements and calibration techniques

The frequency of active microwave sources, which operate in the 220 Mc to 12.4 Gc frequency range, can be measured using the setup shown in Fig. 15·1. The technique can be extended to higher frequency ranges by use of external waveguide mixers and harmonic generators.

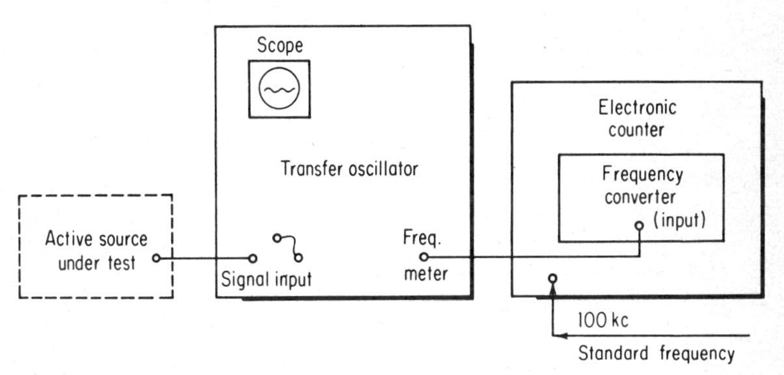

Fig. 15·1 Frequency measurement system.

The transfer oscillator generates a stable frequency which is adjustable over the range from 100 to 220 Mc. The frequency of the transfer oscillator is monitored by the high frequency electronic counter. The signal to be measured is compared with the transfer oscillator harmonics in a mixer. The frequency of the unknown signal is obtained by multiplying the reading on the frequency counter by the proper harmonic number.

The proper harmonics are determined by adjusting the transfer oscillator to locate two adjacent fundamental frequencies (f_1 and f_2) which mix with the input signal to produce beat-frequency indications on the scope. The

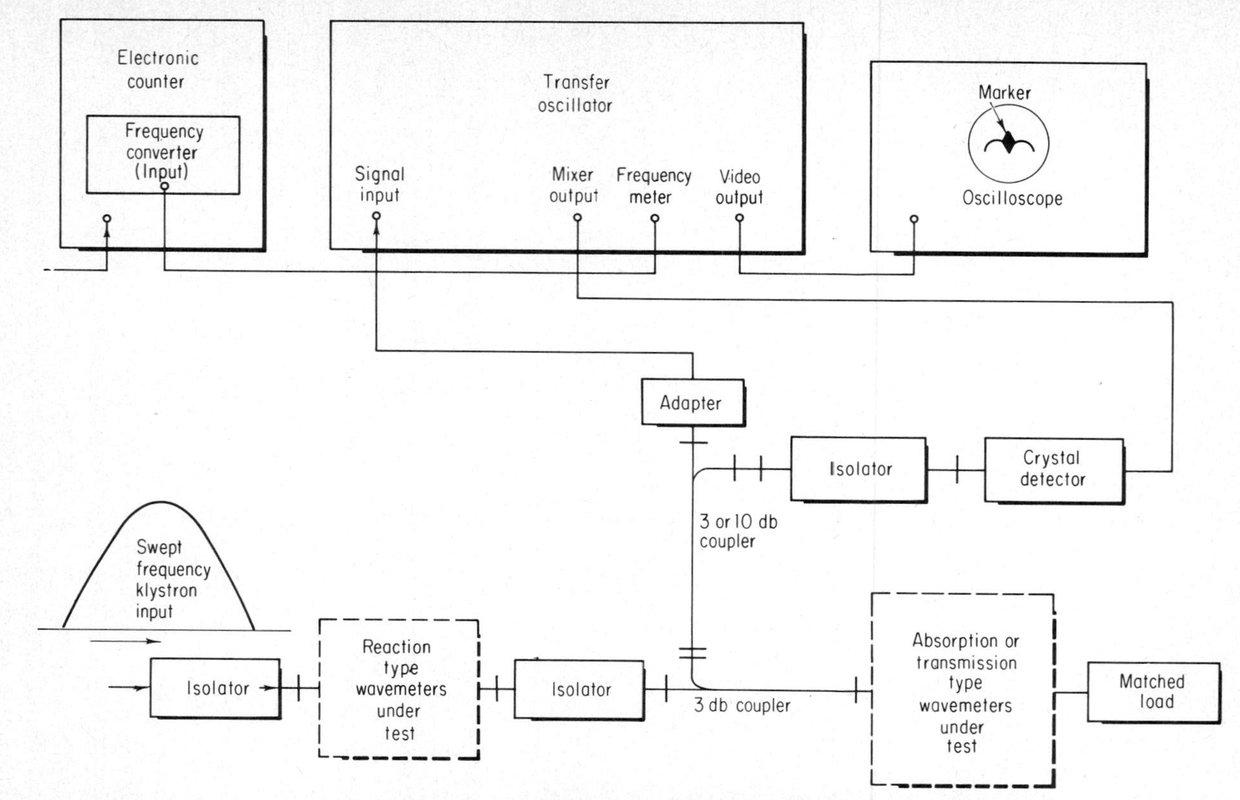

Fig. 15·2 X-band frequency meter calibration system.

harmonics which create the beats, and the exact frequency of the unknown signal are determined from the two fundamental frequencies.

The equations which are used to calculate the unknown frequency and the transfer oscillator harmonics are as follows:

$$f_x = Nf \tag{15·1}$$

$$N_1 = \frac{f_2}{f_1 - f_2} \tag{15·2}$$

$$N_2 = \frac{f_1}{f_1 - f_2} \tag{15·3}$$

where f_x = unknown frequency
N = harmonic number
f = transfer oscillator fundamental frequency for harmonic number N
N_1 = harmonic number of higher fundamental frequency
N_2 = harmonic number of lower fundamental frequency
f_1 = higher of the two fundamental frequencies
f_2 = lower of the two fundamental frequencies

The wavemeter, also referred to as a frequency meter, is a necessary part of every microwave system. Accurate calibration of these devices can be obtained by several techniques.

The calibration system shown in Fig. 15·2 is used to calibrate reaction, absorption, and transmission type wavemeters at X band. The swept-frequency signal from a sawtooth-modulated klystron, or from a backward-wave oscillator, is sent down the main line of the system. The system is described and illustrated in terms of a klystron signal.

The isolator, which is placed in the line between the reaction type frequency meters and the 3-db coupler, is used to isolate the reaction type cavities from the transmission type cavity. The 3-db coupler is connected in the reverse direction in order to obtain a reaction type display of transmission type wavemeter reaction. The swept-frequency signal (klystron mode) is reflected back toward the source, and one-half of the power is coupled into the auxiliary arm of the coupler. Since one-half of the power travels back toward the source, the isolator must present a good match to the reflected signal. Any reflection can cause frequency pulling of the transmission type wavemeter being calibrated.

The swept-frequency input signal is applied to the high frequency mixer in the transfer oscillator. The signal from the internal variable-frequency oscillator is mixed with the swept-frequency signal. Beat notes or markers are obtained when the input signal and harmonics of the variable-frequency oscillator coincide. The beat note or marker is illustrated in Fig. 15·3. As the swept-frequency approaches the frequency of the harmonic, the difference frequency is within the passband of the vertical amplifier in the oscilloscope.

The zero beat appears at the center of the marker, which indicates coincidence of the two frequencies.

The variable-frequency-oscillator output is applied to the frequency converter of the electric counter. The mixing frequency control on the counter frequency converter unit is set to obtain a reading of approximately 10 Mc on the counter. A harmonic marker can be set up for any desired frequency. The oscillator fundamental frequency f is obtained by adding the mixing frequency control reading (Mc) to the counter display. The harmonic number

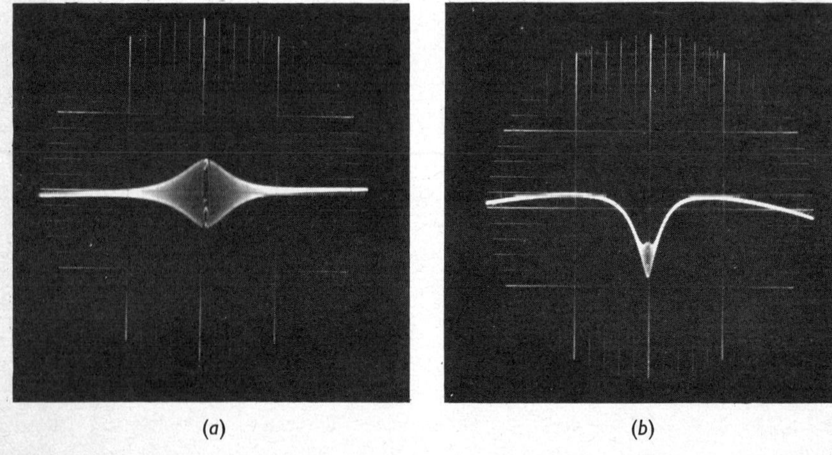

(*a*) (*b*)

Fig. 15·3 (*a*) Beat note. (*b*) Beat note located at the bottom of the frequency meter reaction.

is obtained by adjusting the transfer oscillator to two adjacent zero beats and calculating N_1 from Eq. (15·2).

A harmonic table should be calculated so that mixing frequency control settings and counter readings can be quickly set to obtain a harmonic at any desired frequency.

Markers spaced at 100-Mc intervals can be obtained by setting the counter frequency mixing control to 190 Mc and tuning the transfer oscillator to obtain a frequency of 10 Mc as displayed on the counter. The transfer oscillator frequency of 200 Mc produces large markers spaced 200 Mc apart and smaller markers spaced at 100-Mc intervals.

A part of the signal is detected and applied to the mixer output of the transfer oscillator in order to improve the mode display on the oscilloscope.

The wavemeter calibration is performed by adjusting the frequency meter reaction so that the marker, corresponding to the known frequency f_x, is centered in the frequency meter reaction.

The electronic counter can be gated by a 100-kc standard frequency which is checked and compared with WWV.

The frequency-meter calibration system shown in Fig. 15·4 can be used, where the necessary components are available, up to 40 Gc.

The basic oscillator in the PRD Electronics Model 500 frequency multiplier covers the range from 3.25 to 3.5 Mc. The basic oscillator frequency is multiplied up to the 270 Mc range at output *E*. The output from the frequency multiplier is applied to a crystal harmonic generator at the input of the traveling-wave-tube amplifier operating in the 2 to 4 Gc range. Harmonics generated in the 2 to 4 Gc range are amplified by the TWT amplifier. These amplified harmonics are applied to a crystal harmonic generator which generates the required harmonics up to 40 Gc.

The harmonic number N is referred to output B of the PRD 500 frequency multiplier (9.75 to 10.5 Mc). This frequency is displayed on the counter which is gated by a standard 100-kc signal checked against WWV.

The unknown frequency and harmonic number are determined as follows:

Assume that f_{b2} is greater than f_{b1}.

f_{b2} = frequency at output B of the PRD 500 frequency multiplier corresponding to the marker which identifies f_x

f_{b1} = frequency at output B of the PRD 500 which corresponds to a marker adjacent to the marker which corresponds to f_{b2}

K = the separation, in digits, of the marker adjacent to N

The value of K indicated as being probably 3, 9, or 27 arises since the frequency-multiplier stages following output B of the PRD 500 are triplers. The value of N is established from

$$(N + K)f_{b1} = f_x \tag{15·4}$$

$$N = K\frac{f_{b1}}{f_{b2} - f_{b1}} \tag{15·5}$$

The quantity $f_{b1}/(f_{b2} - f_{b1})$ can have a decimal associated with it because of the loss in precision in obtaining the f_b values and taking the differences $f_{b2} - f_{b1}$. The value is therefore rounded off to the nearest whole number. If the assumed value of K was not correct, the calculated value of f_x will be unreasonable.

Harmonic tables should be calculated for values of f_{b2} corresponding to f_x.

The frequency calibration is performed by tuning the swept-klystron frequency to the approximate frequency to be measured. A precision-calibrated frequency meter is used to obtain the approximate frequency setting. The PRD 500 is tuned to the desired f_{b2} obtained from the tabulated values. The tripler stages are tuned for maximum output, and the marker corresponding to f_x should appear on the swept-frequency mode. The klystron frequency is tuned so that the marker is at the top of the mode. The frequency meter is tuned so that the marker is centered in the reaction. The frequency-meter dial reading corresponds to the known harmonic f_x.

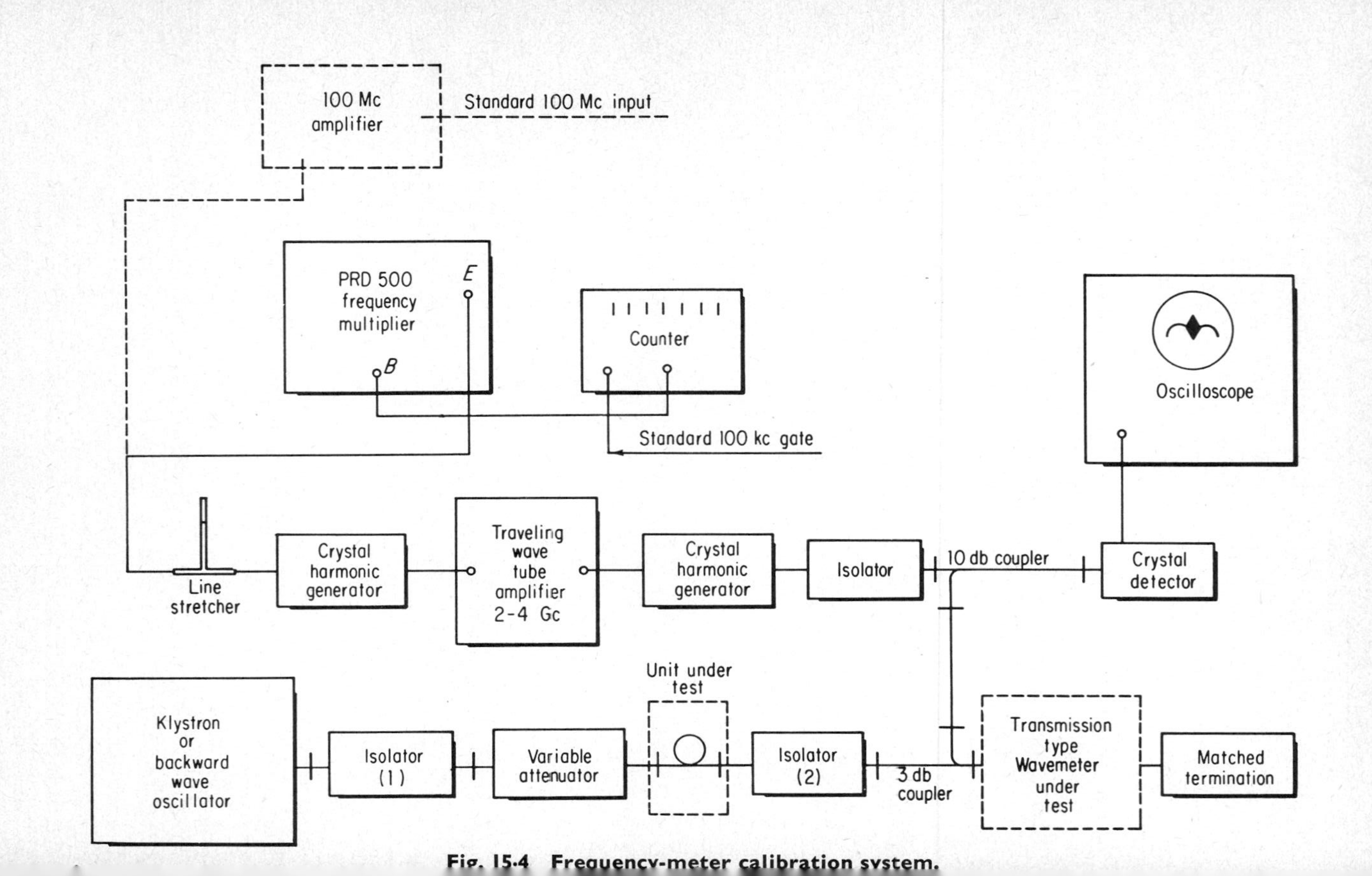

Fig. 15.4 Frequency-meter calibration system.

Rapid frequency calibration can be obtained by setting up a system in which markers are available at 100-Mc intervals.

Frequency calibration markers spaced at 100-Mc intervals can be obtained by amplifying a standard 100-Mc signal and using this amplified signal to generate harmonics as indicated in Fig. 15·4 (dotted block).

A system which can be used to perform frequency-meter calibration in the 40- to 70-Gc range is shown in Fig. 15·5. The swept-frequency klystron mode is detected by the crystal detector and displayed on the oscilloscope. The klystron frequency and wavemeter to be calibrated are tuned to the approximate calibration frequency (a calibrated frequency meter should be used to obtain the proper setting).

The frequency-indicating marker is generated in the harmonic generator. The exact X-band frequency f_{xb}, of which a proper harmonic gives the desired frequency f_v, is set up by use of the transfer oscillator, transfer oscillator synchronizer, and electronic counter. The transfer oscillator is synchronized to the X-band signal. A harmonic table is calculated using the following equations:

$$f_{xb} = \frac{f_v}{N_x} \tag{15·6}$$

$$f_{hi} = \frac{f_{xb} + 30}{N_t} \tag{15·7}$$

$$f_{lo} = \frac{f_{xb} - 30}{N_t} \tag{15·8}$$

where f_v = a frequency in the range from 40 to 70 Gc at which the wavemeter is to be calibrated

f_{xb} = fundamental X-band frequency which is used to generate the appropriate harmonic corresponding to f_v

N_x = harmonic of f_{xb} which corresponds to f_v

f_{hi} = higher of the two transfer oscillator settings which correspond to the selected X-band frequency

f_{lo} = lower of the two transfer oscillator settings which correspond to the selected X-band frequency

N_t = transfer oscillator harmonic which determines f_{hi} and f_{lo}

Procedure

1. For a desired frequency f_v select a harmonic number N_x which results in a fundamental frequency f_{xb} which is within the range of the stabilized X-band source.
2. Set the frequency converter dial and counter controls to read f_{hi} in cycles.
3. Tune the X-band source to the exact required reading as displayed on the counter.

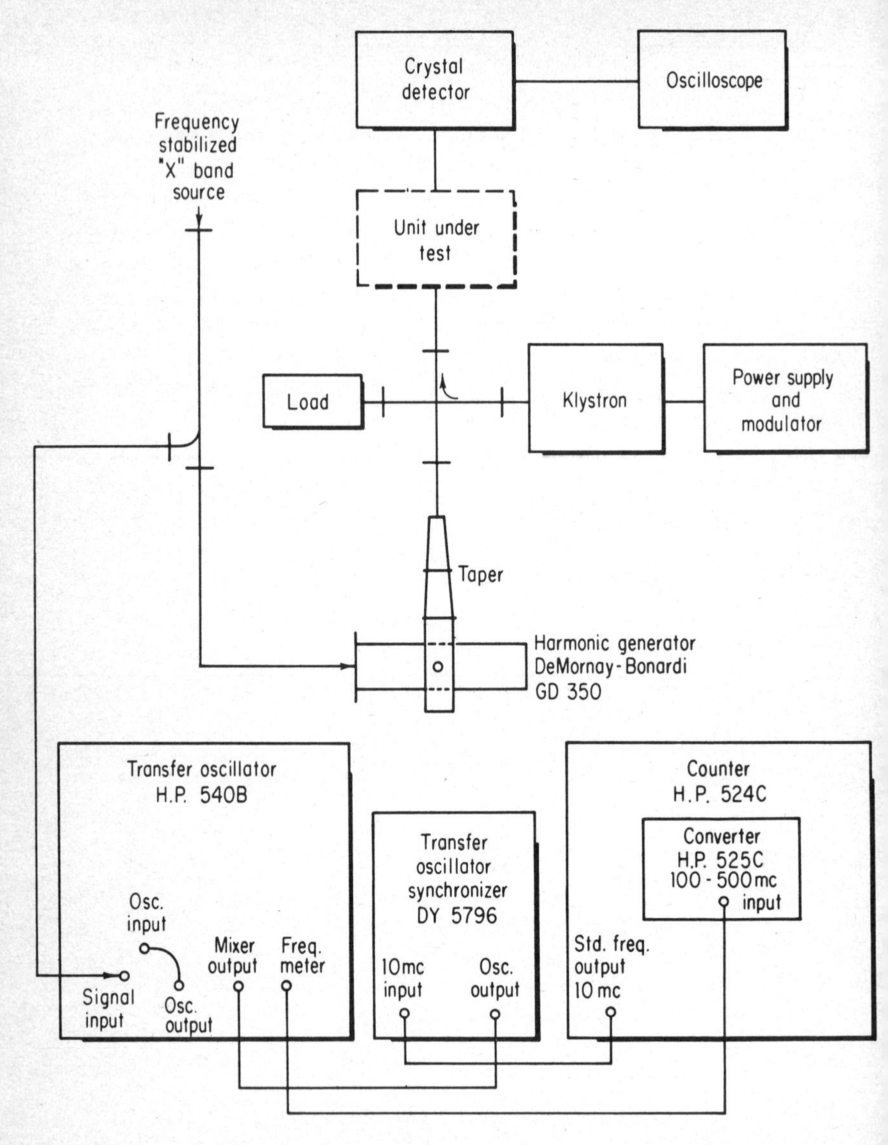

Fig. 15·5 Frequency-meter calibration system in the 40 to 70 Gc range.

4. The harmonic beat note should appear on the oscilloscope display when the swept-frequency klystron is tuned within the range of f_v.
5. If a marker cannot be detected after all efforts have been expended to optimize harmonic generation, select another harmonic and repeat the outlined procedure.

Example

$$f_v = 55 \text{ Gc}$$
$$N_x = 60$$
$$f_{xb} = \frac{55 \cdot 10^9}{60} = 9{,}166.666 \text{ Mc}$$
$$N_t = 44$$
$$f_{hi} = \frac{9{,}166.666 + 30}{44} = 209.015151$$
$$f_{lo} = \frac{9{,}166.666 + 30}{44} = 207.651616$$

The frequency converter dial is set to 200, and the counter controls are set to indicate 9.015151 when the X-band source is tuned to obtain this exact counter reading. The 60th harmonic should appear on the swept-frequency display. The frequency meter reaction is adjusted to center the marker in the reaction. The frequency-meter dial reading corresponds to 55 Gc.

15·2 Phase-shift measurements

Several phase-shift measurement techniques are based on the use of a calibrated short circuit located in a precision waveguide casing. This basic phase standard offers an accuracy approaching a small fraction of a degree. However, large errors in phase-shift measurements can result in certain circuit arrangements.

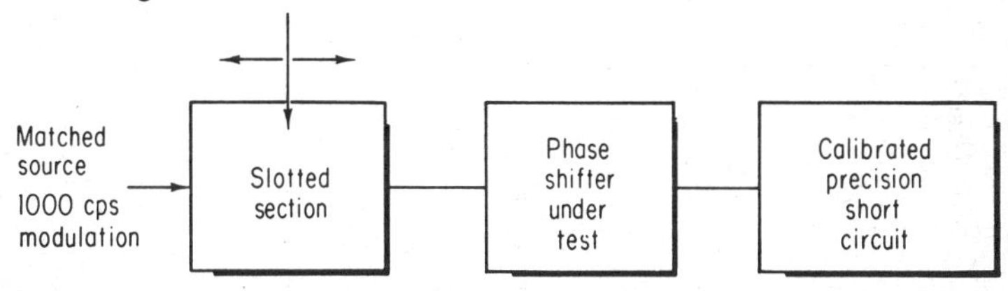

Fig. 15·6 Phase-shift measurement system using the slotted section as a null detector.

One technique for making phase-shift measurements using a precision short circuit is shown in Fig. 15·6.

The phase shifter is calibrated by keeping the slotted section probe fixed in position while introducing phase shift with the test phase shifter and maintaining the null at the probe with the calibrated short circuit. The phase shift is determined from

$$\phi = 360 \frac{d}{\lambda_g} \tag{15·9}$$

where ϕ = phase shift, degrees
λ_g = waveguide wavelength in the calibrated short circuit, in.
d = displacement of the calibrated short circuit, in.

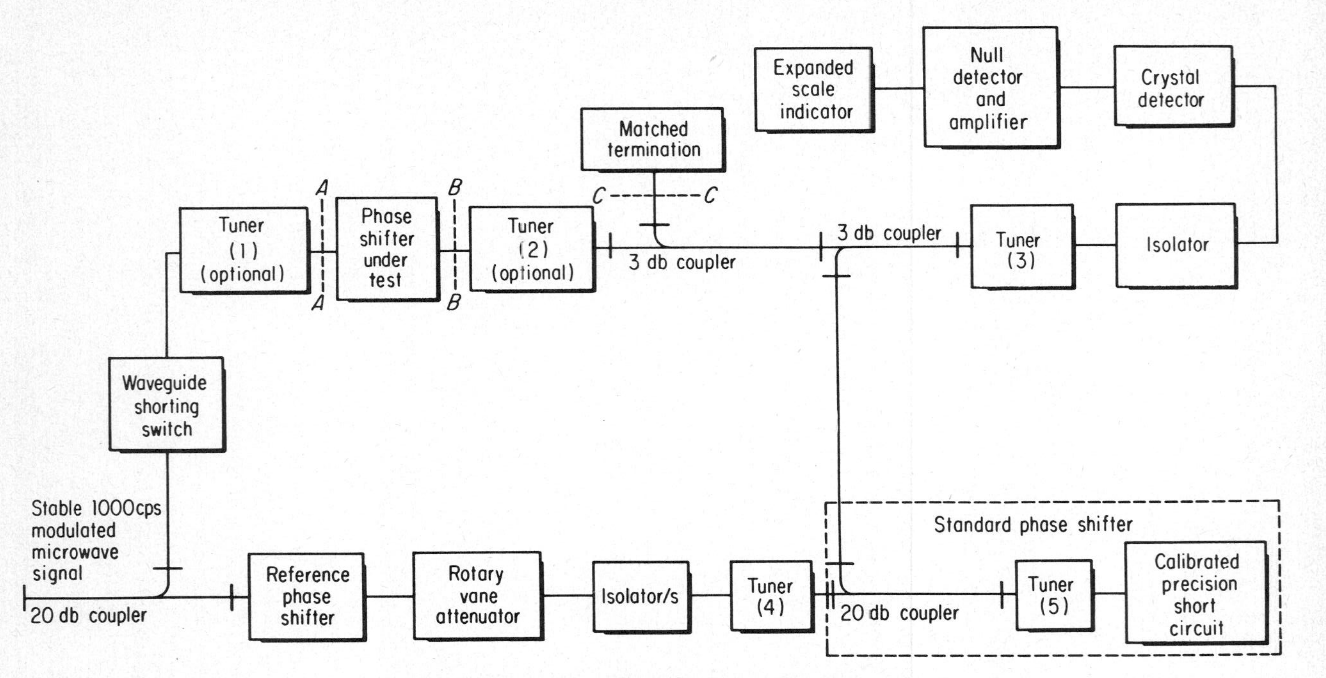

Fig. 15·7 Phase-shift measurement system.

An accurate calibration cannot be obtained using this technique unless the phase shifter under test and the source are perfectly matched. Since phase shifters are not perfectly matched, errors are introduced because of the interaction between the signal reflected from the short circuit and the signal reflected from the mismatch of the phase shifter under test.

The errors associated with the previous measurement technique are eliminated by the use of a two-channel system,[1] or microwave bridge, in which the phase shifter under test is isolated from the standard calibrated short circuit. The two-channel system is shown in Fig. 15·7.

A stabilized microwave signal is supplied to the input 20-db directional coupler. The source is 100 per cent modulated at 1,000 cps. A ferrite type amplitude modulator is used to modulate a stabilized c-w source in order to reduce frequency modulation.

The signal is split into two channels by the input 20-db coupler. Power flows through the two channels, recombines in a 3-db directional coupler, and both signals flow to a matched null detector and amplifier. A null is obtained in the output when the two signals are equal in amplitude and 180° out of phase.

The waveguide-shorting switch is used to shut off power on the upper channel during adjustment of tuners 3, 4, and 5. Tuners 1 and 2 are tuned to present a matched source and a matched load to the phase shifter under test. The first 3-db coupler is used when tuning tuner No. 3. The signals are combined in the other 3-db coupler. Tuner No. 3 is used to obtain maximum isolation between the two channels.

The phase shifter in the lower channel is used to set up reference phase-shift values. The rotary-vane attenuator is used to adjust the signal in the lower channel to equal the signal level in the upper channel. A rotary-vane attenuator is used because it does not exhibit phase changes with changes of attenuation. This is important since it is sometimes necessary to adjust the signal level because of attenuation changes of the unit under test. This is especially true when the system is used to measure phase and gain characteristics of parametric amplifiers or other active devices. If a phase measurement system employs an attenuator which introduces phase shift, the attenuator cannot be changed once the initial signal null has been obtained.

The isolators in the lower channel serve to present a constant impedance to tuner No. 4.

The 20-db directional coupler and calibrated short circuit form the standard phase shifter. Phase shift measurements are made on a differential basis, that is, the phase shift is calculated from a measurement of the difference between two micrometer readings on the calibrated short circuit. Two basic sources of error arise from the use of the directional coupler and calibrated short as a phase standard. The imperfect directivity of the coupler permits a signal to reach the auxiliary arm without being reflected from the calibrated short. Tuner No. 5 is used to eliminate the directivity leakage signal. The source mismatch causes phase errors. The signal from the mismatch returns to the calibrated short to be re-reflected in combination with the reflected signal from the short. The phase addition of the two signals varies as the short is varied. Tuner No. 4 is tuned to eliminate this error.

Tuners 1 and 2 are tuned by channeling the normal input power through a slotted section into the line at *A-A* and *B-B* and performing the necessary tuner adjustments. These two tuners can be eliminated if the system components present a matched load and source to the unit under test.

With power supplied to the input 20-db coupler, the waveguide switch is placed in the "short" position so that no power flows in the upper channel. The probes of tuners 4 and 5 are withdrawn, and a rapid-moving short-circuit termination is placed in the standard phase-shifter casing. The short circuit shown in Fig. 8·2 facilitates this operation since the short can be fast-moving or micrometer-driven.

The signals in the two channels are combined in the second 3-db coupler. Isolation of the two channels is obtained by adjustment of tuner No. 3. A sensitive crystal detector is placed at *C-C* and connected to a standing-wave indicator. A matched termination is connected at *B-B*. This is necessary since tuner No. 3 can be incorrectly tuned if there is a leakage signal through the waveguide switch and tuning is performed with the unit under test in the line. Tuner No. 3 is adjusted to eliminate any variation at the standing-wave indicator when the short circuit in the standard casing is varied over a wavelength. Adjustment of the tuner usually carries the signal down to the noise level. Each parameter is adjusted on either side of noise, and the tuner is locked at the average settings of probe position and probe depth. The crystal detector is then replaced with the matched termination.

Five stub tuners are employed for tuners 4 and 5 if available. The probes of tuner No. 5 must be completely withdrawn prior to adjustment of tuner No. 4. Tuner No. 4 is adjusted to eliminate variations which occur on the expanded scale meter when the moving short circuit is varied over its range. This eliminates the source mismatch.

The moving short is replaced with a broken or mismatched sliding load which has a nominal return loss in the range of 25 to 35 db (corresponding to VSWR values in the range from 1.12 to 1.04). Tuner No. 5 is adjusted to eliminate variations on the expanded scale indicator which result when the mismatched load is varied over its range. This tuning procedure was thoroughly explained in Sec. 12·15. Upon completion of the tuning procedure, the matched termination is removed from *B-B*, and the phase shifter to be tested is placed in the system.

The calibrated micrometer-driven short circuit is placed in the standard phase-shifter casing, and the waveguide shorting switch is placed in the normal "open" position. The phase shifter under test is set to zero phase-shift reading, and the calibrated short circuit is set to a convenient reference micrometer setting. The reference phase shifter and rotary-vane attenuator are adjusted to obtain equal amplitude and 180°-out-of-phase signals as indicated by a null at the output null detector and amplifier.

The performance of the standard phase shifter and the complete measurement system is checked after the system has been completely assembled. The phase-shifter measurements are made on a differential basis (by taking the difference between two micrometer readings). If the tuning procedure has achieved phase-shifter-system linearity, a given phase shift in the test phase shifter will cause the same differential micrometer reading on the standard phase shifter regardless of the portion of the micrometer range used. Therefore, the system can be made self-checking. The system error evaluation is achieved by successive measurements of a chosen standard short-circuit micrometer displacement d when the initial reference point is changed by successive values up to 360°. As an example, assume that the standard phase-shifter micrometer is set to zero, and the phase shifter under test is also set to zero dial reading. The reference phase shifter and rotary-vane attenuator in the lower channel are adjusted to obtain a null at the output detector. The standard phase-shifter micrometer is displaced by $d = 0.1$ in. and the phase shifter under test is adjusted to reestablish the null. The corresponding phase shift is recorded. The test phase shifter is set back to the same zero or initial starting point. The reference phase shifter is changed by, for example, 60°, in such a direction that the standard short-circuit micrometer may be set to give the same displacement d. The standard phase-shifter micrometer is again displaced by $d = 0.1$ in. The test phase shifter is adjusted to reestablish the null, and the phase shift value is recorded. This process is continued until the reference phase shifter has been changed through 360°; it can be repeated for other standard phase-shifter displacements d_1, d_2, etc. The overall system error is given by the overall apparent change in the test phase-shifter readings for the given phase-shift displacements of the standard short.

Calibration of the unit under test is performed by using the test unit to measure fixed short displacements, or by using the standard phase shifter (short) to measure the phase shift associated with particular dial readings of the unit under test. Phase-shift is calculated from

$$\phi = 720 \frac{d}{\lambda_g} \tag{15·10}$$

where ϕ = electrical phase shift, degrees
d = displacement of calibrated short circuit, in.
λ_g = waveguide wavelength, in., as measured with calibrated short circuit in precision waveguide casing after final assembly of system

If the standard short circuit is adjusted from one null to an adjacent null, the standing-wave pattern in the waveguide has been shifted one-half wavelength. However, the phase of the reflected signal has changed 360° because of the two-way travel of the signal. If the standard short circuit is lengthened 0.1 in., the signal path length from the original reference is 0.1 in. down the line in the forward direction and 0.1 in. back to the original reference plane, a total distance of 0.2 in. Therefore, when the waveguide wavelength λ_g is measured as the distance to the second null, the phase change is 720°.

The two-channel (microwave bridge) phase-shift measurement technique can be used in any frequency band where the necessary components are available.

15·3 Measurement of Q

The figure of merit Q of a resonant cavity is defined as the ratio of the resonant frequency to the frequency difference between the half-power frequencies.

$$Q = \frac{f_0}{f_1 - f_2} \tag{15·11}$$

The Q of a cavity can be determined experimentally by many techniques.[2] The measurement of bandwidth between the half-power points is the most frequently used technique. Space limitations do not permit a detailed explanation of the various techniques. The impedance measurement technique is discussed in detail because of the relative simplicity of measurement and minimum equipment requirements.

The *transient decay* or *decrement method* of measuring cavity Q is based on measurement of the decay of energy in a cavity. Application of sharp pulses of r-f power, at the resonant frequency, causes an exponential rise and decay of power in the cavity. The Q can be calculated from

$$Q = 4.343\,\frac{(t_1 - t_2)\omega}{A} \tag{15·12}$$

where A is the decibel attenuation (ratio of powers) of the decay between two points at times t_1 and t_2, and $\omega = 2\pi f_0$ (f_0 is the resonant frequency of the cavity).

In *dynamic measurements* of Q, the cavity response is presented on a swept-frequency display. The half-power points on the response curve are located by frequency markers which are generated by heterodyning techniques as set forth in Sec. 15·1.

Cavity Q can be measured in the basic attenuation system using the *impedance measurement technique*. The output section of the attenuation measurement system is shown in Fig. 15·8.

When tuned off resonance (detuned), the transmission or absorption type cavities present a short circuit to incoming signals. The slotted section probe can be adjusted to locate the minimum of the standing-wave pattern, which represents the *detuned* short position. The source must present a perfect match to the reflected signal.

If the cavity is tuned to exact resonance, the cavity absorbs power and the VSWR in the line decreases. Therefore, the output power level at the standing-wave indicator increases. If the cavity is tuned to cxact resonance and if there are no reflections from the source, the slotted section probe will be at a voltage maximum or voltage minimum on the standing-wave pattern. The cavity presents a pure resistance to the line. If the slotted section probe, at the detuned short position, is at a voltage maximum of the standing-wave

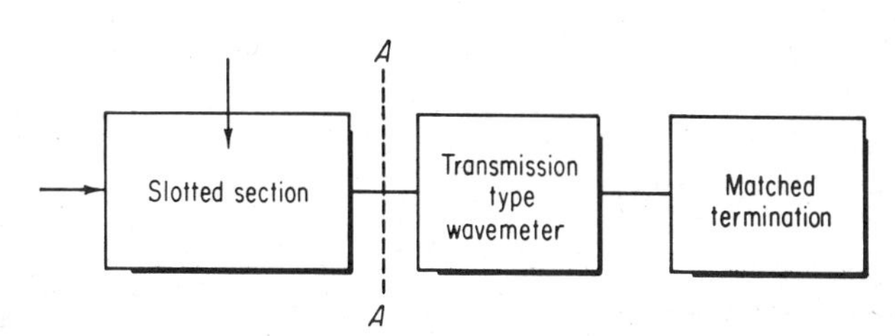

Fig. 15·8 Q measurement using the impedance measurement technique.

pattern, the cavity impedance is $Z_c = Z_0(\text{VSWR})$. The cavity-coupling parameter β is defined as Z_c/Z_0 and is greater than unity. The cavity impedance is a maximum, greater than Z_0, and the cavity is overcoupled. Alternately, if the slotted section is at a minimum of the standing-wave pattern, $Z_c = Z_0/\text{VSWR}$, and the cavity is undercoupled.

Measurement Procedure

1. Tune the cavity off resonance and locate the detuned short position by tuning the slotted section probe to a minimum of the standing-wave pattern.
2. Tune the cavity to resonance as indicated by a maximum output reading on the standing-wave indicator.
3. Set up a reference level on the standing-wave indicator at a power level within the square-law-response region of the probe-detecting element.
4. Tune the cavity to each side of resonance and record the cavity frequencies (f_1 and f_2) at which the standing-wave indicator level drops by 3 db.
5. Tune the cavity back to resonance. Move the slotted section probe and note whether the probe (at the detuned short position) was at a minimum or at a maximum of the standing-wave pattern. If at a maximum, $Z_c/Z_0 = \text{VSWR}$. If at a minimum, $Z_c/Z_0 = 1/\text{VSWR}$.

6. Measure the VSWR and calculate the parameters as follows:

$$Q_L = \frac{f_0}{\Delta f}$$

$$Q_0 = (1 + \beta)Q_L \qquad \text{single-port cavity}$$

$$Q_0 = (1 + 2\beta)Q_L \qquad \text{two-port equal-coupled cavity}$$

where Δf = bandwidth at 3-db points $(f_1 - f_2)$
Z_0 = waveguide characteristic wave impedance
Z_c = cavity impedance
Q_L = loaded Q of the cavity
Q_0 = unloaded Q of the cavity
f_1 = high-frequency 3-db point
f_2 = low-frequency 3-db point
f_0 = cavity resonant frequency

The Q of reaction type frequency meters can be measured using the same technique preceded by the following:

Tune the cavity near resonance; then alternately vary the cavity and short circuit to obtain a minimum VSWR. Repeat the previous measurement procedure. If the probe was not at exact minimum or maximum as noted in step 5, the original adjustments of the cavity and short circuit were not properly tuned for minimum VSWR. The complete procedure must be repeated.

15·4 Noise measurements

The signal level that will produce a satisfactory output signal from a detection system is limited by noise that is presented to the system and/or by noise that is generated within the system. The most important characteristic of a detection or receiving system is its ability to discriminate between noise and the weakest signal which must be detected. Noise signals which are generated by atmospheric or man-made sources enter the detection system via the antenna and are usually beyond our control. Noise that is generated by various electronic phenomena in the detection and amplification process must be minimized. The effects of the internal noise sources can be minimized by proper circuit design and by control of the system operating conditions.

If the instantaneous values of noise voltages and currents are squared, the resulting squared quantity is always positive and has a definite average value over a sufficiently long time. This *average square* value is used as a measure of the magnitude of noise. The average square voltage of the sum of two noise voltages, which are obtained from different sources, is equal to the sum of the average square voltages. Also, the power dissipated in a resistive element is equal to the sum of the combined powers.

The noise level at the input of the system, if there are no external noise signals, is determined by the random motion of electrons in the conductors constituting the input circuit. The random motion of electrons produces small voltage and current fluctuations in linear, passive elements. This fluctuation is referred to as *thermal* or *Johnson*[3] noise. The magnitude and the frequency spectrum of this noise voltage can be considered as having uniform spectral density over the entire frequency band. The magnitude and frequency spectrum of thermal noise has been derived theoretically[4] and has been verified experimentally.[3]

A noise source in a linear system can be represented as an equivalent current source or as an equivalent voltage source. The equivalent voltage source is shown in Fig. 15·9. Maximum noise power is obtained from the

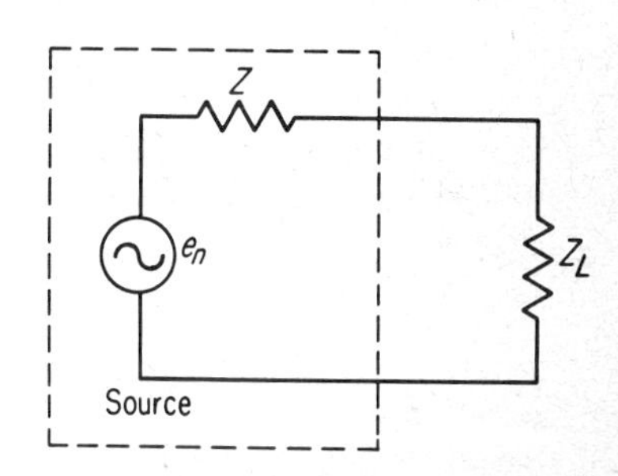

Fig. 15·9 Equivalent voltage source.

source when the load impedance Z_L is the complex conjugate of the source impedance Z. This is the familiar matched system ($R_{\text{source}} = R_{\text{load}}$). The fluctuation voltage across the ends of a resistance is defined by

$$\overline{e_n^2} = 4kTR\,\Delta f \tag{15·13}$$

Equation 15·13 indicates that the average square value of open-circuit voltage is a constant, independent of the center frequency, for a fixed bandwidth Δf.

The *available noise power* is defined as the maximum noise power which can be transmitted from a noise-generating resistance which is connected to a similar resistance. It can be shown that the available noise power of the source shown in Fig. 15·9 is

$$P_n = \frac{|e_n^2|}{4R} \tag{15·14}$$

Since most noise considerations are concerned with finite bandwidths, it is necessary to express the maximum available power in the frequency interval Δf. If the temperature is the same for all components of a linear, passive network, the available noise power is

$$P_n = kT\,\Delta f \tag{15·15}$$

In Eqs. (15·13) to (15·15)

P_n = available noise power

Δf = frequency interval (bandwidth) over which the sinusoidal components of noise are considered

T = temperature of the network, °K. At room temperature $T = 290°K$ (degrees Kelvin = degrees centigrade plus 276)

k = Boltzmann's constant. Boltzmann's constant is a basic physical constant which relates temperature to energy. $k = 1.38 \times 10^{-23}$ joules per degree Kelvin

e_n^2 = fluctuating noise voltage due to thermal agitation of electrons within the resistance R

The quantity kT is an energy. Multiplication of Δf in cycles per second by kT in joules yields P_n in joules per second or watts.

In the practical detection system, the input network or input termination may be an antenna or some type of mixer. If an amplifier is perfectly matched to this input termination, the available signal power is the *actual input signal power* S_i, and the *available signal power output* S_0 is

$$S_0 = G_s S_i \tag{15·16}$$

where G_s is the *signal power gain* of the amplifier. Also, in the case of a conjugate match, the *input noise power* N_i is equal to the available noise power P_n. Therefore, $N_i = kT_0 \Delta f$ if there are no extraneous noise signals. The output signal noise N_0 is given by

$$N_0 = kT_0 \Delta f G_s \tag{15·17}$$

where T_0 is room temperature 290°K.

The *available signal-to-noise ratio* is the ratio of the available signal power to the available noise power. The input and output signal-to-noise ratios become S_i/N_i and S_0/N_0, respectively.

The noise ratio N is defined as the ratio of the available noise power output N_0 to the available noise power input N_i.

$$N = \frac{N_0}{N_i} = \frac{N_0}{kT_0 \Delta f} \tag{15·18}$$

The overall effect of the various sources of noise in a microwave receiver is frequently specified by means of the *noise figure*. This figure of merit of the receiver system takes into account the available power ratio as well as the noise ratio. The noise figure is defined as the ratio of the *available signal-to-noise ratio at the input* to the *available signal-to-noise ratio at the output*.

$$F = \frac{S_i/N_i}{S_0/N_0} \tag{15·19}$$

A noise figure of one (1) would be obtained if the amplifier or receiver did not introduce additional noise to the system. Since $S_0/S_i = G_s$ and $N_i = kT_0 \, \Delta f$, Eq. (15·19) becomes

$$F = \frac{N_0}{kT_0 \, \Delta f \, G_s} \qquad (15\cdot20)$$

N_0 is the measured noise-power output of the system when T_0 is 290°K. *It is important to note that the noise figure is defined as the figure of merit of the receiver at room temperature.* The noise which is contributed by the receiver is obtained by subtracting the total available input noise from the total output noise.

$$N_r = kT_0 \, \Delta f \, G_s F - kT_0 \, \Delta f \, G_s = (F - 1)kT_0 \, \Delta f \, G_s \qquad (15\cdot21)$$

Noise measurements can be made with a calibrated signal generator and power meter. This technique calls for the measurement of the noise bandwidth and gain of the receiver in order to obtain the gain-bandwidth product, which is a measure of the merit of the receiver. This technique is time-consuming and often results in errors.

If a known noise source is used to perform the noise measurement, the problem of measuring the gain-bandwidth product is eliminated. A known noise source is used to produce a known level of broadband noise at the input of the device under test. Each of the available types of noise sources, described below, has certain advantages and disadvantages which are related to the frequency range, the available noise power, and the simplicity of operation.

The temperature-limited diode is a reliable and convenient noise source. The diode is mounted in parallel with a resistive load equal to the value of the source resistance for which the receiver was designed. The diode is operated temperature-limited, and a meter, placed in the plate circuit of the diode, is calibrated directly in noise figure in terms of diode current. The amount of noise obtained from this source can be varied.

The operation of the heated-resistance noise source is based on the facts stated in Eq. (15·14). This type of noise source is limited to measurements of low noise figures since the resistance cannot withstand the high temperatures which are required in order to obtain higher noise power.

The gas-discharge noise source is a reliable source of noise at microwave frequencies. The effective thermal agitation represents an equivalent temperature in the range of 11,500°K. The gas tube is matched to the waveguide structure by mounting the tube so that it extends through the broad sides of the waveguide at an angle of approximately 10°.

The excess noise power generated by the noise source is

$$P_{ns} = k(T_2 - T_0) \, \Delta f \qquad (15\cdot22)$$

where T_2 is the equivalent absolute temperature of the noise source.

A basic noise measurement system which employs gas-discharge noise sources is shown in Fig. 15·10. The excess noise output is constant at a fixed level determined by the gas in the tube. The power detector is used to indicate the noise power output with and without the excess noise.

The output signal noise N_0, measured when the excess noise tube is off, consists of the amplified input termination noise N_i plus the noise generated in the receiver N_r.

$$N_0 = N_i G_s + N_r \tag{15·23}$$

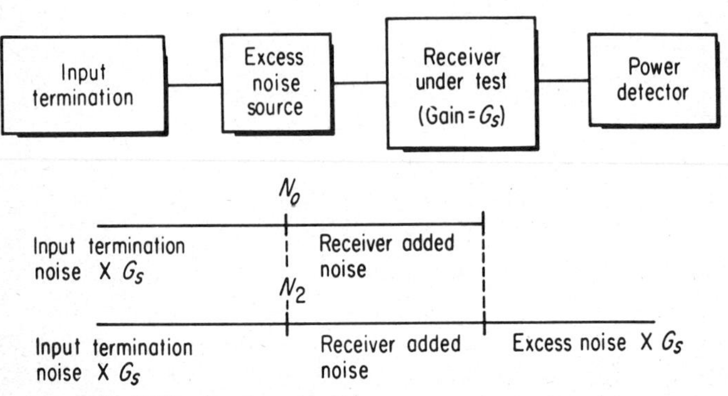

Fig. 15·10 Basic noise-figure measurement system.

When the excess noise tube is turned on, the output noise N_i consists of N_0 plus the amplified excess noise power viewed at the receiver output.

$$N_2 = N_0 + P_{ns}G_s = N_i G_s + N_r + P_{ns}G_s \tag{15·24}$$

If it is assumed that the equivalent noise temperature for the measurement condition is $T = T_0 = 290°\text{K}$, the ratio of N_2 to N_0 is found by substituting from previous equations. The ratio is found to be

$$\frac{N_2}{N_0} = \frac{FT_0 + (T_2 - T_0)}{FT_0} \tag{15·25}$$

Solving for F,

$$F = \frac{T_2 - T_0}{T_0} \times \frac{1}{(N_2/N_0) - 1} \tag{15·26}$$

The noise figure in decibels is

$$F_{db} = 10 \log \frac{T_2 - T_0}{T_0} - 10 \log \left(\frac{N_2}{N_0} - 1\right) \tag{15·27}$$

The ratio $(T_2 - T_0)/T_0$ is a measure of the relative excess noise power available from the noise source. The numerical ratio of N_2/N_0 is called *Y factor*. Extensive tests[5] have indicated that typical average values of excess noise for argon gas tubes are 15.7 db at 9,250 Mc, 15.8 db at 2,295 Mc, 15.9 db at 1,300 Mc, and 16.0 db at 960 Mc. The measurement at 9,250 Mc is in close agreement with measurements performed at the National Bureau of Standards[6] where the service is provided only at X band.

The *twice-power* method of manual noise-figure measurement is illustrated in Fig. 15·11. In this measurement technique, the excess noise source is used to obtain an excess noise power which equals the sum of the amplified input-termination noise plus the receiver noise. That is, N_2 is adjusted for the

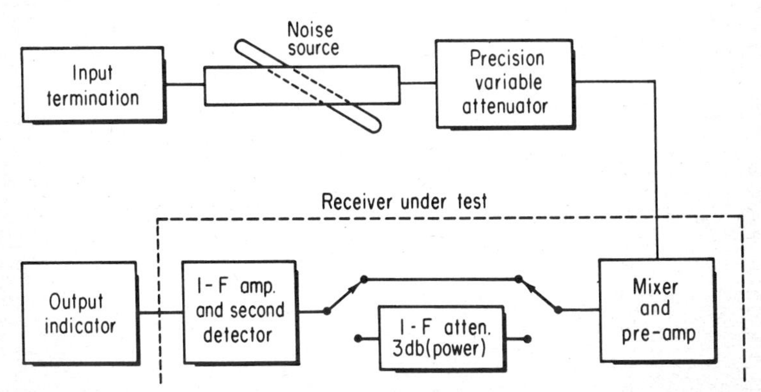

Fig. 15·11 "Twice-power" method of manual noise-figure measurement.

condition, $N_2 = 2N_0$. When this condition is set up, it is noted that Eq. (15·27) reduces to

$$F_{db} = 10 \log \frac{T_2 - T_0}{T_0} - A$$

A is the insertion loss (db) of the precision variable attenuator.

The measurement procedure is as follows:

1. Turn the excess noise source "off" and remove the 3-db attenuator from the system, as indicated in Fig. 15·11.
2. Set up a convenient reference level on the power detector.
3. Insert the 3-db attenuator and turn the excess noise source "on."
4. Adjust the variable attenuator to obtain the original reference. This sets the condition $N_2 = 2N_0$.
5. The attenuated excess noise ratio is the noise figure of the receiver. If the excess noise ratio is known, the attenuator setting is subtracted from the excess noise value to obtain the noise figure of the receiver.

Y-factor Method of Noise-figure Measurement. The excess noise of the noise tube is usually specified by the manufacturer. Since this value is known, it is only necessary to obtain the ratio N_2/N_0 in order to obtain the noise figure. The ratio N_2/N_0 (Y factor) is obtained by using an i-f attenuator to maintain an output reference level. The i-f attenuator, at the output of the receiver, is adjusted to maintain a reference level at the output when the excess noise source is alternately switched "on" and "off."

Equation (15·27) becomes

$$F_{db} = 15.7 - 10 \log \left(\frac{N_2}{N_0} - 1 \right) \tag{15·28}$$

when the measurement is performed using an argon gas tube which has an excess-noise figure of 15.7 db. The ratio N_2/N_0 is measured in decibels using the i-f attenuator. Therefore, in order to solve the equation, it is necessary to solve for the ratio using the equation db $= 10 \log N_2/N_0$.

Example: $N_2/N_0 = 5$ db. $F_{db} = 15.7 - 10 \log (N_2/N_0 - 1)$. The power ratio corresponding to 5 db is 3.16.

$$F_{db} = 15.7 - 10 \log (3.16 - 1) = 15.7 - 3.35 = 12.35 \text{ db}$$

Since noise figure has a limited meaning, it is necessary to be able to obtain the noise figure when the source temperature differs from $T_0 = 290°$K. The extent to which the noise performance is changed is given by

$$F_1 = 1 + (F - 1) \frac{T_0}{T_1} \tag{15·29}$$

Noise-figure Measurements Using Hot-Cold Body Standards.[5] The Y-factor technique can be used to measure the noise figure of a receiver when hot and cold body standard temperatures are supplied as shown in Fig. 15·12. The

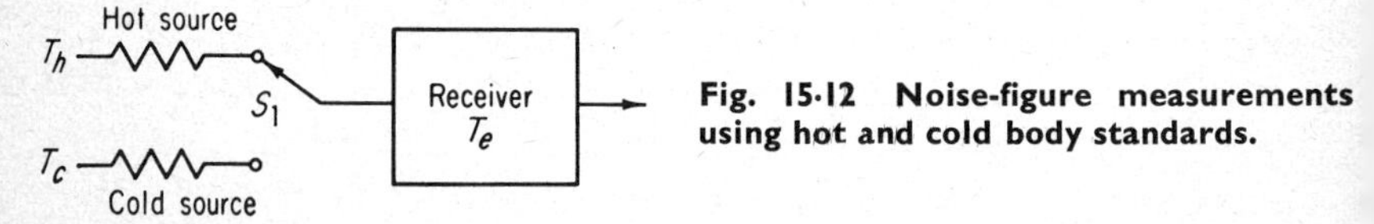

Fig. 15·12 Noise-figure measurements using hot and cold body standards.

general formula which relates noise figure to T_0 when the reference temperature differs from T_0 is given by

$$F = \frac{(T_h - T_0)/T_0 - Y(T_c - T_0)/T_0}{Y - 1} \tag{15·30}$$

where T_h = temperature at receiver input terminal with S_1 connected to hot source

T_c = temperature at receiver input terminal with S_1 connected to cold source

In the above formula, it is assumed that there is no loss in the network between the noise source and the receiver. The effective noise temperatures are lowered when coupled to the receiver through any attenuation.

The Y factor is measured with an i-f attenuator when S_1 is alternately switched to the hot and cold sources. The noise figure is calculated using Eq. (15·30).

Calibration of Gas Tubes with a Cold Body Standard.[5] The basic system shown in Fig. 15·13 can be used to calibrate gas-tube noise sources if the gain of the receiver and the noise factor of the receiver do not change during the measurement.

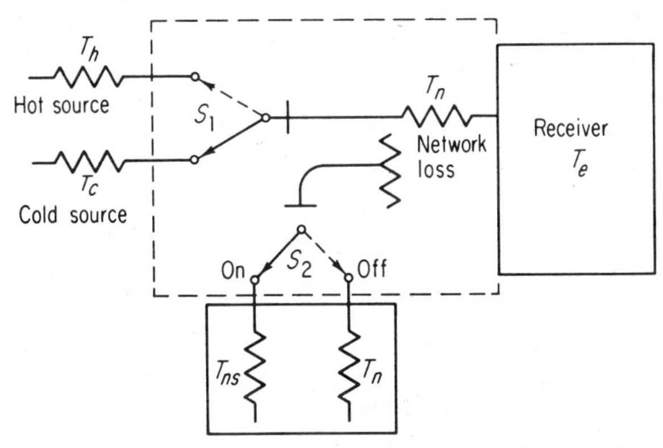

Fig. 15·13 A simplified diagram of system temperatures involved in determining the excess-noise ratio of a gas-tube noise source.

The technique utilized in determining noise figure and excess-noise ratio of gas discharge tubes is the Y-factor method, in which the i-f attenuator in the receiver is used to obtain the Y factor. Two Y-factor measurements are made. The first Y-factor measurement determines the noise figure of the receiver using the hot and cold standard temperatures. The second Y-factor measurement (called Y') determines the excess-noise ratio E_{ns} of the gas tube from the measured noise factor of the receiver.

For the present consideration, $T_h = T_n$, and the excess noise from the hot source is zero. Therefore the noise figure of the receiver is obtained using a cold body standard, and the following equations apply.

$$T_e = \frac{T_n - YT_{1c}}{Y - 1} \tag{15·31a}$$

$$F = \frac{YE_c}{Y - 1} \tag{15·31b}$$

where T_e = effective receiver noise temperature referred to receiver input
T_n = network temperature
T_{1c} = total temperature at receiver input terminal with S_1 connected to cold source at temperature T_c and S_2 remaining at T_n
E_c = excess noise of cold source $= 1 - T_c/290$, where T_c is less than 290°K

The noise figure in decibels is

$$F_{db} = 10 \log Y + 10 \log \left(1 - \frac{T_c}{290}\right) - A - 10 \log (Y - 1) \quad (15\cdot32)$$

where A is the db value of the transmission coefficient (insertion loss of the network).

The first Y-factor measurement is made by keeping the gas tube unfired and adjusting the i-f attenuator for equal i-f output levels when S_1 is switched from the hot to cold load. The noise figure is calculated using Eq. (15·32).

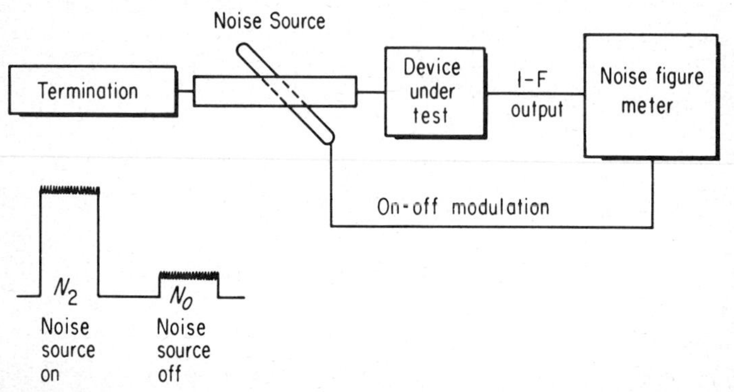

Fig. 15·14 Equipment arrangement for measuring noise figure using the Hewlett-Packard noise-figure meter.

The second Y-factor (Y') measurement is accomplished by connecting S_1 to the standard cold temperature and obtaining equal i-f output by adjusting the i-f attenuator for the fired and unfired conditions of the gas tube. The cold load temperature is used as a standard in order to obtain a larger Y factor which results in smaller errors.

If the receiver gain and system parameters remain constant, the excess noise of the gas tube can be obtained by equating the two noise values. The resulting formula is

$$E_{ns} = C - A + 10 \log (Y' - 1) - 10 \log (Y - 1) + 10 \log \left(1 - \frac{T_c}{290}\right) \quad (15\cdot33)$$

It should be noted that Y and Y' are decibel values and must be converted to power ratio before the equations can be evaluated.

Automatic Noise-figure Measurement. The Hewlett-Packard automatic noise-figure meter[7] is characterized by its ability to automatically display measured values of noise figure. The instrument is designed to operate with a series of gas-discharge noise sources for measurements on microwave devices and with temperature-limited diodes for measurements on i-f amplifiers.

The basic operating principle of the instrument is illustrated in Fig. 15·14. The noise source, connected to the input of the receiver, is automatically modulated on and off. When the noise source is turned on, the receiver output noise N_2 consists of the amplified power from the noise source and the amplified receiver noise. The receiver output noise N_0 consists of the amplified receiver noise and the amplified input termination noise when the noise source is turned off. The output noise N_2 is maintained at a standard level during measurements; thus N_0 is the only variable because the excess noise ratio of the source is known. The instrument performs the ratio of these two noise powers, and the indicating meter is calibrated to display noise figure directly.

15·5 Dielectric measurements

The properties of a material may be specified by two complex constants called *complex dielectric constant* ϵ and *complex permeability* μ. The real and imaginary parts of these constants are indicated in the equations

$$\epsilon = \epsilon' - j\epsilon''$$
$$\mu = \mu' - j\mu'' = \mu_0$$

where ϵ' is the *dielectric constant* and ϵ'' is *the loss factor*. Since the materials are considered to be nonmetallic, $\mu = \mu_0$.

A measure of the energy lost in the form of heat is called the *loss tangent* (tan δ). It is the ratio of the power dissipated to the power stored per cycle.

$$\tan \delta = \frac{\epsilon''}{\epsilon'}$$

The dielectric properties may be determined at microwave frequencies by measuring the propagation characteristics of the electromagnetic wave through the medium. There is a great variety of experimental techniques by which dielectric measurements can be made. The technique used for a particular measurement depends upon the frequency, the dielectric properties of the material, and the amount and form of the available material. This presentation is limited to resonant cavity[8] and shorted-line[9] techniques which are used to determine the characteristics of low loss materials.

In the shorted-line technique, a slotted section is used to measure the shift in minimum of a standing wave and the change in the standing-wave ratio. The minima of the standing-wave pattern occur at intervals of one-half wavelength from the short circuit when the sample is absent. When the sample is inserted in front of the short circuit, the minima shift toward the short circuit, as shown in Fig. 15·15. The shift in minimum is a measure of the dielectric constant. The signal that is lost in the form of heat in the dielectric causes a decrease in the standing-wave ratio. The decrease in standing-wave ratio is a measure of the loss tangent. The sample absorbs maximum power when the wavelength inside the sample is a multiple of a quarter wavelength. This indicates that the approximate value of dielectric constant must be known in order to make the sample one-quarter wavelength long. The thickness of the sample and the wavelength inside the material determine the shift in the minimum.

Measurement Procedures. The measurements can be performed using the attenuation measurement system previously discussed. The tuning procedures are performed in order to assure a matched source as seen by the slotted section. The section of waveguide which is used to enclose the dielectric sample under test is connected to the slotted section. The sample is formed to

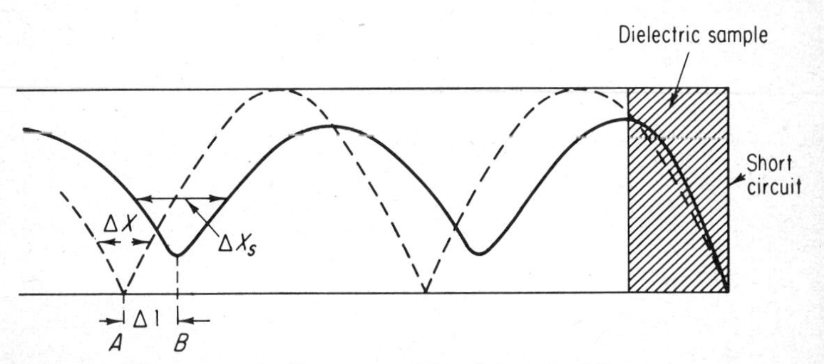

Fig. 15·15 Standing waves in the waveguide with and without the sample.

fit tightly into the waveguide, and the length of the sample d conforms to the dimensions corresponding to the approximate value of dielectric constant. A precision dial gauge is mounted on the slotted section in order to obtain precision distance measurements. All measured distances are in centimeters.

1. Terminate the empty waveguide section with the short circuit and measure the waveguide wavelength λ_g.
2. Measure the distance Δx, indicated on the diagram, using the "twice-minimum" method for measuring high standing-wave ratio. Note: The distances which are to be measured are indicated on the diagram. The measurements are performed with a slotted section connected to the waveguide section which contains the sample.
3. Record the position of the minimum (A).
4. Place the dielectric sample in the waveguide so that it is against the short circuit.
5. Measure Δx_s by the "twice-minimum" method and record the position of the minimum (B).
6. Record the shift in the minimum Δl.
7. Calculate:

$$\frac{\tan x}{x} = \frac{\lambda_g}{2\pi d} \tan \frac{2\pi(\Delta l + d)}{\lambda_g}$$

The unknown x is multivalued (theoretically infinite number of values). Therefore, it is necessary to know the approximate value of the dielectric constant so that the correct value of x can be chosen from the tables of $(\tan x)/x$. The approximate value of x is found using the estimated value of dielectric constant to solve for x in the following equation:

8. Calculate the dielectric constant:

$$\epsilon' = \left(\frac{x\lambda}{2\pi d}\right)^2 + \left(\frac{\lambda}{\lambda_c}\right)^2$$

where λ/λ_c is the waveguide proportionality constant which is a function of the waveguide dimensions and the mode of propagation.

9. Calculate loss tangent:

$$\tan\delta = \frac{\Delta x_1 - \Delta x}{\epsilon' d}\left(\frac{\lambda}{\lambda_g}\right)^2$$

Resonant Cavity Method of Measuring Dielectric Constant. This method of measuring the properties of materials is based on perturbation theory. In general, perturbation methods are utilized in studying the effects of small changes. It is assumed that the solution deviates little from the ideal. The resultant changes caused by the perturbation are calculated from the ideal solution. The present problem involves placement of a dielectric material inside a resonant cavity and equating the frequency shift from that of the unperturbed mode. The change in frequency is a function of the dielectric constant, and the change in cavity Q is a function of the loss factor of the material.

The quantities which must be measured are the resonance frequency of the cavity with and without the dielectric sample, the loaded Q of the cavity with and without the sample, and the cavity dimensions.

The results of the perturbation theory are expressed in the form

$$\frac{\Delta f}{f} = \frac{-(\epsilon - 1)\int_{V_s} E_0 E_s \, dv}{2\int_{V_0} E_0^2 \, dv} \tag{15·34}$$

where Δf = difference in cavity frequency with and without sample
f' = resonant frequency of empty cavity
$f = f' - jf'/2Q$
V_s = volume of sample
V_0 = volume of cavity
E_0 = microwave field strength in empty cavity
E_s = microwave electric field strength inside sample
Q = loaded Q of cavity

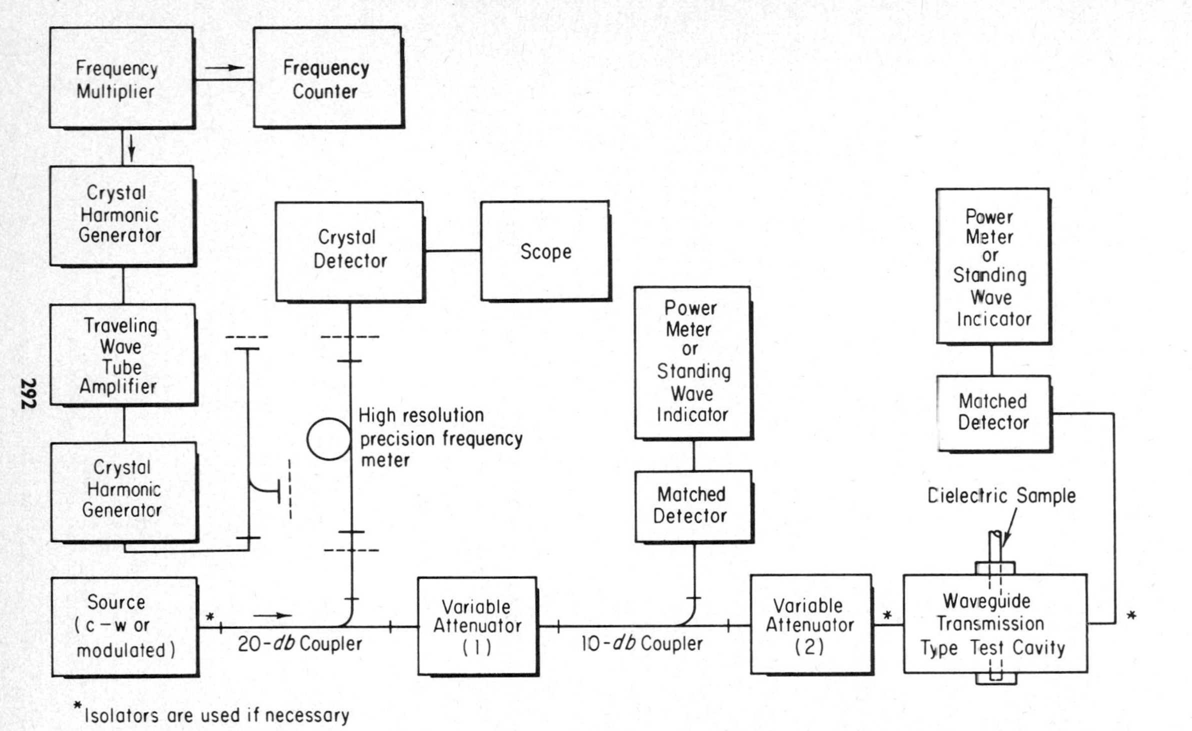

Fig. 15·16 Block diagram of the dielectric constant measurement system.

For a rectangular cavity operating in the TE_{10n} mode where n is odd and the dielectric rod is across the cavity at the center, Eq. (15·34) can be reduced to

$$\frac{\Delta f'}{f'} = -2(\epsilon' - 1)\frac{V_s}{V_0} \tag{15·35}$$

$$\Delta\frac{1}{Q} = 4\epsilon''\frac{V_s}{V_0} \tag{15·36}$$

$\Delta(1/Q)$ is the difference in the reciprocal Q of the cavity with and without the sample. At X band, a sample 0.04-in.-diameter rod is inserted in the transmission type cavity so that the axis is along constant microwave electric field and zero magnetic field. The Q of the cavity must be greater than 2,000. The ends of the rod pass through 0.42-in. holes in both cavity walls.

A block diagram of a measurement system is shown in Fig. 15·16. The indicating meters can be standing-wave indicators if the source is amplitude modulated.

Measurement Procedure

1. Tune the frequency of the c-w or modulated source to the resonant frequency f' of the empty transmission type test cavity. Resonance is indicated by maximum output reading on the standing-wave indicator connected at the output of the test cavity.
2. Set up reference power levels on both power indicators. The power level at the first power monitor is a reference which must be maintained throughout a given measurement. If power levels change when the source frequency is varied, then level set attenuator (1) is adjusted to obtain the original reference level.
3. Vary the source frequency and measure the two frequencies at which the output level drops 3 db from the reference set up at cavity resonance. The frequency is measured using a high Q, high resolution frequency meter or by heterodyning techniques. The frequency difference is designated $\Delta f'$.
4. Place the dielectric rod in the cavity and repeat steps 1 through 3. The frequencies obtained using steps 1 and 3 are designated f_c' and $\Delta f_c'$, respectively.
5. Calculate the dielectric constant from the following equation which was obtained by substitution from Eq. (15·35).

$$\epsilon' = \frac{2lw}{\pi d^2}\frac{f' - f_c'}{f'} + 1$$

6. Calculate the loss factor from the following equation which was obtained by substitution from Eq. (15·36).

$$\epsilon'' = \frac{lw}{\pi d^2}\left(\frac{\Delta f_c'}{f_c'} - \frac{\Delta f'}{f'}\right)$$

l and w are the length and width of the cavity, and d is the diameter of the dielectric sample rod.

7. Calculate the loss tangent:

$$\tan \delta = \frac{\epsilon''}{\epsilon'}$$

8. Calculate the wavelength inside the sample. The diameter of the sample must be small compared to one-quarter wavelength of the input frequency.

$$\lambda = \frac{1}{f'(\epsilon')^{1/2}}$$

15·6 Ferrite measurements

The measurements of fundamental resonance line width ΔH, temperature and frequency dependence of ferrite properties, and power absorption as a

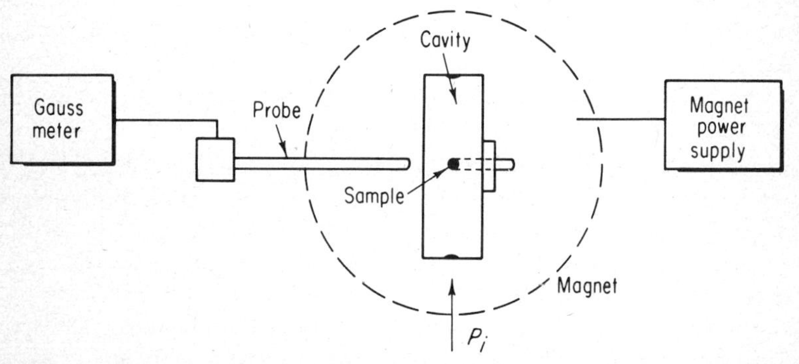

Fig. 15·17 Cavity and magnet arrangement used in measurements of gyromagnetic ratio and resonance line width.

function of d-c magnetic field are performed using cavity techniques. The ferrite-loaded cavity problem was solved in 1953.[10] When a small amount of ferrite material is placed in a resonant cavity, a shift in cavity resonance and a change in cavity Q occur when an external magnetic field is adjusted for ferromagnetic resonance in the ferrite.

The ferromagnetic resonance linewidth and the gyromagnetic ratio of ferrite materials are measured utilizing a cavity perturbation technique. The sample is in the form of a small sphere (about 0.040 in. at X band). The sample is positioned in a transmission type cavity at a point of *minimum microwave* electric field and *maximum microwave magnetic field.* The sample is mounted on a quartz, fused silica, or other dielectric rod and is inserted into the cavity through a small hole located in the narrow wall of the cavity. The sample must be small compared to one-quarter of a wavelength.

The wavelength of microwave radiation in the sample is given by

$$\lambda_s = \frac{3 \times 10^4}{f\sqrt{\epsilon'}}$$

The measurements are performed using the dielectric constant measurement system shown in Fig. 15·16. The cavity and magnet arrangement shown in Fig. 15·17 replaces the transmission type test cavity in the dielectric constant measurement system.

The absorption in the sample is measured by determining the changes in power required to maintain a fixed level at the output of the cavity.

Procedure

1. The modulated or c-w source is tuned to the resonant frequency of the empty cavity. Resonance is indicated by the maximum output at the standing-wave indicator or power meter.
2. Precision-calibrated attenuator No. 2 is set to a convenient reference A_0 (5 or 6 db is usually sufficient). Level set attenuator No. 1 is adjusted to obtain convenient output reference levels on the monitoring power meter or standing-wave indicator and the output power meter or standing-wave indicator.
3. The sample is inserted in the cavity, and the magnetic field is adjusted to obtain the maximum absorption of the microwave signal as indicated by minimum transmission through the cavity. The microwave frequency must be tuned to cavity resonance for all measurements. The frequency f and the magnetic field $\mathbf{H}_r$ are recorded.
4. Attenuator No. 2 is adjusted to obtain the original reference level. The attenuator reading in decibels is labeled A_r.
5. The attenuation corresponding to a sample absorption of half the resonance value is calculated from

$$A_{1/2} = A_0 + 20 \log 2 - 20 \log (10^{(A_0 - A_r)/20} + 1)$$

6. Attenuator No. 2 is set to the $A_{1/2}$ db value, and the magnetic field is determined at the two points at which the output is returned to the original reference value. When the magnetic field is varied on either side of resonance, the absorption decreases and the output level increases. The difference in the two magnetic fields at these two points is the ferrimagnetic resonance linewidth $\Delta\mathbf{H}$.

The sphere diameter must be reduced if the values of $\Delta\mathbf{H}$ and $\mathbf{H}_r$ do not satisfy the following equation:

$$A_0 - A_r \leq 20 \log \left(1 + \frac{120\,\Delta\mathbf{H}}{\mathbf{H}_r}\right)$$

Measurements are made for different orientations of the sphere because the values of $\mathbf{H}_r$ and $\Delta\mathbf{H}$ should not vary as a function of sample rotation.

15·7 Cable characteristics

The three quantities, characteristic impedance Z_0, velocity of propagation v, and attenuation constant α, are used for calculations in transmission line applications.

The attenuation of a given length of cable is measured in the manner described in previous discussions of attenuation measurements. The resistive component of the attenuation varies as the square root of the frequency and is predominant at lower frequencies. The conductance component, which is directly proportional to frequency, is predominant at high frequencies and accounts for the increase in attenuation with frequency.

In low frequency applications the capacitance C is often used in order to calculate the characteristic impedance. Generally, Teflon or polystyrene is used as the insulating material in the cable, and the capacitance per unit length is essentially constant at all frequencies.

The inductance L varies because of the change in current penetration (skin effect) with changes in frequency. The frequency characteristic of the characteristic impedance varies similarly to the inductance. The characteristic impedance is constant at very low frequencies, decreases somewhat in the medium-frequency range, and is again constant (at a lower value) at high frequencies (above 40 Mc).

At low frequencies, the characteristic impedance is measured with an impedance bridge. The characteristic impedance is computed from the open-circuit and short-circuit input impedance measurements.

$$Z_0 = \sqrt{Z_{os}Z_{ss}} \tag{15·37}$$

The characteristic impedance can be measured by an indirect method[11] since the capacitance is independent of frequency.

$$Z_0 = \frac{101{,}600}{v_r C} \tag{15·38}$$

where v_r = relative velocity of propagation, per cent
C = capacitance, $\mu\mu$f per ft

The per cent velocity of propagation is

$$v_r = \frac{v_{ca}}{v} 100 = \frac{f\lambda_{ca}}{9.84} \tag{15·39}$$

where v_{ca} = velocity in cable
λ_{ca} = wavelength inside cable, ft

The number of quarter wavelengths contained in the sample of physical length l is designated k.

$$k = \frac{l}{\lambda_{ca}/4} \quad \text{or} \quad \lambda_{ca} = \frac{4l}{k}$$

$$v_r = \frac{fl}{2.46k} \tag{15·40}$$

The value of k is determined by adjusting the source frequency to obtain a null at some fixed point on the line. This frequency setting is designated f_1.

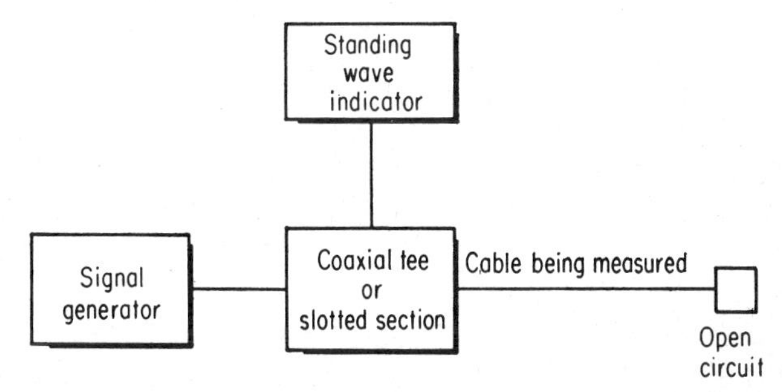

Fig. 15·18 System used to measure velocity of propagation.

The source frequency is increased until the next successive null occurs. This frequency is designated f_2.

$$k = \frac{2f_1}{f_2 - f_1} \tag{15·41}$$

A block diagram of the system used to measure the velocity of propagation is shown in Fig. 15·18. The signal source frequency is varied to obtain successive nulls on the standing-wave indicator. The physical length (ft) of the cable is measured and the velocity is calculated from Eq. (15·40). The characteristic impedance is calculated from Eq. (15·38).

The time delay and the electrical length of a transmission line can be measured using a phase-interference technique. A block diagram of the system is shown in Fig. 15·19. The input signal is split into two channels and then recombined in a crystal detector. The attenuator in one channel is used to obtain the best presentation on the oscilloscope. As the source is swept over the desired range, the cycles of the phase-interference pattern appear as illustrated on the diagram. The number of cycles presented on the display is determined by the sweep frequency range and the difference in path length between the two channels.

The measurement is performed by measuring the number of cycles of the interference pattern with and without the delay unit (transmission line) in the line. The frequency meter reaction is varied over the range to obtain the number of cycles of the phase-interference pattern and the total change in frequency. Increased accuracy of measurement is obtained when many cycles of the interference pattern are present. The number of cycles of the phase-interference pattern is increased when the line in which the unit under test is placed is made considerably longer than the opposite arm. When the unknown is placed in the line, the number of cycles over the frequency range from f_1 to f_2 increases.

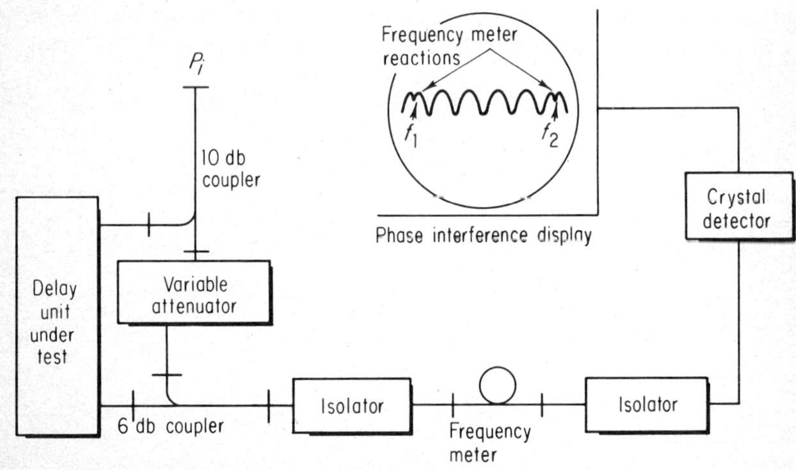

Fig. 15·19 System used to measure delay and electrical length.

The electrical length l_e is calculated as

$$l_e = \frac{v(N_2 - N_1)}{f_2 - f_1} = \frac{2.998N}{f_2 - f_1} \tag{15·42}$$

and

$$t_d = \frac{l_e}{v} = \frac{N_2 - N_1}{f_2 - f_1} \tag{15·43}$$

where t_d = time delay
l_e = electrical length, cm
f_2 = frequency, Gc, at upper limit of chosen frequency range
f_1 = frequency, Gc, at lower limit of chosen frequency range
N_1 = number of cycles of the interference pattern over frequency range $f_2 - f_1$ *without* the unit under test connected in the line
N_2 = number of cycles of the interference pattern over frequency range $f_2 - f_1$ *with* the unit under test connected in the line

It is convenient to make N_1 an integer by choice of f_1 and f_2 and accept N_2 to its nearest integer. However, N_1 and N_2 are not necessarily whole integers.

After establishing the time delay, the physical length can be measured, and the actual velocity in the delay unit can be established using Eq. (15·43).

15·8 Antenna measurements

The field pattern is a good indication of the characteristics of the antenna. Basic pattern measurements and gain measurements are considered in this section.

The space around the antenna consists of the *antenna region* and the *outer region*. The boundary between the two regions is a sphere whose center is at the middle of the antenna and whose surface passes across the ends of the antenna. The outer region is divided into two regions called the *near-field*, or Fresnel region, and the *far-field*, or Fraunhofer region. In the far-field, contributions to the intensity at a point on the axis arrive with essentially

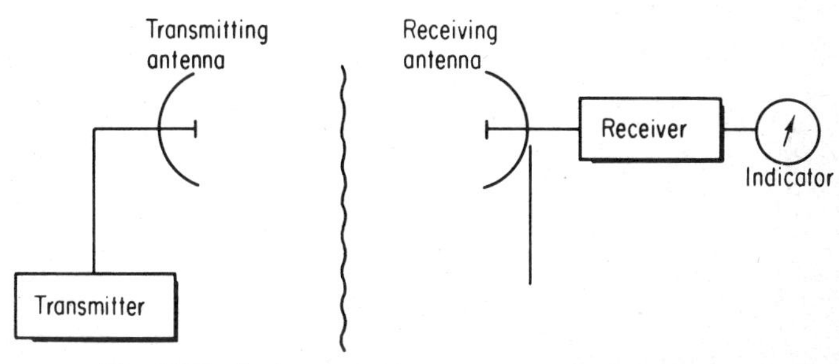

Fig. 15·20 Antenna-pattern measurement system.

their initial phase relationships. Antenna measurements are performed under the far-field conditions which can be established by an approximation of the minimum range R for a given pair of antennas as expressed by

$$R = \frac{(D_1 + D_2)^2}{\lambda} \quad \text{or} \quad \frac{(2D)^2}{\lambda}$$

where D_1 and D_2 are the aperture diameters (D is diameter of larger aperture).

The antennas being measured are usually placed on towers in order to minimize ground reflections.

Pattern intensity measurements can be performed by measuring the relative power received by an antenna at various points of interest. A pattern-measuring arrangement is illustrated in Fig. 15·20. Usually the transmitting antenna is fixed in position, and the antenna under test is rotated on the desired axis. Rapid automatic-measurement plotting systems are usually incorporated when large numbers of patterns are required.

Since gain is measured with respect to some reference antenna, a simple comparison measurement can be performed using the measurement system shown in Fig. 15·20. The indicator can be calibrated in relative voltage, and the gain is calculated from

$$G = \left(\frac{V_1}{V_r}\right)^2$$

where V_1 and V_r are the voltages received from the test antenna and reference antenna, respectively. A direct decibel reading can be obtained using a calibrated precision attenuator to adjust the radiated power at the transmitter so that the received indication is the same for both antennas.

Antenna-gain Measurements. A practical measurement system which can be used to obtain antenna-gain characteristics is shown in Fig. 15·21. A monitored square-wave-modulated source supplies power to the antenna through a precision calibrated attenuator. A matched detector and standing-wave indicator are connected to the receiving antenna. A light-weight

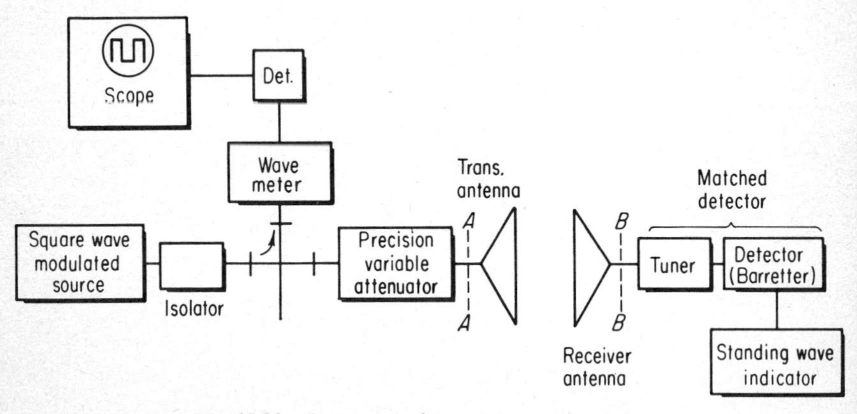

Fig. 15·21 Antenna-gain measurement system.

battery-powered indicator is desirable, as evidenced by this particular measurement technique.

The transmitting antenna is oriented to transmit directly toward the receiving antenna, which is located at an appropriate distance in the far-field region. The receiving antenna is positioned to obtain maximum received power, as indicated on the standing-wave indicator. The precision attenuator is set to a convenient reference level (preferably zero db), and a convenient reference level is set on the standing-wave indicator, using the range switch and gain controls. When this reference level is set up, the gain controls of the standing-wave indicator cannot be changed during the following operations. The matched detector is removed from *B-B*, and the standing-wave indicator and matched detector are carried to the transmitter site. The precision attenuator is set to a high value of attenuation in order to prevent bolometer (barretter) burnout; then the matched detector is connected at *A-A*. The precision attenuator is adjusted to obtain the original reference-level setting on the standing-wave indicator. The difference in attenuator settings is a measure of the familiar power ratio (10 log P_t/P_r). This attenuation or power ratio value is used to calculate the antenna gain according to the case in question as determined by one of the three conditions which will now be discussed.

The gain of an antenna can be calculated using the following formula.[12]

$$G_1G_2 = \left(\frac{4\pi R}{\lambda}\right)^2 \frac{P_r}{P_t} \tag{15·44}$$

where G_1 and G_2 are the gains of the transmitting and receiving antennas, respectively. P_t and P_r are the transmitted and received powers. λ is the free-space wavelength in feet, and R is the range of separation of the antennas in feet. If we consider the electromagnetic horns, R is the range of separation of the apices of the two horns measured in feet.

The gain of either antenna can be computed if the gain of one antenna is known. If the antenna gain is expressed in decibels, then from Eq. (15·44)

$$g_1 + g_2 = 20 \log 4\pi + 20 \log R - 20 \log \lambda - 10 \log \frac{P_t}{P_r} \tag{15·45}$$

It is noted that the measured attenuation value is the last term of this equation. If the gain of the transmitting antenna is known, the receiving antenna gain is found from

$$g_2 = 20 \log 4\pi + 20 \log R - 20 \log \lambda - 10 \log \frac{P_t}{P_r} - g_1 \tag{15·46}$$

If the two antennas are identical, then Eq. (15·44) becomes

$$G^2 = \left(\frac{4\pi R}{\lambda}\right)^2 \frac{P_r}{P_t}$$

$$G = \frac{4\pi R}{\lambda}\sqrt{\frac{P_r}{P_t}} \tag{15·47}$$

The gain in decibels is

$$g = 10 \log 4\pi + 10 \log R - 10 \log \lambda - \frac{1}{2}\left(10 \log \frac{P_t}{P_r}\right) \tag{15·48}$$

In general, the gain of any antenna may be measured by using three unknown antennas and the following formulas.[12]

$$G_1G_2 = \left(\frac{4\pi R}{\lambda}\right)^2 \frac{P_{12}}{P_t} \tag{15·49a}$$

$$G_1G_3 = \left(\frac{4\pi R}{\lambda}\right)^2 \frac{P_{13}}{P_t} \tag{15·49b}$$

$$G_2G_3 = \left(\frac{4\pi R}{\lambda}\right)^2 \frac{P_{23}}{P_t} \tag{15·49c}$$

As an example, consider two identical antennas spaced 60 ft apart. The gain is to be measured at 9.0 Gc ($\lambda = 0.1093$ ft). The reading obtained on the attenuator during the power ratio measurement is 28 db. From Eq. (15·48)

$$g = 10 \log 4\pi + 10 \log 60 - 10 \log 0.1093 - (\tfrac{1}{2})28$$

$$g = 11.0 + 17.8 - (-9.6) - (\tfrac{1}{2})28 = 24{\cdot}4 \text{ db}$$

PROBLEMS

15·1 The distance between two adjacent nulls, measured with the standard short circuit in the phase-shift measurement system of Fig. 15·7 operated at X band (WR 90), is 0.956 in.
a. What is the waveguide wavelength?
b. What is the operating frequency?
c. What value of phase shift is obtained if the short circuit is displaced 0.2 in.?

15·2 At a frequency of 9.0 Gc, the VSWR of an absorption type cavity is 3.6 and the frequency difference between 3-db power points is 0.6 Mc. The voltage standing wave is a minimum at the detuned short position when the cavity is tuned to resonance.
a. Calculate the loaded and unloaded Q.
b. If the characteristic impedance is 460 ohms, what is the cavity impedance at resonance?
c. Repeat *a* and *b* if the voltage standing wave is maximum at the detuned short position when the cavity is tuned to resonance.

15·3 Derive Eq. (15·2) in Sec. 15·4.

15·4 Calculate the noise power at room temperature for a 1-Mc bandwidth. What is the corresponding level below 1 watt?

15·5 An argon gas tube is used in the Y-factor method of noise measurement. The excess noise of the gas tube is 15.8 db and the measured Y factor is 3.6 db. Calculate the noise figure in decibels.

15·6 Calculate the noise figure of the receiver in Fig. 15·12 if $T_n = 400°$K, $T_c = 78°$K, and $Y = 4$ db.

15·7 Calculate the receiver noise figure and the excess noise of the gas tube as measured in the system shown in Fig. 15·13 if $T_c = 78°$K, $Y = 1.95$ db, $Y' = 5.8$ db, $C = 10.8$ db, and $A = 0.5$ db.

15·8 Two antennas are placed 54 ft apart. The operating frequency is 9.5 Gc. If the gain of the transmitting antenna is 20 db, and the measured power ratio is 24 db, what is the gain (in decibels) of the receiving antenna?

15·9 Identical antennas are spaced 46 ft apart. The frequency of operation is 12.4 Gc, and the measured power ratio is 18 db. Calculate the gain (in decibels) of the antennas.

REFERENCES

1. M. Magid, Precision Microwave Phase Shift Measurements, *IRE Trans. Instr.*, vol. 1–7, no. 3–4, December, 1958.
2. Edward L. Ginzton, "Microwave Measurements," chap. 9, McGraw-Hill Book Company, New York, 1957.

   M. Wind and H. Rapaport, "Handbook of Microwave Measurements," Polytechnic Institute of Brooklyn, New York, 1954.
3. J. B. Johnson, Thermal Agitation of Electricity in Conductors, *Phys. Rev.*, vol. 32, pp. 97–109, July, 1928.
4. H. Nyquist, Thermal Agitation of Electric Charge in Conductors, *Phys. Rev.*, vol. 32, pp. 110–113, July, 1928.
5. T. Mukaihata, B. L. Walsh, M. F. Bottjer, and E. B. Roberts, Subtle Differences in System Noise Measurements and Calibration of Noise Standards, *IRE Trans. Microwave Theory Tech.*, vol. MTT-10, no. 6, pp. 506–516, November, 1962.
6. J. S. Wells, W. C. Daywitt, and C. K. S. Miller, Measurement of Effective Temperatures of Microwave Noise Sources, *IRE Intern. Conv. Record*, part 3, pp. 220–230, 1962.
7. Howard C. Poulter, An Automatic Noise Figure Meter for Improving Microwave Device Performance, *Hewlett-Packard Journal*, vol. 9, no. 5, January, 1958.

   Marco R. Negrete, Additional Conveniences for Noise Figure Measurements, *Hewlett-Packard Journal*, vol. 10, no. 6–7, February–March, 1959.
8. G. Birnbaum and J. Franeau, Measurement of the Dielectric Constant and Loss of Solids and Liquids by a Cavity Perturbation Method, *J. Appl. Phys.*, vol. 20, pp. 817–818, August, 1949.
9. A. Von Hippel, "Dielectric Materials and Applications," The Technology Press of the Massachusetts Institute of Technology, Cambridge, Mass., John Wiley & Sons, Inc., New York, 1958.
10. B. Lax and A. D. Berk, Resonance in Cavities with Complex Media, *IRE Natl. Conv. Record*, vol. 1, part 10, p. 70, 1953.
11. "The Measurement of Cable Characteristics," General Radio Company, Reprint E-104, February, 1958.

    "Electronic Test Instruments," Hewlett-Packard Company, 1959.
12. C. G. Montgomery, "Technique of Microwave Measurements," M.I.T. Radiation Laboratory Series, McGraw-Hill Book Company, New York, 1948.

    J. D. Kraus, "Antennas," McGraw-Hill Book Company, New York, 1950.

INDEX